计算机技术开发与应用丛书

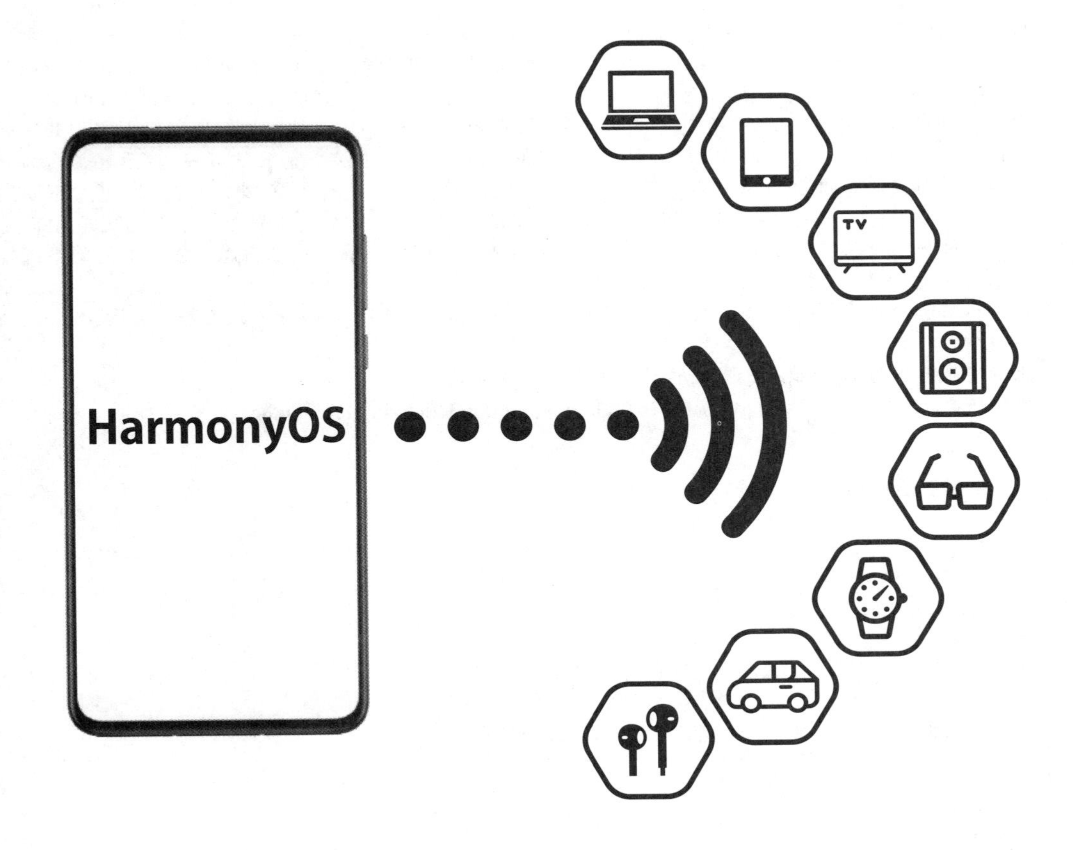

鸿蒙操作系统开发入门经典

徐礼文 ◎ 著
Xu Liwen

清華大学出版社
北京

内容简介

本书涵盖鸿蒙北向应用开发全部内容和南向硬件开发基础内容。

本书共5篇14章，开发准备篇(第1章和第2章)分别是鸿蒙操作系统简介和鸿蒙应用开发准备。第1章总体介绍鸿蒙操作系统的重要特性及技术架构，第2章介绍鸿蒙应用开发环境搭建，鸿蒙应用开发IDE、真机调试证书申请、真机模拟器使用。基础知识篇(第3～5章)系统地讲解鸿蒙两大应用开发框架的使用：Java UI和JavaScript UI框架，以及鸿蒙面向Ability开发技术。分布式开发篇(第6～8章)分别介绍鸿蒙分布式任务调度、分布式数据服务和分布式文件服务。应用实战篇(第9～12章)通过不同语言框架开发不同平台应用，最后介绍如何申请鸿蒙发布证书，以及将鸿蒙应用发布到华为官方应用市场的流程。硬件开发篇(第13章和第14章)介绍开源鸿蒙源代码编译环境搭建和海思开发版的鸿蒙系统烧录及嵌入式开发入门。

本书适合鸿蒙系统应用开发爱好者、Java及JavaScript开发者，以及嵌入式爱好者阅读。

图书在版编目(CIP)数据

鸿蒙操作系统开发入门经典/徐礼文著.—北京：清华大学出版社，2021.6(2022.12重印)
(计算机技术开发与应用丛书)
ISBN 978-7-302-58200-7

Ⅰ.①鸿…　Ⅱ.①徐…　Ⅲ.①移动终端－应用程序－程序设计　Ⅳ.①TN929.53

中国版本图书馆CIP数据核字(2021)第099085号

责任编辑：赵佳霓
封面设计：吴　刚
责任校对：李建庄
责任印制：沈　露

出版发行：清华大学出版社
网　　址：http://www.tup.com.cn，http://www.wqbook.com
地　　址：北京清华大学学研大厦A座　　**邮　　编**：100084
社 总 机：010-83470000　　**邮　　购**：010-62786544
投稿与读者服务：010-62776969，c-service@tup.tsinghua.edu.cn
质量反馈：010-62772015，zhiliang@tup.tsinghua.edu.cn
课件下载：http://www.tup.com.cn，010-83470236
印 装 者：三河市君旺印务有限公司
经　　销：全国新华书店
开　　本：186mm×240mm　　**印　　张**：30.75　　**字　　数**：691千字
版　　次：2021年7月第1版　　**印　　次**：2022年12月第3次印刷
印　　数：3501～4300
定　　价：119.00元

产品编号：091752-01

前 言

PREFACE

2019年8月9日，华为公司在华为开发者大会上正式发布了鸿蒙操作系统(HarmonyOS) 1.0，同时宣布该操作系统源代码开源。2020年9月10日，鸿蒙操作系统2.0正式发布，与鸿蒙1.0版本相比，鸿蒙2.0在分布式软总线、分布式数据管理、分布式安全等分布式能力上进行了升级，同时发布了自适应的UX(用户体验)框架，让开发者能够非常简单且快速地开发鸿蒙应用程序。

鸿蒙操作系统，打破了硬件间各自独立的边界，融入了全场景智慧生态，鸿蒙操作系统不局限于手机，还包括可穿戴设备、智能汽车等物联网，创造一个超级智能终端互联的世界，将人、设备、场景有机地联系在一起。鸿蒙操作系统作为面向物联网时代的操作系统，将有望重塑物联网生态，将芯片、系统、人工智能等技术分享给全球，推动全社会数字化转型，继而进入智能社会新时代。对此，围绕系统构建庞大软硬生态，将带来万物智能的全场景生活生态。

鸿蒙的问世，恰逢中国整个软件业亟须补短，对国产软件的全面崛起能起到刺激和战略性带动作用。鸿蒙是时代的产物，它代表中国高科技必须开展的一次战略突围，是中国解决诸多“卡脖子”问题的一个带动点。

本书适合于有一定基础的JavaScript和Java开发者，可帮助读者快速掌握鸿蒙应用开发的技巧，本书案例涉及华为智能手表、智慧屏、手机及第三方开发板的开发，通过多个游戏案例帮助开发者快速掌握鸿蒙开发。

本书内容

本书共5篇14章。开发准备篇(第1和2章)、基础知识篇(第3～5章)、分布式开发篇(第6～8章)及应用实战篇(第9～12章)全面讲解北向应用开发的两大UI框架：Java UI和JavaScript UI框架，通过3个案例介绍鸿蒙应用开发的基本步骤和技巧。硬件开发篇(第13和14章)讲解南向硬件开发的基础入门，以及开源鸿蒙系统的编译和开发板烧录鸿蒙系统的步骤。本书提供了大量代码示例，读者可以通过这些例子理解知识点，也可以直接在开发实战中稍加修改应用这些代码。

本书读者对象

学习本书内容需要具备一定的 Java、HTML、CSS、JavaScript 基础知识，希望本书能够对读者学习使用鸿蒙开发者框架构建美观、快速、跨终端的移动应用程序有所帮助，并恳请读者批评指正。

配套资源

扫描下面二维码可获取本书教学课件(PPT)及源代码。

徐礼文

2021 年 4 月

教学课件(PPT)

本书源代码

目录

CONTENTS

第一篇　开发准备篇

第二篇　基础知识篇

第三篇 分布式开发篇

第四篇 应用实战篇

第一篇　开发准备篇

学习目标

本篇为鸿蒙开发的准备篇，第1章主要介绍鸿蒙操作系统的技术特性和技术架构，第2章介绍如何配置鸿蒙应用开发的环境、下载安装鸿蒙开发工具DevEco Studio、申请华为开发者账号以及真机测试证书的步骤和流程。

主要内容如下：

- 介绍鸿蒙操作系统的技术特性和技术架构；
- 鸿蒙应用开发环境搭建、模拟器及真机调试。

如何学习本篇：

- 通过本篇的学习，读者将对鸿蒙操作系统的特性、技术架构有比较全面的了解，为学习后续章节奠定基础；
- 学习第2章，按照本书介绍的鸿蒙应用开发环境搭建的流程进行开发前的环境准备。

第1章 鸿蒙操作系统简介

2019年8月9日，华为公司在华为开发者大会上正式发布了鸿蒙操作系统（HarmonyOS）1.0，同时宣布该操作系统源代码开源。2020年9月10日，鸿蒙操作系统2.0正式发布，与鸿蒙1.0版本相比，鸿蒙2.0在分布式软总线、分布式数据管理、分布式安全等分布式能力上进行了升级，同时发布了自适应的UX（用户体验）框架，让开发者能够非常简单且快速地开发鸿蒙应用程序。

鸿蒙操作系统是一款面向全场景分布式操作系统。鸿蒙操作系统不同于现有的Android、iOS、Windows、Linux等操作系统，它设计的初衷是解决在5G万物互联时代，各个系统间的连接问题。鸿蒙操作系统面向的是1＋8＋N的全场景设备，能够根据不同内存级别的设备进行弹性组装和适配，并且实现跨设备交互信息。

1.1 鸿蒙全场景战略

鸿蒙操作系统，打破了硬件间各自独立的生态边界，融入了全场景智慧生态，鸿蒙操作系统不局限于手机，还包括可穿戴设备、智能汽车等，创造一个超级智能终端互联的世界，将人、设备、场景有机地联系在一起。作为面向物联网时代的操作系统，将有望重塑物联网生态，对于华为来讲，将芯片、系统、人工智能等技术分享给全球，推动全社会数字化转型，继而进入智能社会新时代。对此，围绕系统构建庞大软硬生态，将带来万物智能的全场景生活生态。

华为的“1＋8＋N”这个产品战略是为了打造未来5G全场景智慧生活而制定的，面向5G高品质全场景的智慧生活，生态在各领域都可以体现出它的存在和价值，如图1-1所示。

在华为的1＋8＋N战略中，1和8都是华为自己构建的。1是指手机，8是指平板、PC、眼镜、智慧屏、AI音箱、耳机、手表、车机。从手机的优势向外围延展，N是指由生态系统合作伙伴提供的智能设备，基于用户为中心的家庭场景，提供全场景的视听、娱乐、社交、教育和健康等解决方案，从而很好地迎合时代更新换代的消费升级。

在华为的1＋8＋N生态中有一个非常重要的应用就是Huawei Share。最早是在手机

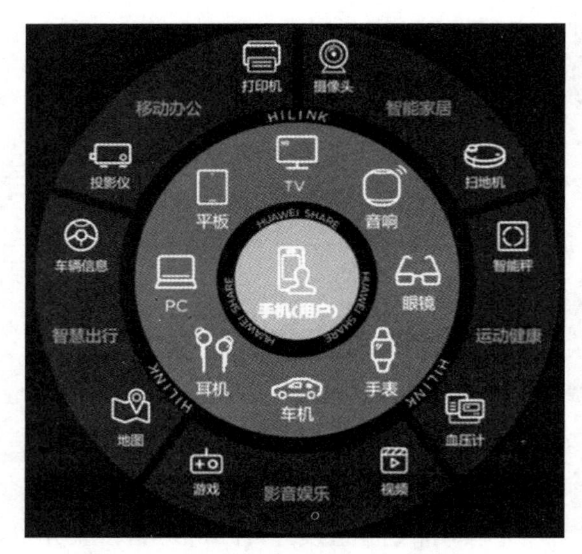

图 1-1　华为“1+8+N”生态

与 PC 之间可以实现“一碰传”,后来 Huawei Share 在华为的 1+8 中实现更多的连接。通过 Huawei Share,在华为自有的 1+8 中,可以实现一碰传文件、一碰传音、一碰联网、多屏协同等创新体验;通过 HUAWEI HiLink,华为 1+8 设备可同海量的 N 设备之间智慧互联,设备一键操控、语音交互、场景联动等极致体验被实现。

1.2　鸿蒙操作系统技术特性

相对于市面上已有的操作系统,尤其是相对于安卓系统,鸿蒙操作系统具备如下 3 个主要特征:①以分布式为基础的多终端屏幕共享,跨屏交互;②系统与硬件解耦,弹性部署;③应用一次开发,多端部署。

1.2.1　分布式设计

随着智能设备越来越多,为每个设备设计一个独立的操作系统几乎是不可能的,而鸿蒙操作系统的目的是解决多设备如何共享一个操作系统的问题。

鸿蒙操作系统是将设备的硬件能力拆散,当成一个个共享资源,当用户需要某个能力时,就可以将它从硬件库里提取出来跨界使用。它系统层级更清晰,也更加模块化,可以依照不同的处理器性能让开发商去调整系统模块,从而运用于更多的智能设备。

这就意味着,将来洗衣机、电视机、冰箱、空调器,甚至小到灯泡或者门锁智能化之后,它们都能搭载鸿蒙操作系统,这有一个好处,如果你有一个鸿蒙操作系统的便携设备,例如手机,你就可以无须任何有线连接,实现和所有设备的联动。

鸿蒙不是手机系统的简单替代,而是全场景的底座,其三大分布式能力分别是分布式软总线、分布式数据管理和分布式安全的解决方案，如图 1-2 所示。

图 1-2　鸿蒙全场景的底座：三大分布式能力

1. 分布式软总线

鸿蒙分布式软总线技术是基于华为多年的通信技术积累，在 1+8+N 设备间搭建一条“无形”的总线，具备自发现、自组网、高带宽、低时延的特点，如图 1-3 所示。

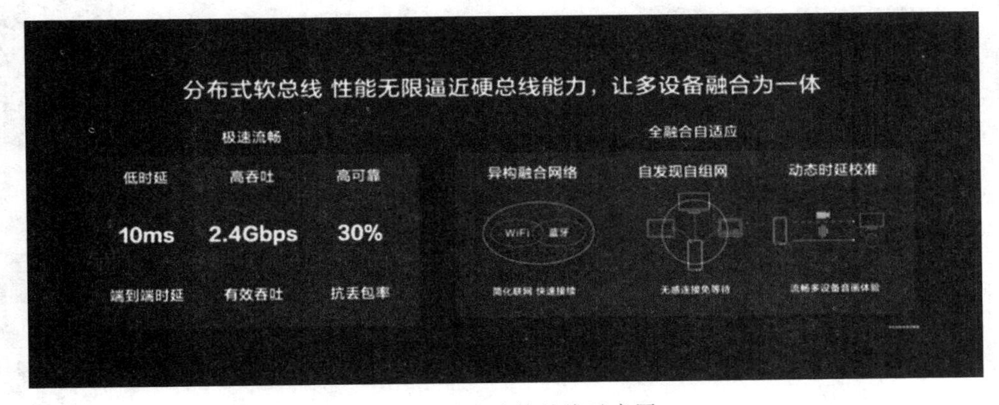

图 1-3　分布式软总线示意图

设备通信方式多种多样（USB、WiFi、BT 等），不同通信方式使用差异很大且烦琐，同时通信链路的融合共享和冲突无法处理，通信安全问题也不好保证。分布式软总线致力于实现近场设备间统一的分布式通信能力管理，提供不区分链路的设备发现和传输接口。目前实现能力包含：①服务发布，服务发布后周边的设备可以发现并使用服务；②数据传输，根据服务的名称和设备 ID 建立一个会话，这样就可以实现服务间的传输功能；③安全，提供通信数据的加密能力。

分布式软总线为设备之间的互联互通提供了统一的分布式通信能力，为设备之间的无感发现和零等待传输创造了条件。开发者只需聚焦于业务逻辑的实现，无须关注组网方式与底层协议。

2. 分布式数据管理

现在，每个人拥有的设备越来越多，大家经常需要将某一设备的数据导入其他设备上，这使得数据在不同设备之间流转越来越频繁。假设我们有一组照片需要在手机、平板、智慧屏和 PC 之间共享和编辑，此时就需要考虑这些照片如何在这些不同的设备上进行存储，以及如何相互访问。HarmonyOS 分布式数据管理的目标就是为开发者在系统层面解决这些问题，让应用开发变得简单，它能够保证多设备间的数据安全，解决多设备间数据同步、跨设备查找和访问等很多关键技术问题。

分布式数据管理，基于分布式软总线的能力，实现应用程序数据和用户数据的分布式管理。用户数据不再与单一物理设备绑定，业务逻辑与数据存储分离，应用跨设备运行时数据无缝衔接，为打造一致、流畅的用户体验创造了基础条件。分布式数据管理优势如图 1-4 所示。

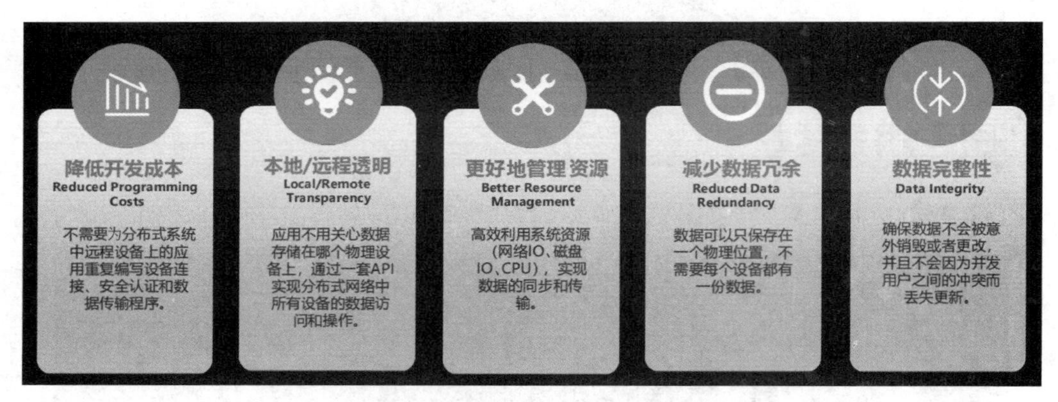

图 1-4　分布式数据管理优势

分布式数据管理可以让跨设备数据处理像本地一样方便快捷，其中，鸿蒙的分布式系统比微软 Samba 软件的远程读写性能快 4 倍，分布式文件系统的远程读写性能是 Samba 的 4 倍；分布式数据库 OPS（Operation Per Second 每秒操作次数）性能是 ContentProvider 的 1.3 倍。ContentProvider 是 Android 四大组件之一，其作用是为不同的应用之间数据共享提供统一的接口。

分布式检索方面，检索性能是 iOS Core Spotlight（为 iOS 的搜索提供一些 App 内部的数据，便于用户在 iPhone 上下拉出现的搜索框中，搜索我们所使用的 App 中的内容）的 1.2 倍，如图 1-5 所示。

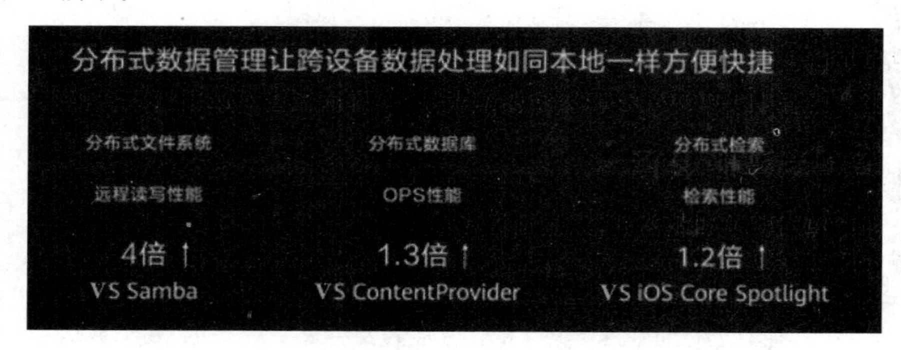

图 1-5　分布式数据管理的方便快捷性

3. 分布式安全

目前华为是业界第一家在微内核领域通过 CC EAL5+安全认证的厂商。分布式安全确保正确的人用正确的设备访问正确的数据。

CC EAL 安全证书,目前国际范围内最受普遍认可的信息安全评价标准是 CC(即 Common Criteria),其中共定义了由低到高 EAL1 到 EAL7 共 7 个等级,可以简单理解为等级越高,消费者使用这款产品时,对它的安全性越有信心。

确保正确的人。如图 1-6 所示,当用户进行解锁、付款、登录等行为时系统会主动拉出认证请求,并通过分布式技术的可信互联能力,完成多设备协同身份认证,确保使用者是正确的人。

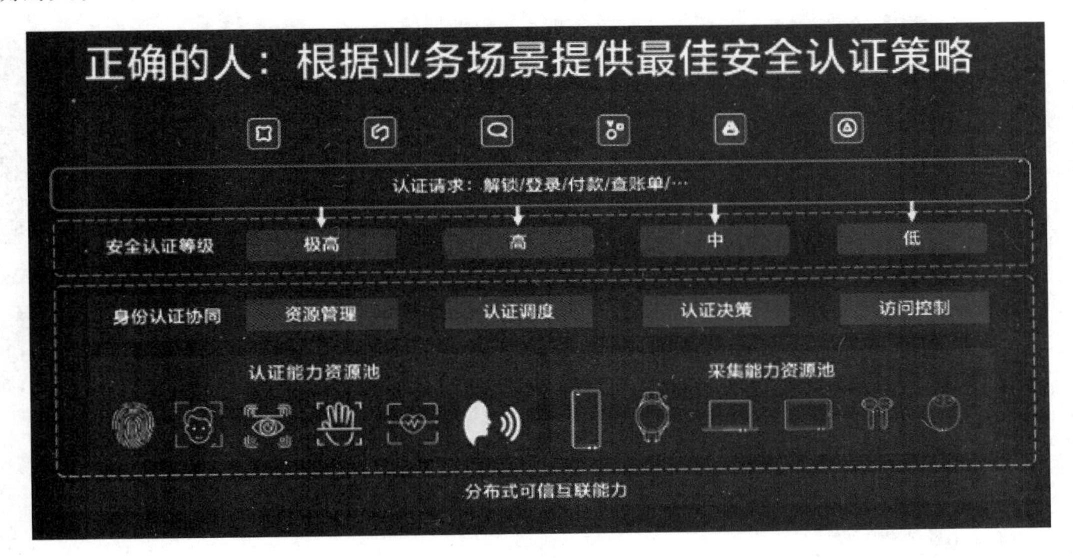

图 1-6 正确的人

用正确的设备。早在 EMUI 10,TEE OS(TEE 即可信执行环境,该环境可以保证不被常规操作系统干扰计算,因此称为“可信”)安全内核就获得了全球商用 OS 内核最高安全等级的 CC EAL5+安全等级认证,而该 TEE OS 可以弹性地部署到任何一个 IoT 设备上。

在多设备融合的情况下,通过 HarmonyOS,每个设备都会获得所有链接在一起的设备的安全能力加持。当单一设备受到外部攻击时,完全可以调用其他设备上的安全能力进行共同防御,如图 1-7 所示。

访问正确的数据。如图 1-8 所示,HarmonyOS 会根据安全等级的不同,对数据和设备进行分类分级保护,敏感数据只能保存在高安全等级设备中。在数据流通中,只有高安全等级设备可以访问低安全等级设备而低安全等级设备不能访问高安全等级设备,由此确保数据流通安全可信。

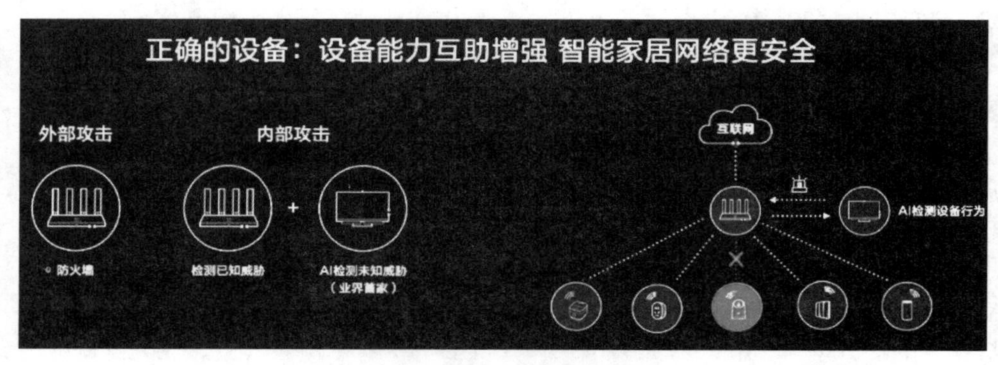

图 1-7 安全能力进行共同防御

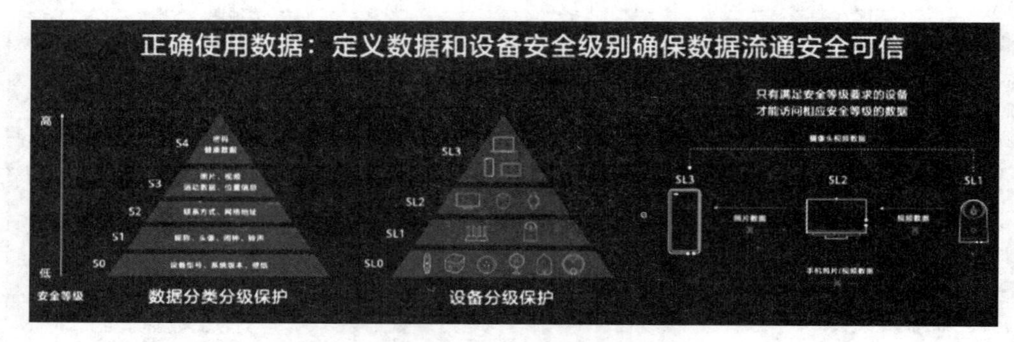

图 1-8 访问正确的数据

1.2.2 一次开发，多端部署

HUAWEI DevEco 2.0 集成开发工具(IDE)，如图 1-9 所示，为开发者提供了多语言统一的开发编译环境，分布式架构 Kit 提供屏幕布局控件及交互的自动适配，支持控件拖曳，面向预览的可视化编程，从而使开发者可以基于同一工程高效构建多端自动运行 App，实现真正的一次开发，多端部署，在跨设备之间实现共享生态。

图 1-9 HUAWEI DevEco 2.0 集成开发工具(IDE)构建全场景应用

华为方舟编译器是首个取代 Android 虚拟机模式的静态编译器，可供开发者在开发环境中一次性将高级语言编译为机器码。此外，方舟编译器未来将支持多语言统一编译，可大幅提高开发效率。

1.2.3 系统与硬件解耦，弹性部署

鸿蒙操作系统分布式将硬件能力虚拟化，将硬件能力与终端解耦，并将多终端硬件能力融合成能力资源池。如图 1-10 所示，能力资源池包括显示、摄像头、扬声器、话筒、通信、传感、计算。

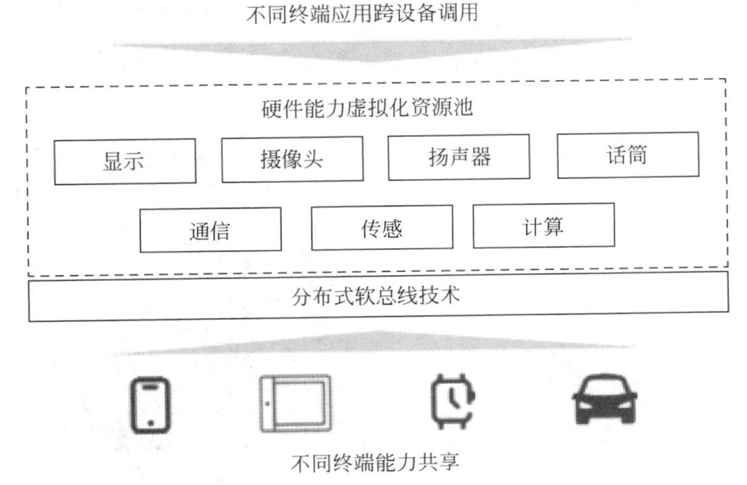

图 1-10 操作系统与硬件能力解耦，硬件能力虚拟化的资源池

HarmonyOS 通过组件化和小型化等设计方法，支持多种终端设备按需弹性部署，能够适配不同类别的硬件资源和功能需求。支撑通过编译链关系自动生成组件化的依赖关系，形成组件树依赖图，支撑产品系统的便捷开发，从而降低硬件设备的开发门槛。

鸿蒙操作系统设计上支持根据硬件的形态和需求，可以选择所需的组件；支持根据硬件的资源情况和功能需求，可以选择配置组件中的功能集，例如，选择配置图形框架组件中的部分控件；支持根据编译链关系，可以自动生成组件化的依赖关系，例如，选择图形框架组件，将会自动选择依赖的图形引擎组件等。

HarmonyOS 组件化设计实现了内存从 512KB 级别到 4GB 级别都有合适的裁剪方案，这样就让鸿蒙能够支持各种各样的终端设备，从 IoT（Internet of Things，物联网）到可穿戴、摄像头、VR、音箱、行车记录仪、电视机、PC、平板、手机等各种终端设备。

1.3 鸿蒙操作系统技术架构

目前鸿蒙操作系统是基于 Linux 系统来开发自研操作系统的。这样有两大好处：一是可以很好地兼容安卓系统的 App，毕竟安卓系统是基于 Linux 系统来开发的，这样在生态上

的问题就解决了很大一部分了。

另外，鸿蒙是一个集计算机、手机、汽车等设备于一体的大一统的系统，目前 Linux 系统在计算机领域的应用生态也是不错的，基于 Linux 系统来开发，在计算机领域的应用生态，也解决了很大一部分。

鸿蒙操作系统整体遵从分层设计，如图 1-11 所示，从下向上依次为内核层、系统服务层、框架层和应用层。在多设备部署场景下，支持根据实际需求裁剪某些非必要的子系统、功能或者模块。

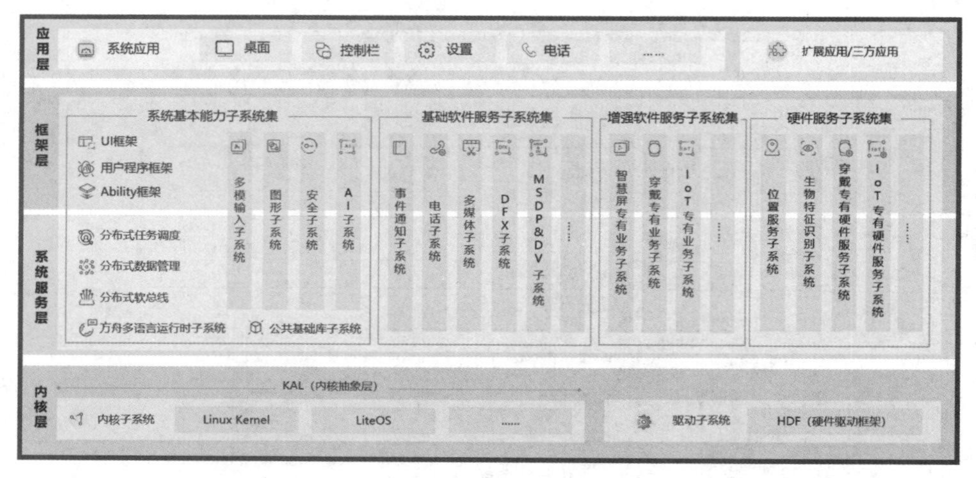

图 1-11　鸿蒙操作系统技术架构

1.3.1　内核层

鸿蒙内核层由鸿蒙微内核、Linux 内核、Lite OS 组成，未来将发展为完全的鸿蒙微内核架构，如图 1-12 所示。

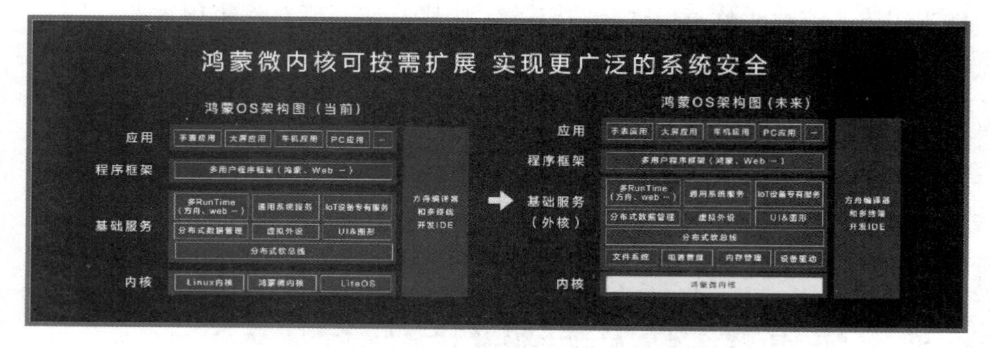

图 1-12　鸿蒙操作系统未来将采用微内核

HarmonyOS 目前采用多内核设计，支持针对不同资源受限设备选用不同的 OS 内核。内核抽象层(KAL，Kernel Abstract Layer)通过屏蔽多内核差异，对上层提供基础的内核能力，包括进程/线程管理、内存管理、文件系统、网络管理和外设管理等。

（1）内核子系统：内核抽象层（Kernel Abstract Layer，KAL）通过屏蔽多内核差异，对上层提供基础的内核能力，包括进程/线程管理、内存管理、文件系统、网络管理和外设管理等。

（2）驱动子系统：HarmonyOS 驱动框架（HDF）是 HarmonyOS 硬件生态开放的基础，提供统一外设访问能力和驱动开发、管理框架。

1.3.2　系统服务层

系统服务层是 HarmonyOS 的核心能力集合，通过框架层对应用程序提供服务。

该层包含以下几部分。

（1）系统基本能力子系统集：为分布式应用在 HarmonyOS 多设备上的运行、调度、迁移等操作提供了基础能力，由分布式软总线、分布式数据管理、分布式任务调度、公共基础库、多模输入、图形、安全、AI 等子系统组成。

（2）基础软件服务子系统集：为 HarmonyOS 提供公共的、通用的软件服务，由事件通知、电话、多媒体、DFX、MSDP（组播源发现协议）&DV 等子系统组成。

（3）增强软件服务子系统集：为 HarmonyOS 提供针对不同设备的、差异化的能力增强型软件服务，由智慧屏专有业务、穿戴专有业务、IoT 专有业务等子系统组成。

（4）硬件服务子系统集：为 HarmonyOS 提供硬件服务，由位置服务、生物特征识别、穿戴专有硬件服务、IoT 专有硬件服务等子系统组成。

根据不同设备形态的部署环境，基础软件服务子系统集、增强软件服务子系统集、硬件服务子系统集内部可以按子系统粒度裁剪，每个子系统内部又可以按功能粒度裁剪。

1.3.3　框架层

框架层为 HarmonyOS 的应用程序提供了 Java、C、C++、JavaScript 等多语言的用户程序框架和 Ability 框架，如图 1-13 所示，以及各种软硬件服务对外开放的多语言框架 API；

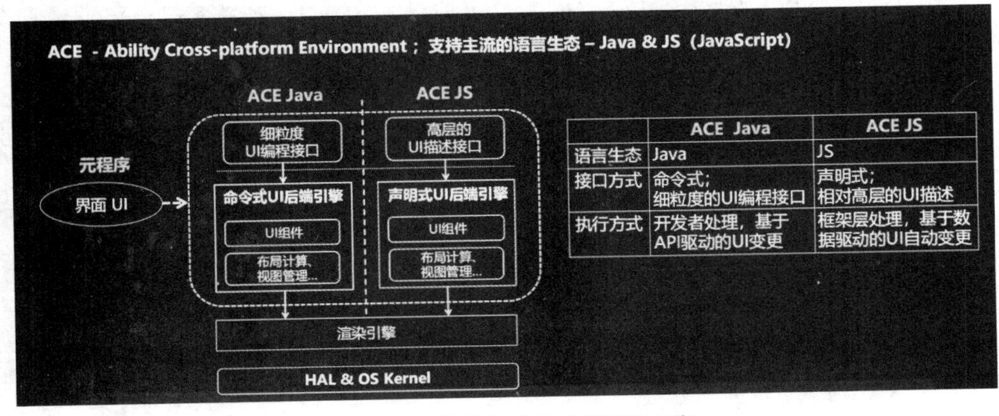

图 1-13　鸿蒙操作系统应用开发框架

同时为采用 HarmonyOS 的设备提供了 C、C++、JavaScript 等多语言的框架 API，不同设备支持的 API 与系统的组件化裁剪程度相关。

1.3.4 应用层

应用层包括系统应用和第三方非系统应用。HarmonyOS 的应用由一个或多个 FA(Feature Ability)或 PA(Particle Ability)组成。其中，FA 有 UI 界面，其提供与用户交互的能力，而 PA 则无 UI 界面，提供后台运行任务的能力及统一的数据访问抽象。基于 FA/PA 开发的应用，能够实现特定的业务功能，支持跨设备调度与分发，为用户提供一致、高效的应用体验，如图 1-14 所示。

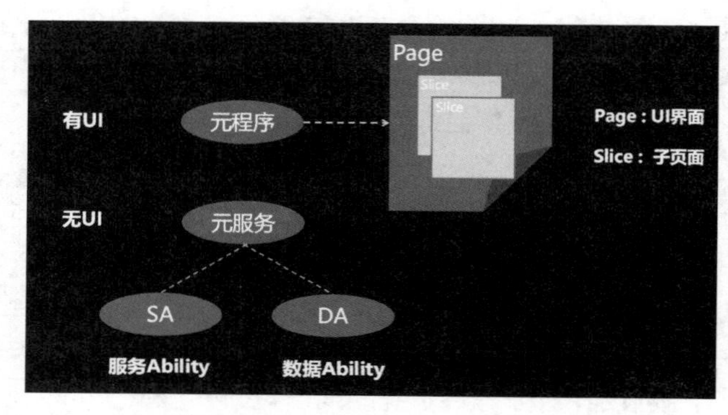

图 1-14 基于 FA/PA 应用开发

1.4 本章小结

这里引用华为消费者业务 CEO 余承东先生的一句话："没有人能够熄灭满天星光，每一位开发者，都是华为要汇聚的星星之火，星星之火可以燎原。"华为鸿蒙操作系统作为底层，为企业数字化赋能，其芯片、人工智能等技术通过云端输出，为全球数字化带来创新活力，而基于鸿蒙操作系统打造的 HMS 移动生态，将是继安卓、iOS 后全球第三大移动应用生态，而这仅用了一年时间。

鸿蒙操作系统的到来，对我国软硬件生态意义重大，在万物互联时代掌握话语权。那么要发挥鸿蒙价值，无论是移动 HMS 生态，亦是数字化生态，关键还在应用。当然，吸引开发者对华为来讲也面临诸多挑战，与此同时，如果鸿蒙操作系统能成功，则将会创造一个千亿，甚至万亿的生态大市场。截至目前，华为 HMS 移动生态有 1840 亿次应用下载和分发量，有 4.9 亿活跃用户，对于积极拥抱鸿蒙生态的开发者来讲，也带来新一波红利。

第2章 鸿蒙应用开发准备

本章介绍如何配置鸿蒙应用开发环境、下载安装集成开发工具 DevEco Studio、使用鸿蒙模拟器、申请真机调试证书,以及如何使用 Scrcpy Android 投屏软件进行真机测试。通过本章的学习,读者可以逐步搭建好鸿蒙应用开发所需的相关环境,为后续章节学习做好准备工作。

2.1 鸿蒙应用开发环境搭建

鸿蒙应用开发环境搭建分两步,分别是下载安装 Node.js 和下载安装 DevEco Studio。

2.1.1 下载和安装 Node.js

Node.js 发布于 2009 年 5 月,由 Ryan Dahl 开发,是一个基于 Chrome V8 引擎的 JavaScript 运行环境,使用了一个事件驱动、非阻塞式 I/O 模型,并可让 JavaScript 运行在服务器端的开发平台。

Node.js 应用于开发鸿蒙 JavaScript 应用程序和运行鸿蒙预览器功能,是开发 HarmonyOS 应用过程中必备的软件。

下载安装 Node.js,可选择 LTS 版本 12.0.0 及以上,Windows 64 位对应的软件包,如图 2-1 所示。Node.js 安装包及源代码下载网址为 https://nodejs.org/en/download/。

双击下载后的软件包进行安装,根据安装向导完成 Node.js 的安装。Mac 系统在安装软件过程中,需要输入用户系统密码来授权系统运行安装新软件。

Windows 系统具体的安装步骤如下。

步骤 1:双击下载后的安装包 node-v14.16.0-x64.msi 进行安装,如图 2-2 所示。

步骤 2:勾选接受协议选项,单击 Next 按钮,如图 2-3 所示。

步骤 3:Node.js 默认安装目录为 C:\Program Files\Node.js\,安装过程中可以修改目录,并单击 Next 按钮,如图 2-4 所示。

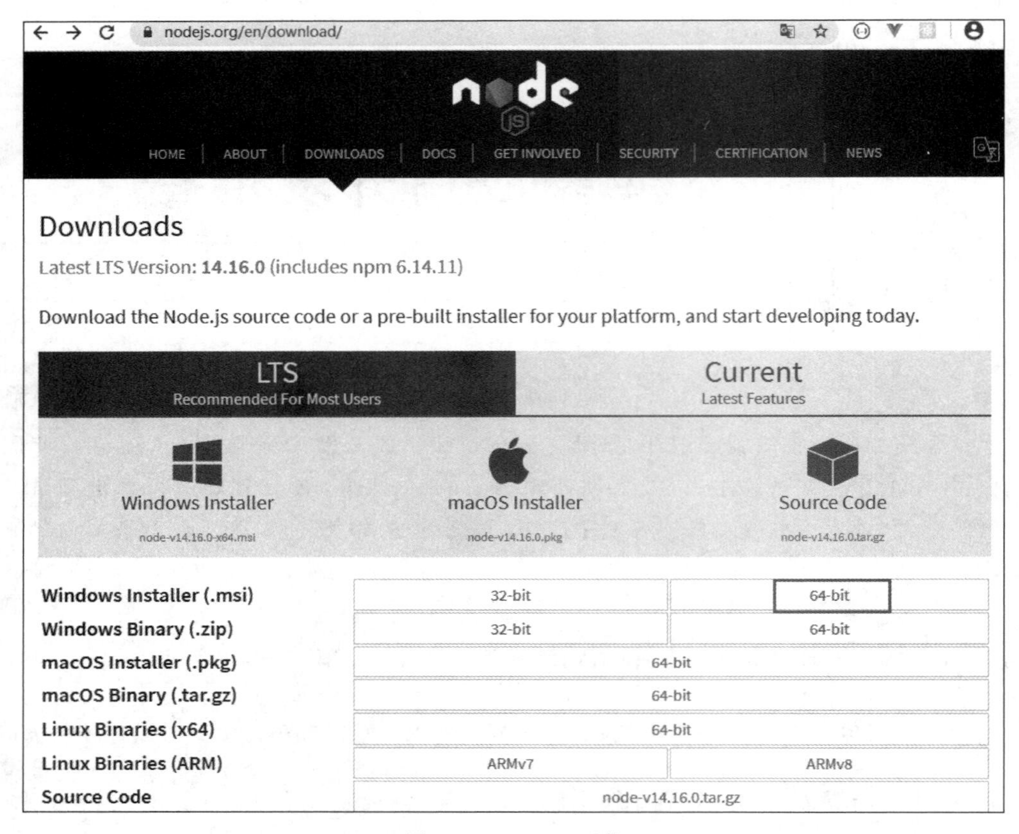

图 2-1　Node.js 下载

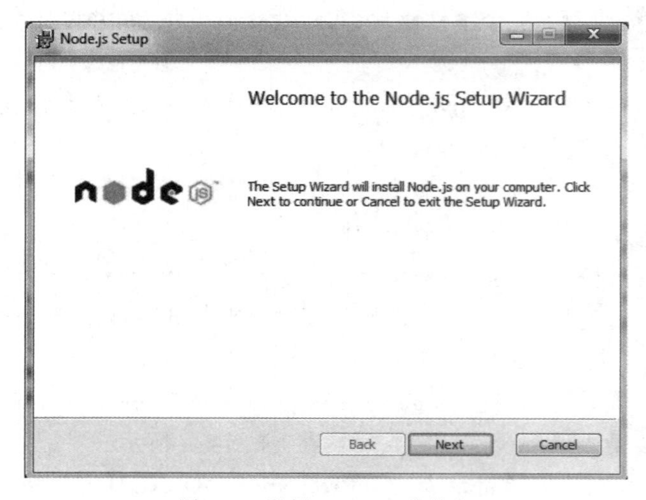

图 2-2　运行 Node.js 安装包

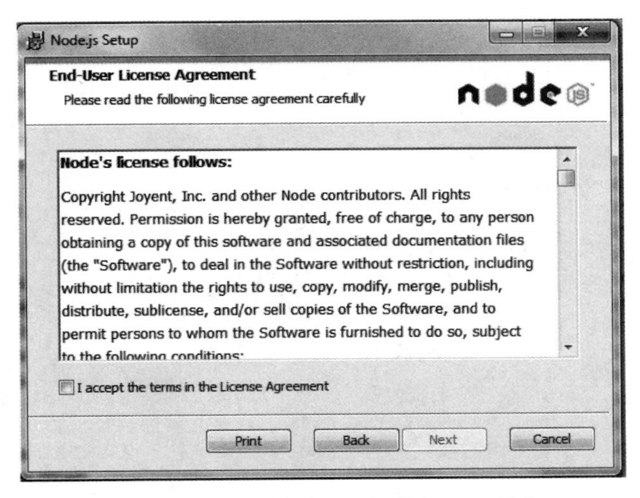

图 2-3 勾选接受协议选项，单击 Next 按钮

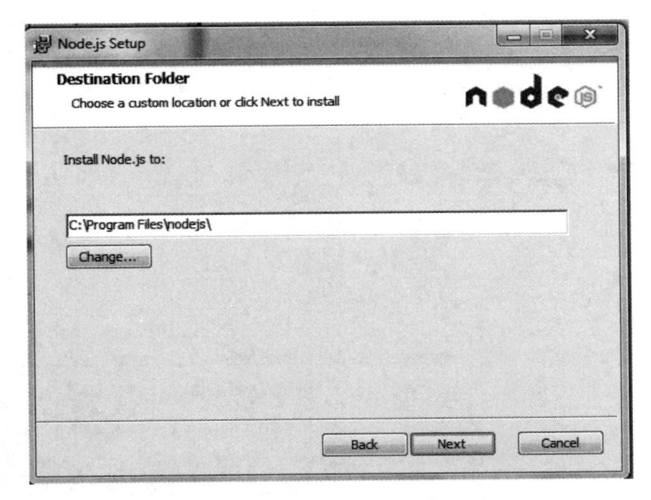

图 2-4 修改 Node.js 安装目录

步骤 4：单击树形图标来选择需要的安装模式，一般选择默认即可，然后单击 Next 按钮，如图 2-5 所示。

步骤 5：单击 Install 按钮开始安装 Node.js，如图 2-6 所示。也可以单击 Back 按钮来修改先前的配置，接下来单击 Next 按钮。

步骤 6：单击 Finish 按钮退出安装向导，如图 2-7 所示。

步骤 7：打开 cmd 输入 node -v 查看 Node.js 的版本号，如图 2-8 所示，表示 Node 安装成功了。

macOS 上安装 Node 可以通过以下两种方式实现。

(1) 在官方下载网站下载 pkg 安装包，直接单击安装即可。

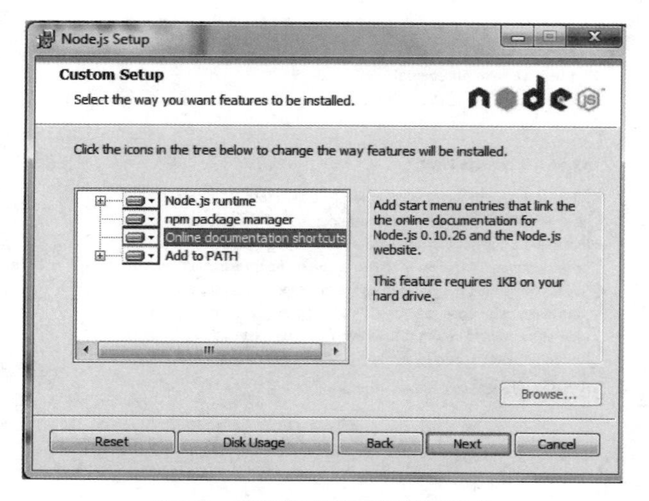

图 2-5　安装模式(选择默认即可)

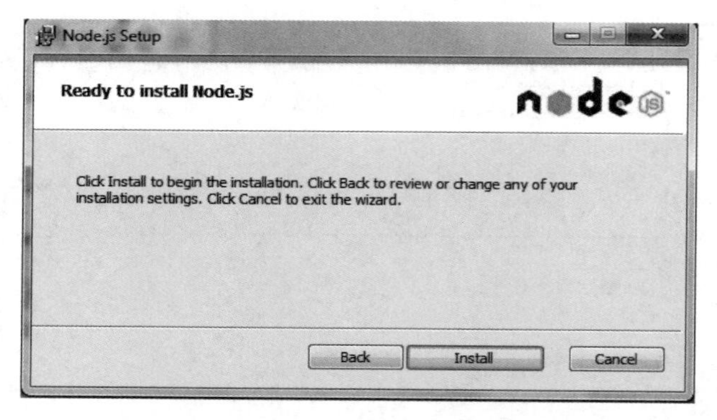

图 2-6　单击 Install 按钮开始安装 Node. js

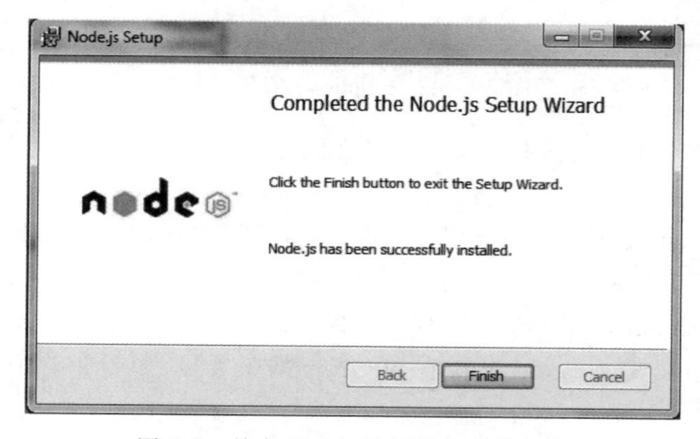

图 2-7　单击 Finish 按钮退出安装向导

图 2-8　查看 Node.js 版本号

(2) 使用 brew 命令来安装，命令如下：

```
brew install node
```

2.1.2　下载和安装 DevEco Studio

HUAWEI DevEco Studio(以下简称 DevEco Studio)是基于 IntelliJ IDEA Community 开源版本打造的，面向华为终端全场景多设备的一站式集成开发环境(IDE)，为开发者提供工程模板创建、开发、编译、调试、发布等 E2E 的 HarmonyOS 应用开发服务。通过使用 DevEco Studio，开发者可以更高效地开发具备 HarmonyOS 分布式能力的应用，进而提升创新效率。

下面介绍如何下载安装 DevEco Studio 开发工具，具体的步骤如下：

(1) 登录 HarmonysOS 应用开发门户，单击右上角注册按钮，注册开发者账号。可以访问如下网址：https://id1.cloud.huawei.com/CAS/portal/login.html，登录成功后，再访问 HUAWEI DevEco Studio 产品页，下载 DevEco Studio 安装包，如图 2-9 所示。

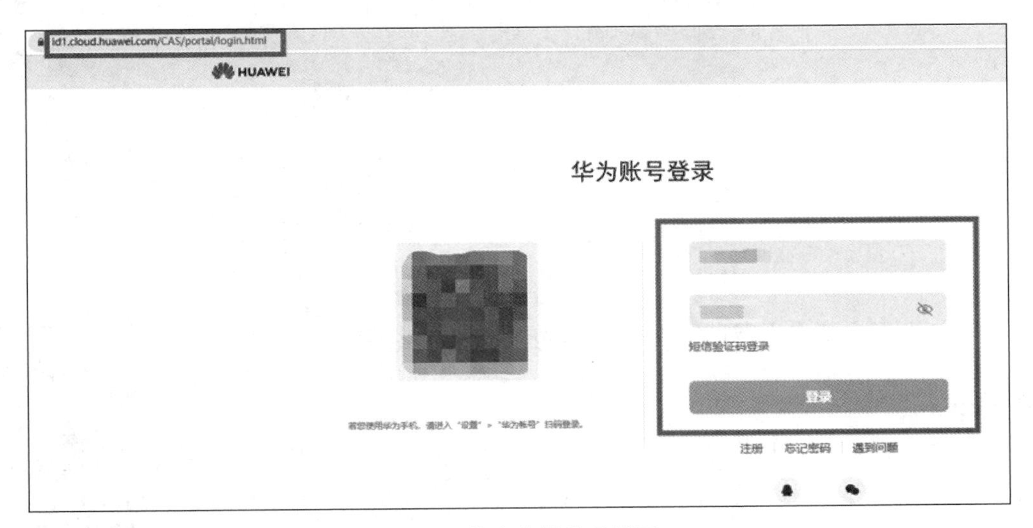

图 2-9　华为账号登录页面

(2) 进入 HUAWEI DevEco Studio 产品页，下载 DevEco Studio 安装包，如图 2-10 所示。

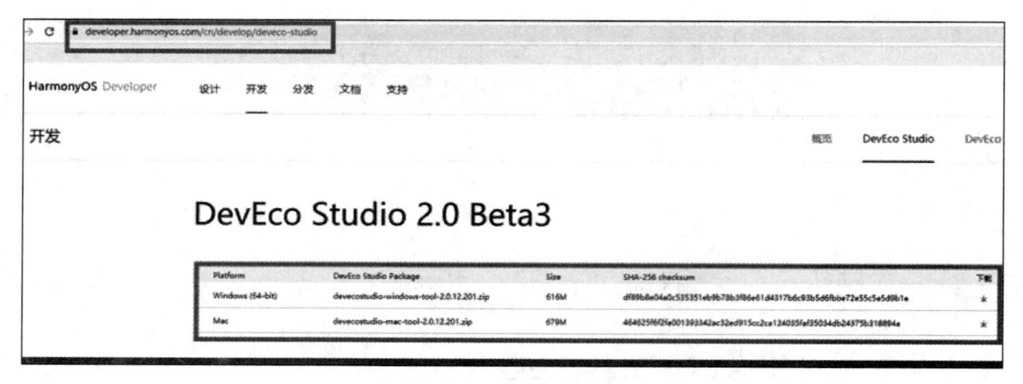

图 2-10 DevEco Studio 2.0 下载

(3) Windows 用户双击下载的安装文件 deveco-studio-xxxx.exe，进入 DevEco Studio 安装向导，在如下安装选项界面勾选 DevEco Studio launcher 后，单击 Next 按钮，如图 2-11 所示，直至安装完成。

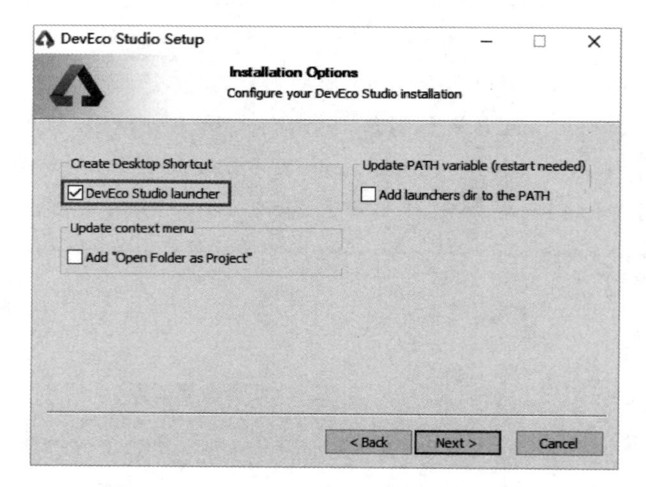

图 2-11 DevEco Studio 2.0 Windows 安装

(4) Mac 用户双击下载的 deveco-studio-xxxx.dmg 软件包。

(5) 在安装界面中，将 DevEco-Studio.app 拖曳到 Applications 中，如图 2-12 所示，等待安装完成。

(6) 安装完成后，先不要勾选 Run DevEco Studio 选项，接下来需要根据需要配置开发环境，如图 2-13 所示，检查和配置开发环境。

(7) DevEco Studio 的编译会构建依赖 JDK，DevEco Studio 预置了 Open JDK，版本为 1.8，安装过程中会自动安装 JDK。

图 2-12 DevEco Studio 2.0 Mac 安装

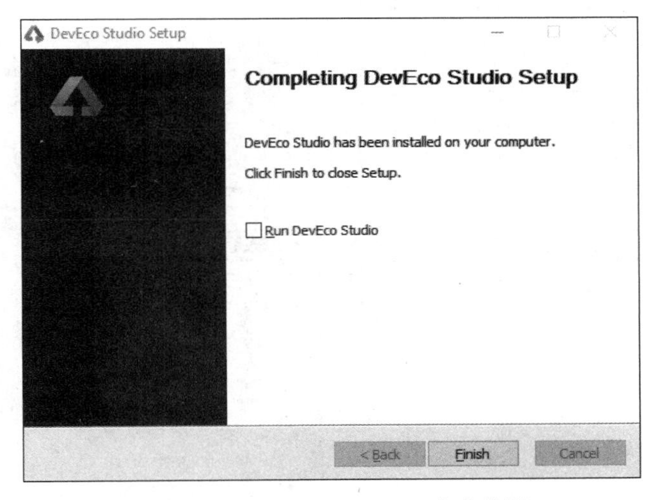

图 2-13 DevEco Studio 2.0 安装检测

2.1.3 运行 Hello World

DevEco Studio 开发环境配置完成后，可以通过运行 Hello World 工程来验证环境设置是否正确。以 Wearable 工程为例，在 Wearable 远程模拟器中运行该工程。

我们来按步骤完成并运行一个 Hello World 程序，步骤如下。

步骤 1：打开 DevEco Studio，在欢迎页单击 Create HarmonyOS Project，如图 2-14 所示，创建一个新工程。

步骤 2：选择设备类型和模板，以 Wearable 为例，选择 Empty Feature Ability(Java)，单击 Next 按钮，如图 2-15 所示。

步骤 3：填写项目相关信息，保持默认值即可，单击 Finish 按钮。

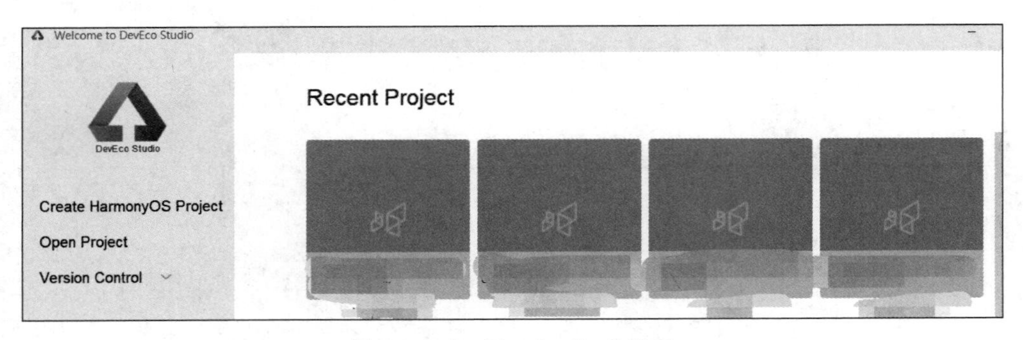

图 2-14　DevEco Studio 欢迎页

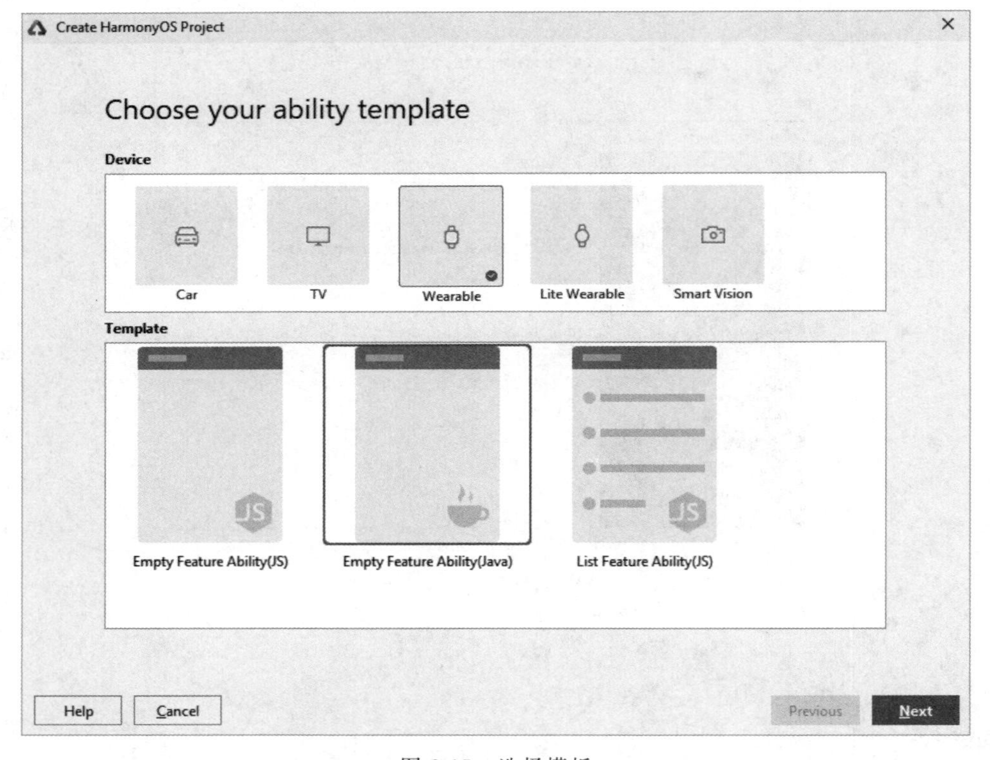

图 2-15　选择模板

步骤 4：工程创建完成后，DevEco Studio 会自动进行工程同步，同步成功后如图 2-16 所示。首次创建工程时，会自动下载 Gradle 工具，时间较长，需耐心等待。

步骤 5：在 DevEco Studio 菜单栏，单击 Tools→HVD Manager。首次使用模拟器需下载模拟器相关资源，单击 OK 按钮，等待资源下载完成后，单击模拟器界面左下角的 Refresh 按钮，如图 2-17 所示。

步骤 6：在浏览器中弹出华为开发者联盟账号登录界面，输入已实名认证的华为开发者联盟账号的用户名和密码进行登录，如图 2-18 所示。

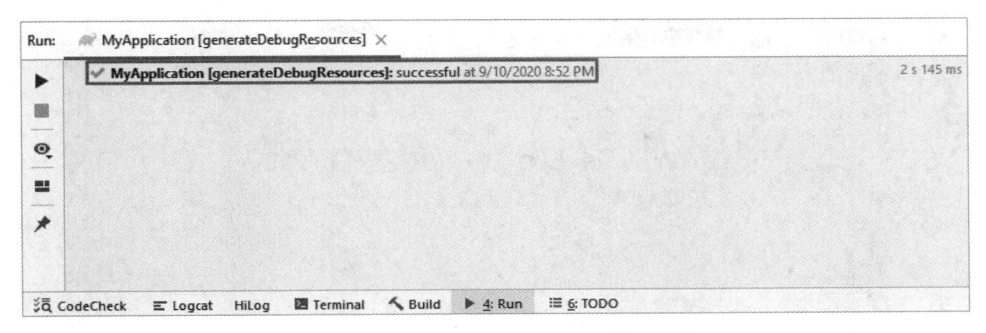

图 2-16　安装 DevEco Gradle 编译环境

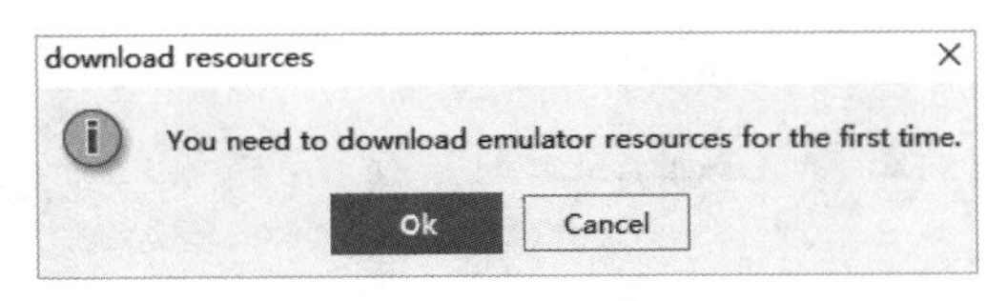

图 2-17　下载安装模拟器

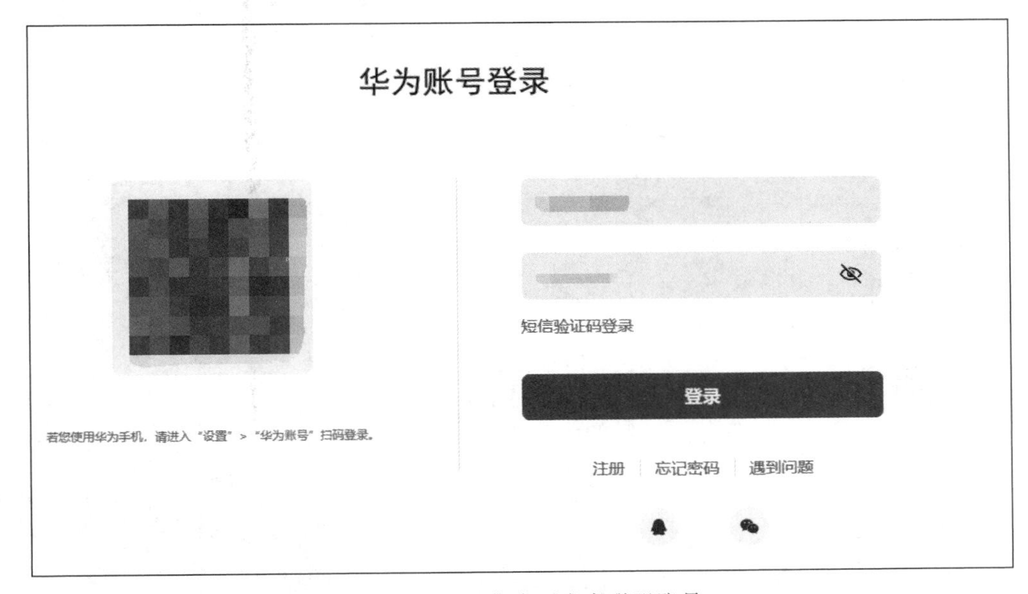

图 2-18　登录华为开发者联盟账号

步骤 7：登录后，单击界面的“允许”按钮进行授权，如图 2-19 所示。

步骤 8：在设备列表中，选择 Wearable 设备，并单击放大按钮，运行模拟器，如图 2-20 所示。

步骤 9：单击 DevEco Studio 工具栏中的单击放大按钮运行工程，或使用默认快捷键 Shift+F10(Mac 系统为 Control+R)运行工程。

图 2-19　登录华为账号授权

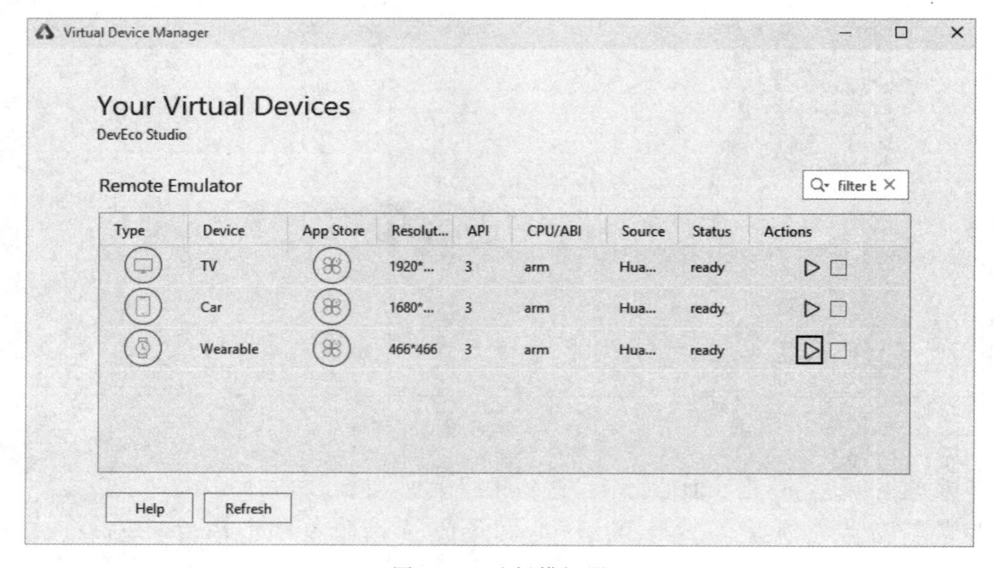

图 2-20　选择模拟器

步骤 10：在弹出的 Select Deployment Target 界面选择已启动的模拟器，单击 OK 按钮，如图 2-21 所示。

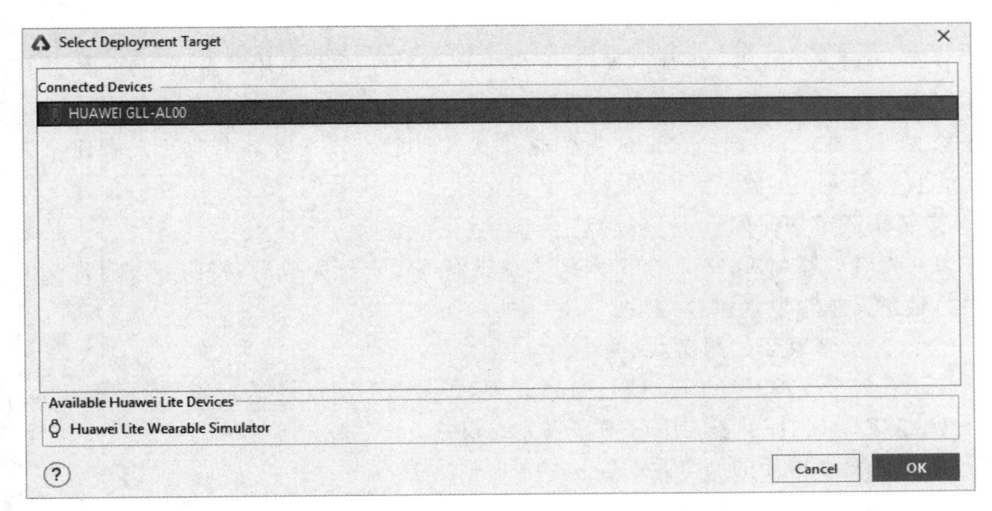

图 2-21 启动模拟器

步骤 11：DevEco Studio 会启动应用的编译构建，完成后应用即可运行在模拟器上，如图 2-22 所示。

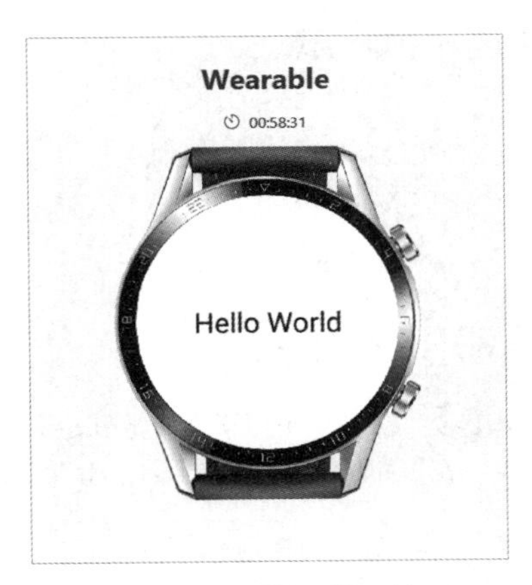

图 2-22 模拟器运行

注意：下载 JavaScript SDK 时，JavaScript 依赖通常下载缓慢。对于国内用户，可以将 npm 仓库设置为华为云仓库。在命令行工具中执行如下命令，重新设置 npm 仓库网址后，再执行 JavaScript SDK 的下载。npm config set registry https://mirrors.huaweicloud.com/repository/npm/

2.2 华为开发者账号申请

安装好 DevEco Studio 开发工具后，开发者还需要注册华为开发者联盟账号，并且实名认证才能享受联盟开放的各类能力和服务。

账号注册完成后，可选择认证成为企业开发者或个人开发者。企业开发者比个人开发者享受的服务更多。

1. 实名认证审核时间

华为开发者联盟将在 1～3 个工作日内完成审核。审核完成后，将向开发者提交的认证信息中的联系人邮箱发送审核结果。

2. 个人开发者享受的服务/权益

个人开发者享受应用市场、主题、商品管理、账号、PUSH（推送）、新游预约、互动评论、社交、HUAWEI HiAI、手表应用市场等服务或权益。

3. 企业开发者享受的服务/权益

企业开发者享受应用市场、主题、首发、支付、游戏礼包、应用市场推广、商品管理、游戏、账号、PUSH（推送）、新游预约、互动评论、社交、HUAWEI HiAI、手表应用市场、运动健康、云测、智能家居等服务或权益。

2.3 鸿蒙应用程序运行调试

鸿蒙应用程序运行调试，可以通过 DevEco Studio 提供的模拟器供开发者运行和调试鸿蒙应用，对于 Phone、Tablet、Car、TV 和 Wearable 可以使用 Remote Emulator 运行应用，对于 Lite Wearable 和 Smart Vision 可以使用 Simulator 运行应用。

2.3.1 在远程模拟器中运行应用

在 DevEco Studio 菜单栏，单击 Tools→HVD Manager。首次使用 Remote Emulator 需下载相关资源，如图 2-23 所示，单击 OK 按钮，等待资源下载完成后，需重新单击 Tools→HVD Manager。

华为允许开发者每次使用 Remote Emulator 的时长为 1h，到期后会自动释放资源，因此需要及时完成 HarmonyOS 应用的调试。如果 Remote Emulator 到期后被释放，则开发者可以重新申请资源。

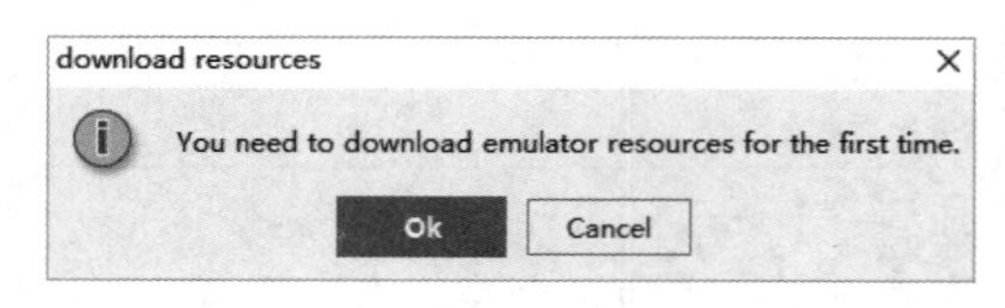

图 2-23 等待资源下载

在浏览器中弹出华为开发者联盟账号登录界面，输入已实名认证的华为开发者联盟账号的用户名和密码进行登录。

注意：使用 DevEco Studio 远程模拟器需要华为开发者联盟账号进行实名认证，建议在注册华为开发者联盟账号后，立即提交实名认证审核，认证方式包括“个人实名认证”和“企业实名认证”。

登录后，单击界面的“允许”按钮进行授权，如图 2-24 所示。

图 2-24 开发者单击允许按钮进行授权

单击已经连接的 Remote Emulator 设备并运行按钮 ▷，如图 2-25 所示，启动远程模拟设备(同一时间只能启动一个设备)。

单击 DevEco Studio 的 Run→ Run‘模块名称’或 ▶，或使用默认快捷键 Shift＋F10 (Mac 系统为 Control＋R)。

在弹出的 Select Deployment Target 界面选择已启动的 Remote Emulator 设备，如图 2-26 所示，单击 OK 按钮。

DevEco Studio 会启动应用的编译构建，完成后应用即可运行在 Remote Emulator 上，如图 2-27 所示。

Remote Emulator 侧边栏按钮的作用如下。

× ：释放当前正在使用的 Remote Emulator，单次使用时长为 1h。

⧉ ：设置 Remote Emulator 设备分辨率。

○ ：返回设备主界面。

◁ ：后退按钮。

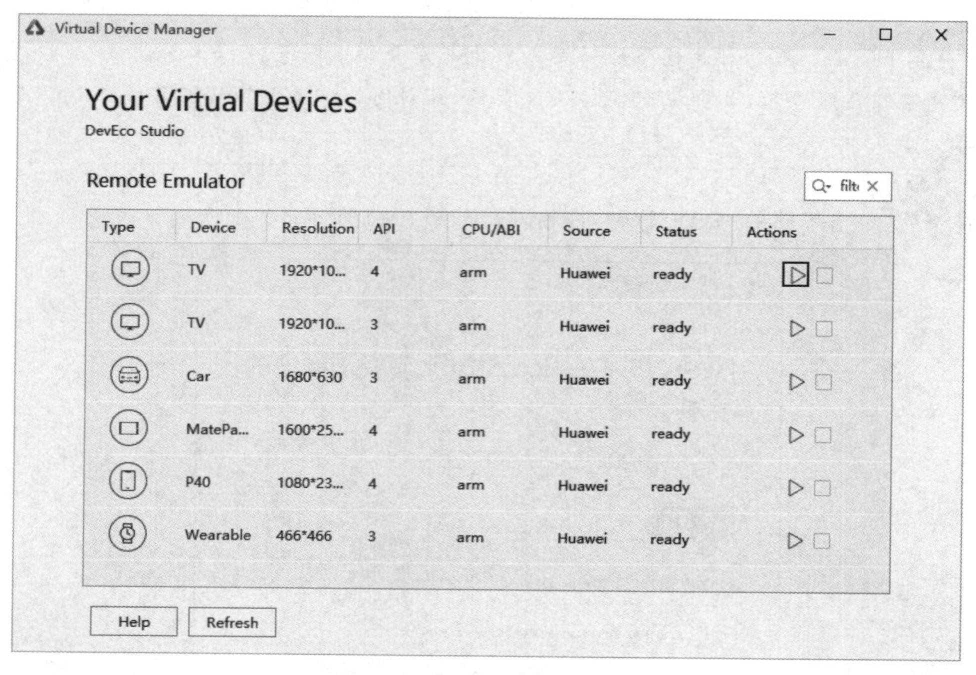

图 2-25　选择远程虚拟设备

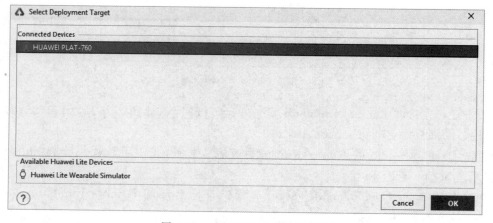

图 2-26　选择已识别的设备名称

2.3.2　在 Simulator 中运行应用

DevEco Studio 提供的 Simulator 可以运行和调试 Lite Wearable 和 Smart Vision 设备的 HarmonyOS 应用。在 Simulator 上运行应用兼容签名与不签名两种类型的 HAP。

单击 DevEco Studio 的 Run→Run'模块名称'或 ▶，或使用默认快捷键 Shift＋F10（Mac 系统为 Control＋R）。

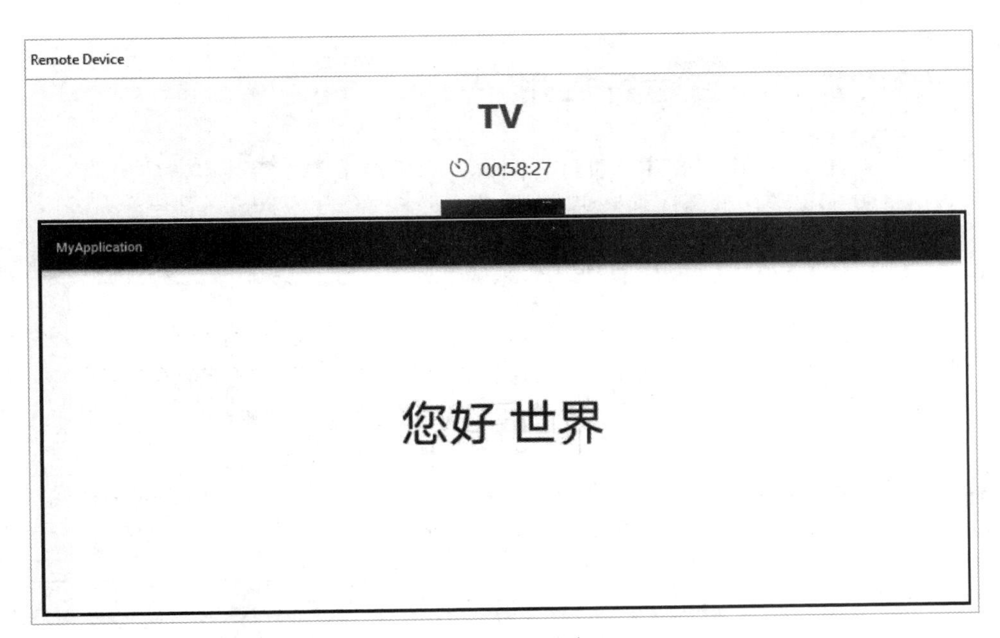

图 2-27 模拟器界面

在弹出的 Select Deployment Target 界面的 Available Huawei Lite Devices 设备列表中，选择需要运行的设备，如图 2-28 所示，单击 OK 按钮。

DevEco Studio 会启动应用的编译构建，完成后应用即可运行在 Simulator 上，如图 2-29 所示。

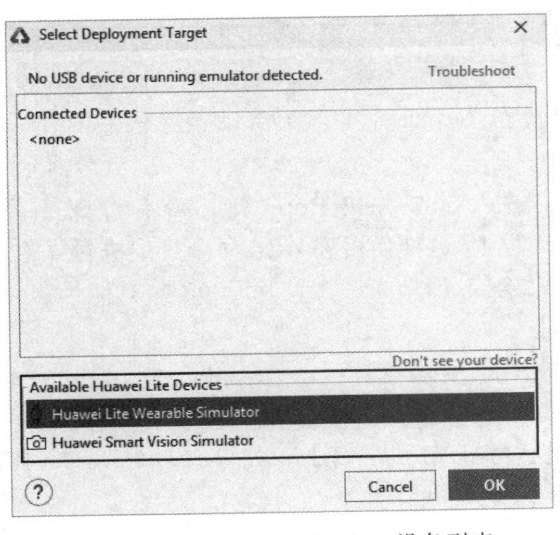

图 2-28 Huawei Lite Devices 设备列表

图 2-29 Huawei Lite Devices 设备模拟器

2.4 使用真机设备运行应用

使用模拟器测试后，还需要在真机进行测试。真机测试首先需要申请应用调试证书，下面将讲解申请真机测试证书的详细流程。

2.4.1 安装真机投屏软件

scrcpy Android 投屏是通过 adb 调试的方式将手机屏幕投到计算机上，并可以通过计算机控制 Android 设备。它可以通过 USB 连接，也可以通过 WiFi 连接(类似于隔空投屏)，而且不需要任何 root 权限，不需要在手机里安装任何程序。scrcpy 同时适用于 GNU/Linux、Windows、macOS 和鸿蒙操作系统。

首先需要下载 scrcpy，目前最新版本为 v1.10，如图 2-30 所示。下载网址为 https://github.com/Genymobile/scrcpy/releases。

图 2-30　下载 scrcpy

在解压后的目录中，adb.exe 为 adb 调试程序，scrcpy.exe 为启动投屏软件，如图 2-31 所示。

打开 cmd 并定位到此目录(在网址栏中输入 cmd 并按 Enter 键)，或者将该目录，如 D:\test\scrcpy-win64-v1.10 加入系统环境变量中，这样便可在 cmd 命令行中进行操作。

大多数手机默认禁止 ADB 通过网络与之连接，因此第一次使用 ADB 时只能通过 USB 数据线连接，如图 2-32 所示。

有些手机还需要再打开一个关于调试模式的附加选项——USB 调试(安全设置)，如图 2-33 所示，否则将无法在计算机上操控手机。

在命令行中运行 scrcpy，手机屏幕即可投射到计算机屏幕上。如果有多个设备，则需要指定序列号，序列号可以通过 adb devices 命令获得，代码如下：

```
scrcpy -s a1171b8    #此处的序列号已通过 adb devices 命令获得
```

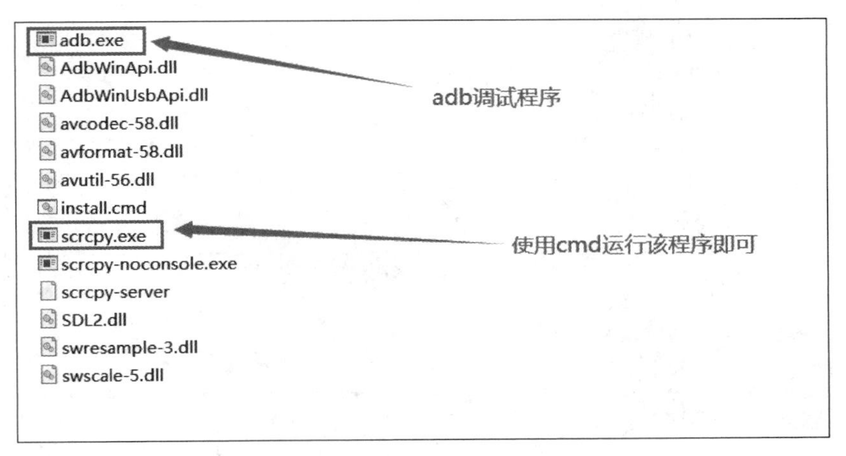

图 2-31 解压后的目录

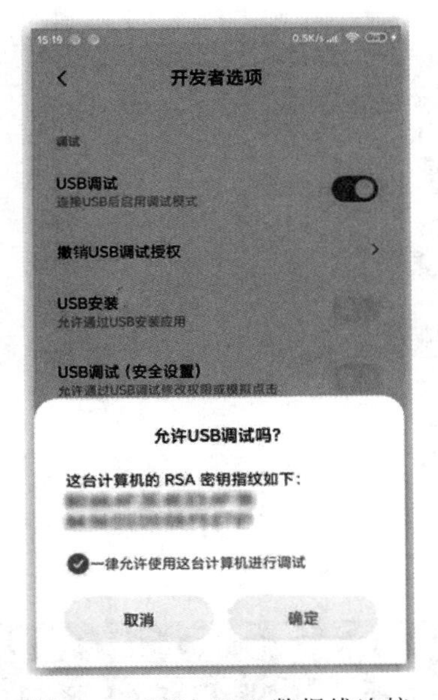

图 2-32 通过 USB 数据线连接

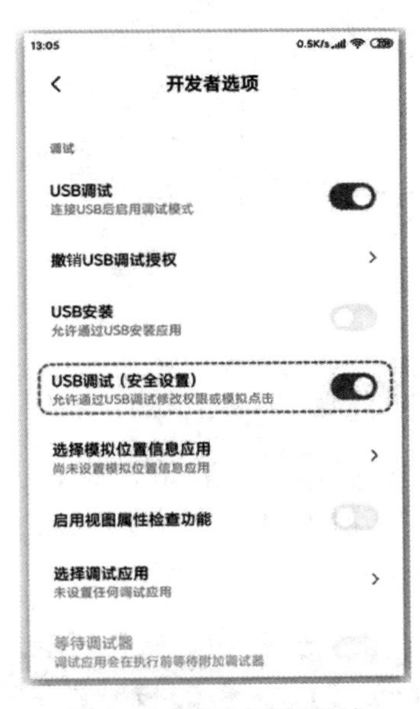

图 2-33 开启 USB 调试

2.4.2 真机设备测试流程

使用真机进行项目测试，需要申请应用调试证书，具体申请流程如图 2-34 所示。

注意：目前只有受邀请开发者才能访问 HarmonyOS 应用相关菜单，如果 AGC 页面未展示文档中的菜单，需联系华为运营人员（邮箱：agconnect@huawei.com）。

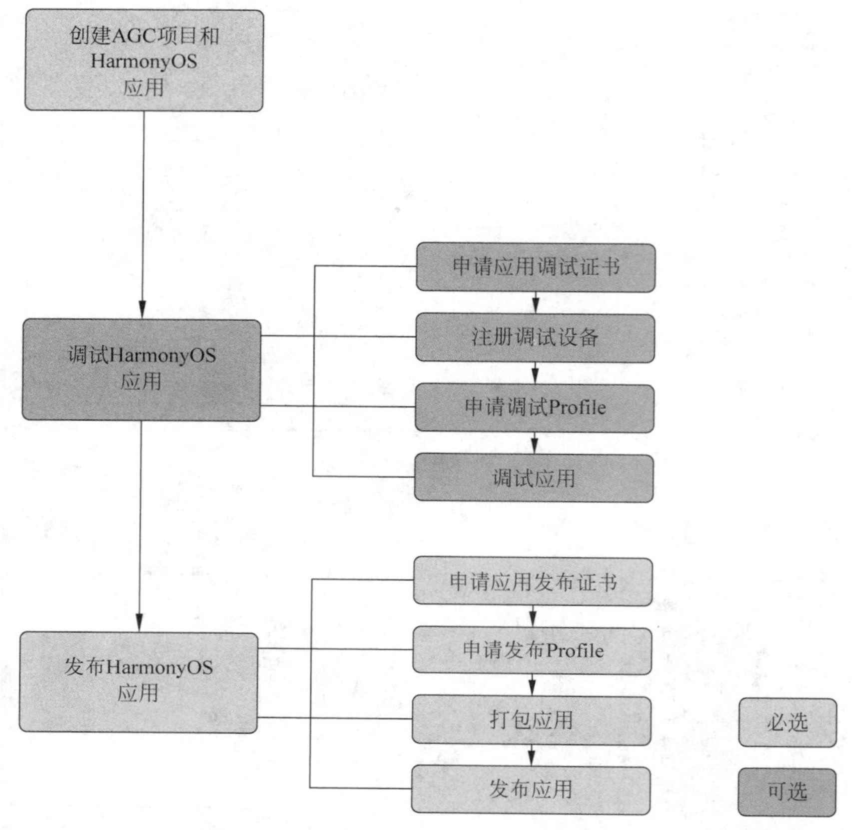

图 2-34　真机设备测试流程

申请真机测试的流程相对比较复杂，开发人员可按照下面的步骤完成申请流程。

步骤 1：创建 HarmonyOS 应用项目。首先需要创建一个鸿蒙应用项目，通过 DevEco Studio 创建，如图 2-35 所示。

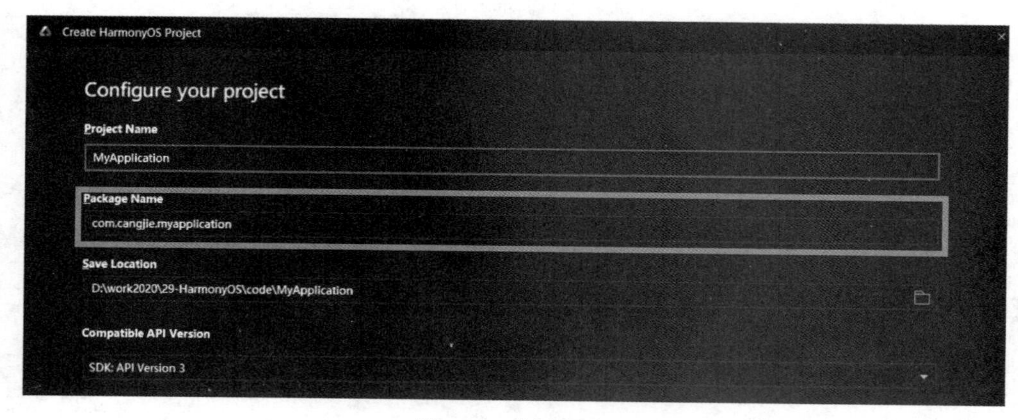

图 2-35　创建项目

这里需要注意 Package Name，Package Name 框所填信息用于生成应用签名信息。

步骤 2：使用 DevEco Studio 生成证书请求文件。在主菜单栏单击 Build→Generate Key，如图 2-36 所示。

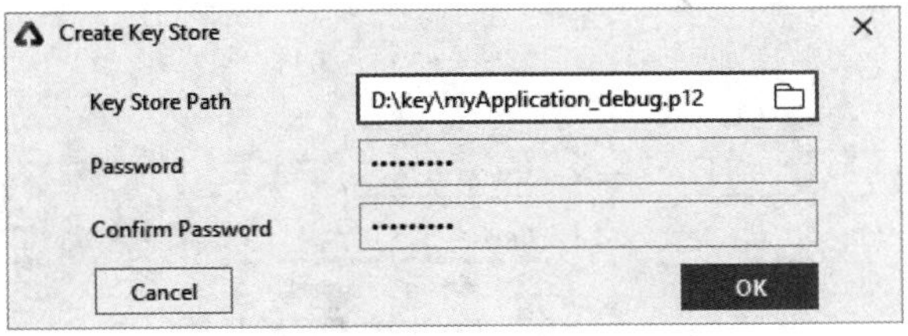

图 2-36 生成 p12 文件

在 Generate Key 界面中，继续填写密钥等信息，填写完毕后单击 Generate Key and CSR 按钮，如图 2-37 所示。

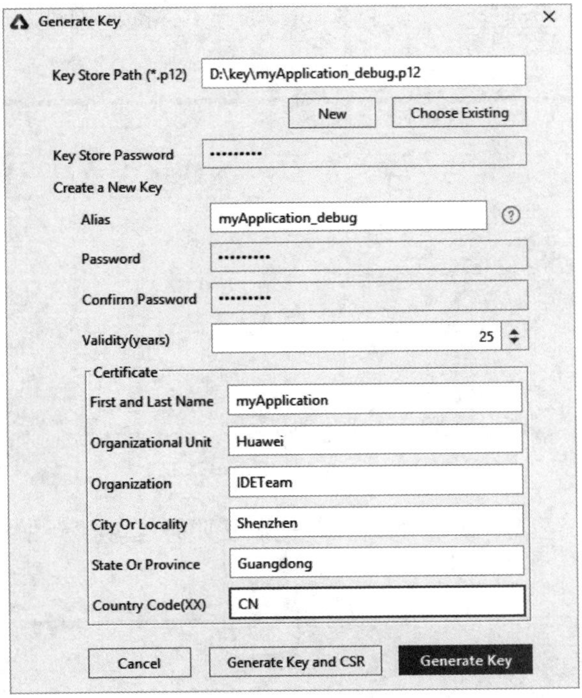

图 2-37 生成 csr 文件

在弹出的窗口中，单击 CSR File Path 对应的图标，选择 CSR 文件存储路径，如图 2-38 所示。

单击 OK 按钮，创建 csr 文件成功，工具会同时生成密钥文件(.p12)和证书请求文件(.csr)，如图 2-39 所示。

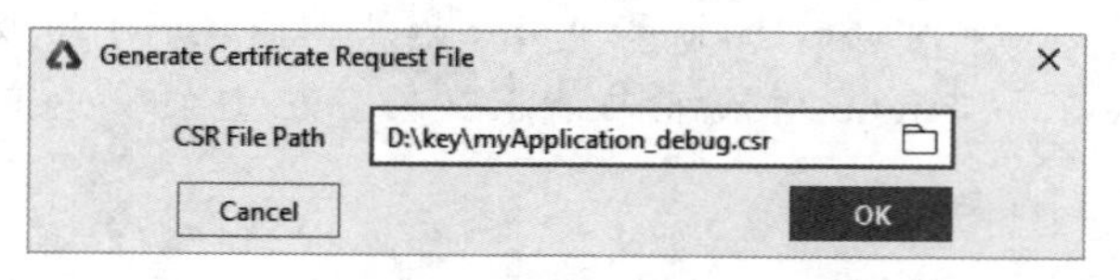

图 2-38　保存 csr 文件

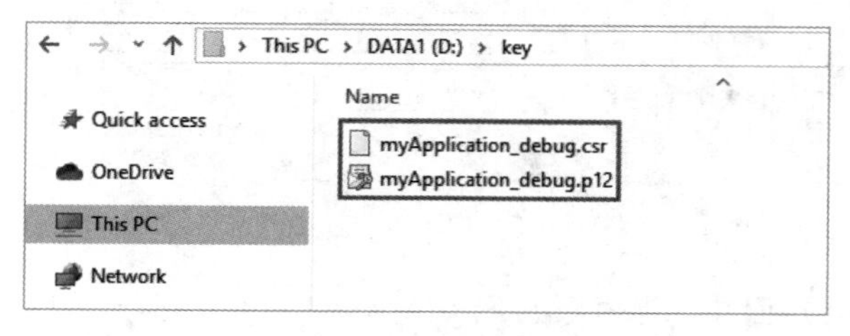

图 2-39　生成的(.p12)和证书请求文件(.csr)

步骤 3：申请应用调试证书。登录 AppGallery Connect 网站，选择“用户与访问”，如图 2-40 所示。

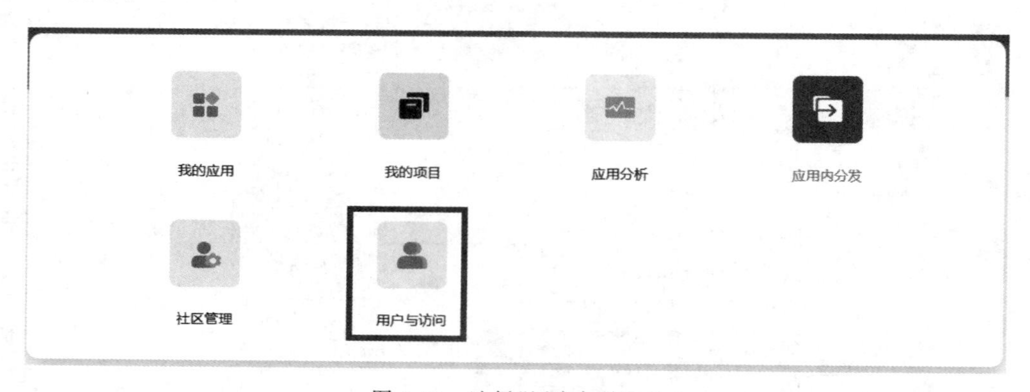

图 2-40　选择“用户与访问”

在左侧导航栏选择“证书管理”，进入证书管理页面，单击“新增证书”按钮，如图 2-41 所示。

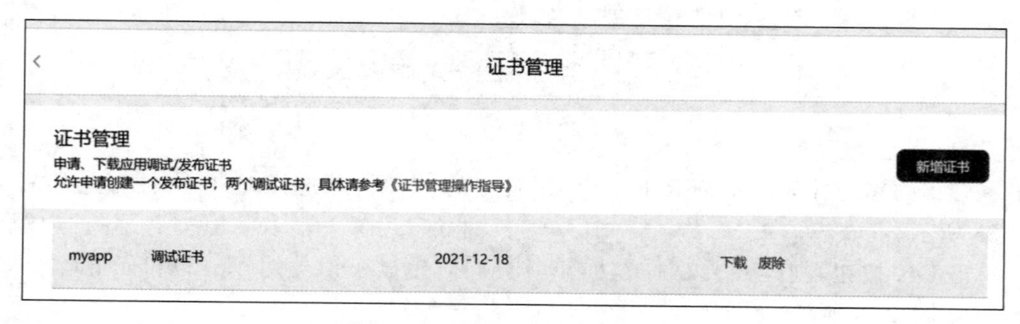

图 2-41　选择“新增证书”按钮

在弹出的"新增证书"窗口，填写要申请的证书信息，填写完毕后单击"提交"按钮，如图2-42所示。

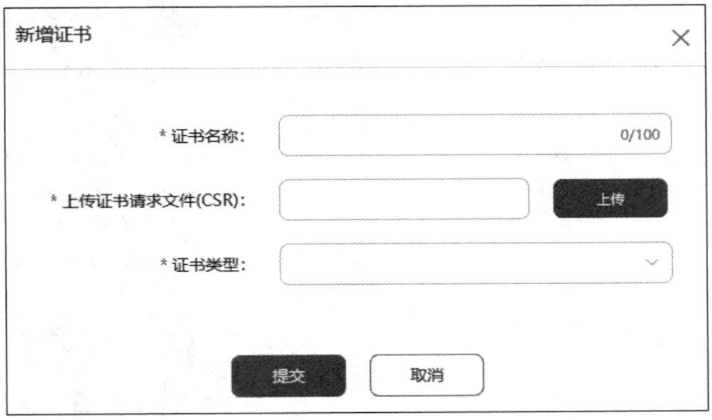

图2-42 填写证书信息

在左侧导航栏选择"设备管理"，进入设备管理页面，单击右上角的"添加设备"按钮，如图2-43所示。

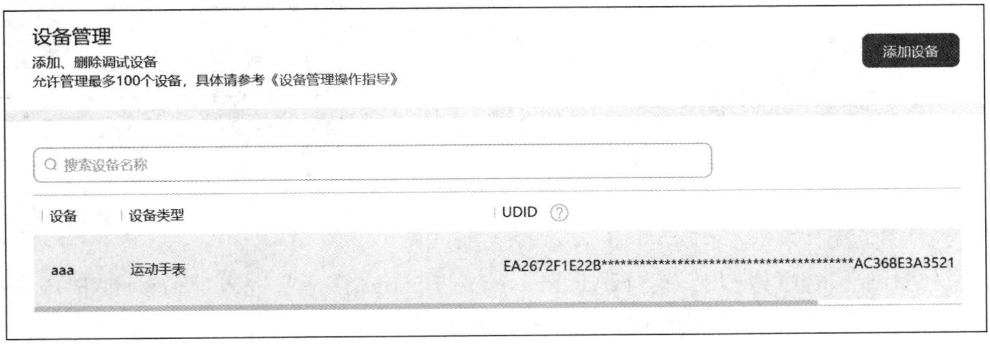

图2-43 选择"添加设备"

在弹出的窗口中填写设备信息，填写完毕后需单击"提交"按钮，如图2-44所示。

图2-44 填写设备信息

通过 adb 命令查看 UDID 信息，命令如下：

```
adb shell dumpsys DdmpDeviceMonitorService
```

步骤 4：申请应用调试证书和 Profile。登录 AppGallery Connect 网站，选择“我的项目”，如图 2-45 所示。

图 2-45 选择“我的项目”

提示：当前在同一个项目下可以创建多个应用，这样就可以共用之前生成的 csr 文件和密钥文件(.p12)，新的应用只需生成 Profile 文件就可以了。

找到你的项目，单击所创建的 HarmonyOS 应用。

选择“HarmonyOS 应用”→“HAP Provision Profile 管理”，进入“管理 HAP Provision Profile”页面，单击右上角“添加”按钮，如图 2-46 所示。

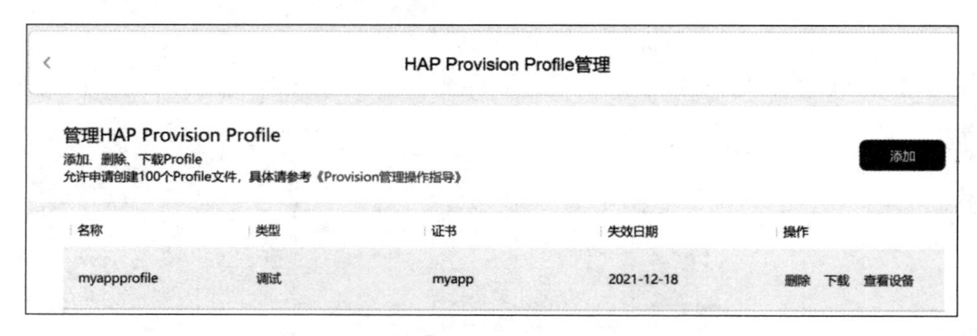

图 2-46 选择并添加 HAP Provision Profile

在弹出的 HarmonyAppProvision 信息窗口添加调试 Profile，如图 2-47 所示。

调试 Profile 申请成功后，管理 HAP Provision Profile 页面会展示 Profile 名称、Profile

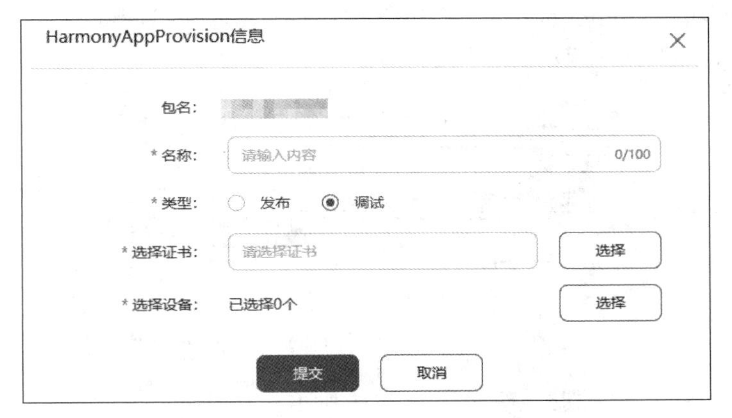

图 2-47　添加调试 Profile

类型、添加的证书和失效日期。下载生成的 Profile 文件，如图 2-48 所示。

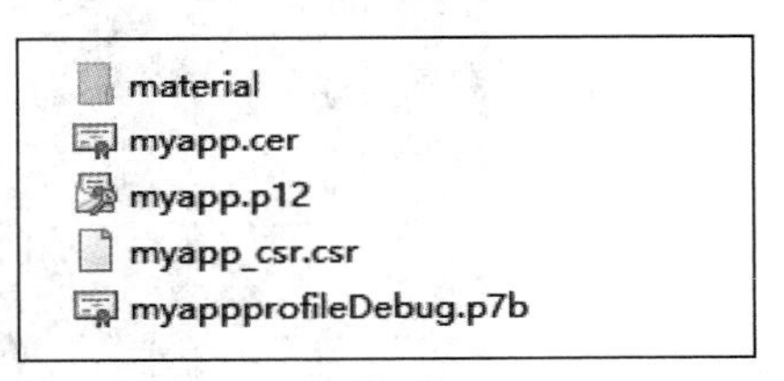

图 2-48　所有的证书文件列表

步骤 5：构建类型为 Debug 的 HAP(带调试签名信息)。打开 File→Project Structure，在 Modules→entry→Signing Configs→debug 窗口中，配置指定模块的调试签名信息，如图 2-49 所示。

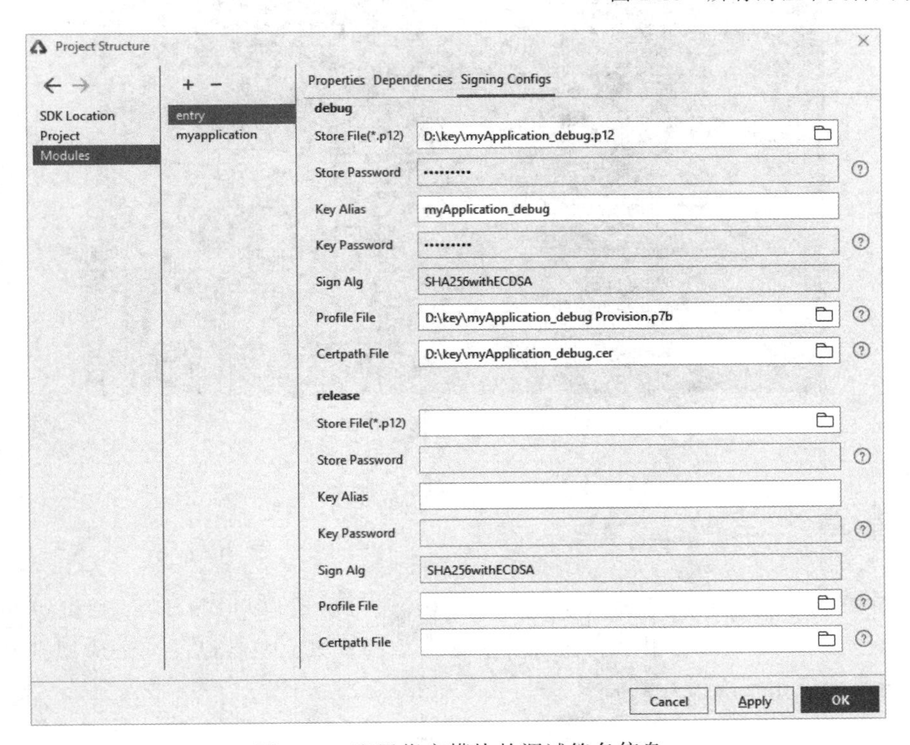

图 2-49　配置指定模块的调试签名信息

在主菜单栏，单击 Build→Build APP(s)/Hap(s) →Build Debug Hap(s)，生成已签名的 Debug HAP，如图 2-50 所示。

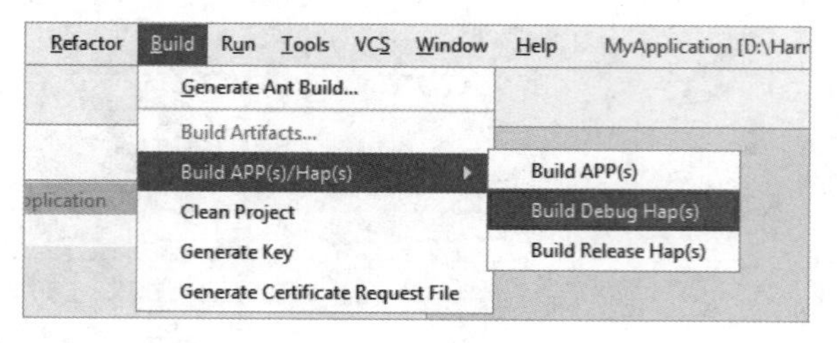

图 2-50 生成已签名的 Debug HAP

步骤 6：运行程序并在真机查看，如图 2-51 所示。

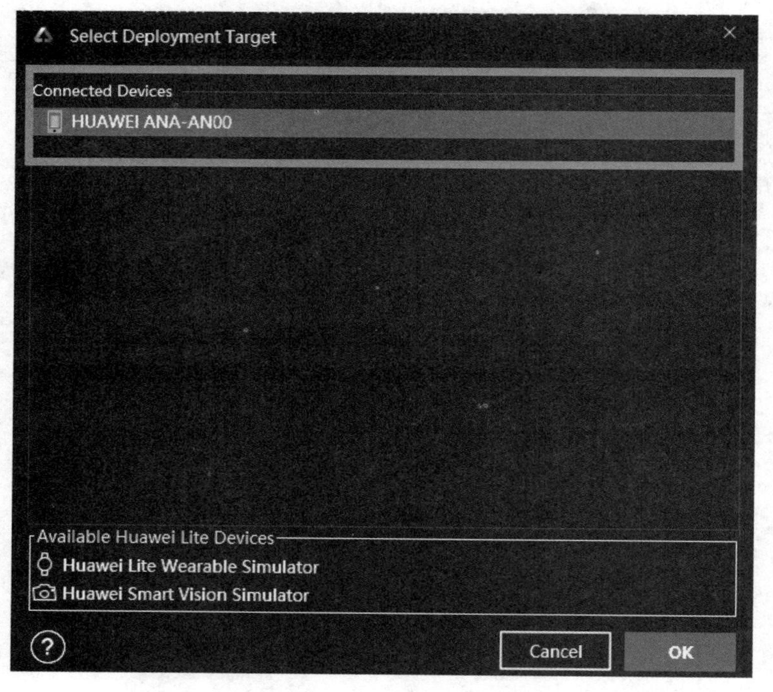

图 2-51 通过选择真机运行

2.5 本章小结

本章介绍鸿蒙应用开发环境搭建、华为开发者账号申请、模拟器使用、真机投屏软件安装及华为真机设备测试证书申请。读者可根据本章的步骤完成自己的鸿蒙应用开发环境搭建，为后面鸿蒙应用开发做好准备。

第二篇　基础知识篇

学习目标

通过本篇的学习,你将可以基于鸿蒙 ACE 框架开发鸿蒙的应用程序。

主要内容如下:

- 鸿蒙 ACE Java UI 框架的详细使用;
- 鸿蒙 ACE JavaScript UI 框架的详细使用;
- 鸿蒙分布式 Ability 开发;
- 鸿蒙的媒体及硬件开发。

如何学习本篇:

- 学习本章的读者可结合每节中的案例反复编码练习。

第3章

鸿蒙 ACE Java 应用框架

本章将详细讲解鸿蒙 ACE Java UI 框架的使用，读者通过本章的学习将对鸿蒙 Java UI 框架有深入的了解。ACE Java UI 框架是鸿蒙 ACE(Ability Cross-platform Environment，鸿蒙面向元能力(Ability)的跨平台开发运行环境)的重要组成部分，用来开发鸿蒙富设备应用程序。

3.1 ACE 运行时简介

ACE 全称为 Ability Cross-platform Environment，是鸿蒙面向元能力(Ability)开发的跨平台运行环境，如图 3-1 所示。

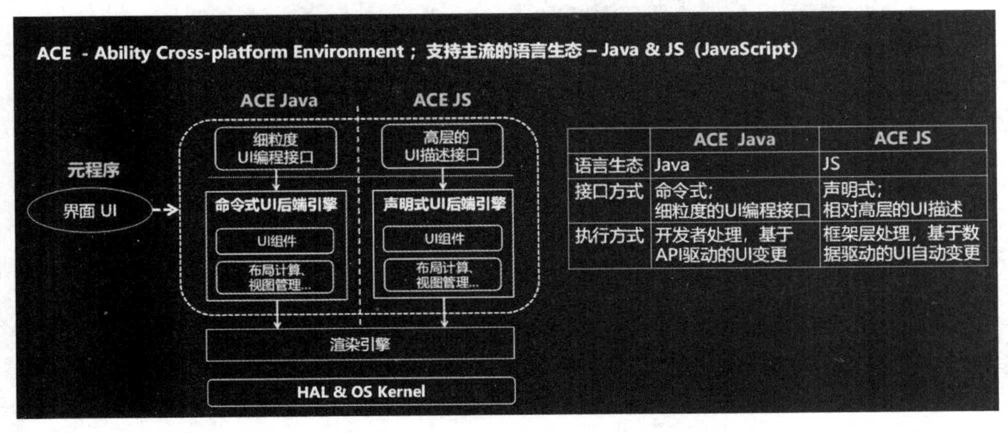

	ACE Java	ACE JS
语言生态	Java	JS
接口方式	命令式；细粒度的UI编程接口	声明式；相对高层的UI描述
执行方式	开发者处理，基于API驱动的UI变更	框架层处理，基于数据驱动的UI自动变更

图 3-1　鸿蒙 ACE Run-time 架构图

ACE 是为华为鸿蒙操作系统定制设计的一套针对全场景环境的跨平台应用开发框架。ACE 支持目前两种主流语言：Java 和 JavaScript。未来将支持华为自己的开发语言“仓颉”。

3.1.1 ACE针对全场景开发

鸿蒙应用开发针对的是1+8+N的全场景设备：1指的是手机；8代表车机、音箱、耳机、手表/手环、平板、智慧屏、PC、AR/VR；N泛指其他IoT设备。

这里可以把这些IoT终端设备分为富设备和轻设备，对于富设备开发鸿蒙提供了基于Java、C++语言的开发框架，对于一些内存有限，功能有限的轻设备则提供了基于JavaScript的开发框架，这样可以让一些低端硬件配置的IoT设备运行鸿蒙应用程序。

3.1.2 ACE支持的两种UI框架

鸿蒙跨平台运行时(ACE)为Java开发者和JavaScript开发者提供了一站式的开发环境。

1. 基于Java的应用开发框架(ACE Java UI)

鸿蒙的应用开发可以基于Java语言+XML的方式进行开发，此方式和安卓类似，如果开发者熟悉Java语言或者Android开发，那么使用此方式可以很快进行应用开发。ACE Java UI框架的架构图如图3-2所示。

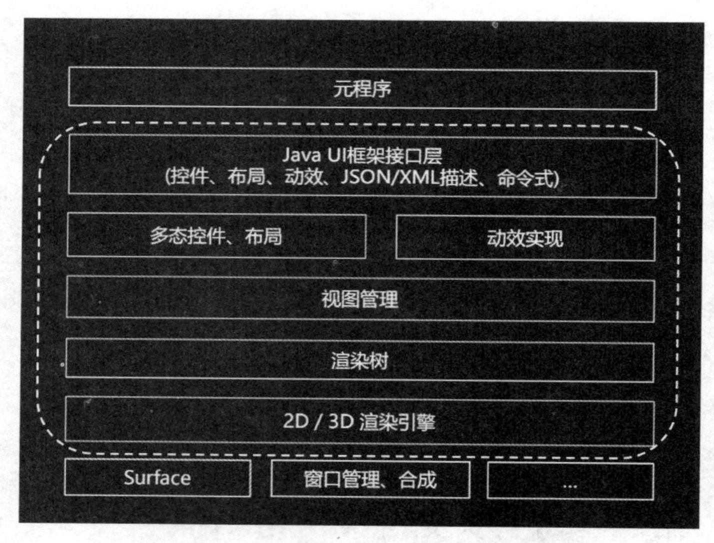

图3-2 鸿蒙ACE Java总体架构图

2. 基于JavaScript的应用开发框架(ACE JavaScript UI)

鸿蒙开发同时还提供了JavaScript形式的开发。JavaScript的语法规则基本遵循ES6的语法规范。如果是前端开发人员，则可以使用HTML+CSS+JavaScript的形式进行应用的开发，但是对于后端一些服务能力，目前只能使用Java语言进行开发，所以对于JavaScript的开发者，依然需要Java语言的辅助。ACE JavaScript UI框架的架构图如图3-3所示。

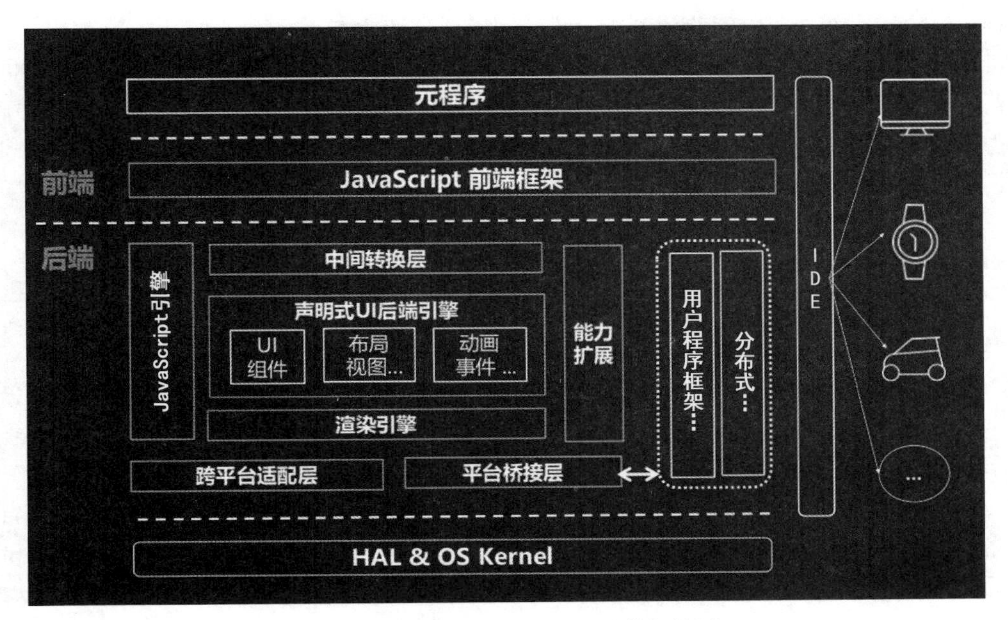

图 3-3 鸿蒙 ACE JavaScript 总体架构图

3.2 ACE 开发中的核心概念

在学习 ACE 开发框架前，需要了解 ACE 中的一些核心概念。

3.2.1 Ability 和 Slice

Ability 是 HarmonyOS 开发的核心，首先了解什么是 Ability，我们用生活中的例子来做一个类比，Ability 就好像是一个画板，初始什么都没有，一片空白，我们可以找一张画纸进行绘画，画完的画纸夹在画板上，画板上就会有我们创作的图画了。

HarmonyOS 中的 Ability 其实就相当于画板，它用来承载画纸，那么画纸又是什么？标题中还有一个 Slice，它就相当于画纸，我们写应用的 UI，写完后通过 Slice 加载布局，就相当于在画纸上绘画，完成后把画纸放在画板上进行展示。

将对应的概念引申到程序中，我们有一个 Ability（画板）用来向用户展示新闻，同时这个 Ability 拥有两个 Slice（画纸），一个用来展示新闻标题，另一个用来展示新闻详情。那么我们就需要在这两个 Slice 中分别画上新闻的标题和详情，然后通过单击或其他交互方式来切换两个 Slice，相当于给画板换另一张画纸。

3.2.2 Ability 分类

有了上述概念，我们基本知道 Ability 能做些什么了，但是还要具体了解一下这个核心组件，在 HarmonyOS 中，Ability 可以分为两类，一类叫 FA(Feature Ability)，另一类叫作 PA(Particle Ability)，这两类 Ability 最直观的区别是 FA 有界面，可以和用户交互，而 PA 无界面，它提供一些应用运行需要的能力，例如读取数据。这两类 Ability 在创建的时候都有相应的模板。

Ability 的模板种类。

(1) 对于 FA 类型的 Ability，Page 类型的模板是其唯一的模板类型，这种类型的模板提供了和用户交互的能力。

(2) 对于 PA 类型的 Ability，它有两类模板。

- Service 模板：用于提供后台运行任务的能力。
- Data 模板：用于对外部提供统一的数据访问抽象。

3.2.3 鸿蒙应用包结构

鸿蒙应用软件包以 App Pack(Application Package)形式发布，它是由一个或多个 HAP (HarmonyOS Ability Package)及描述每个 HAP 属性的 pack.info 组成。HAP 是 Ability 的部署包，HarmonyOS 应用代码围绕 Ability 组件展开。一个 HAP 是由代码、资源、第三方库及应用配置文件组成的模块包，可分为 entry 和 feature 两种模块类型，如图 3-4 所示。

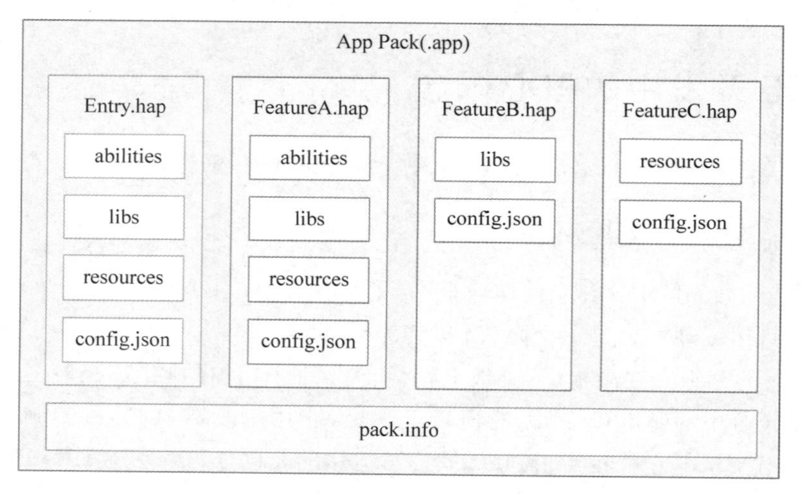

图 3-4 鸿蒙应用包结构

entry：应用的主模块。在一个 App 中，对于同一设备类型必须有且只有一个 entry 类型的 HAP，可独立安装运行。

feature：应用的动态特性模块。一个 App 可以包含一个或多个 feature 类型的 HAP，也可以不含。只有包含 Ability 的 HAP 才能够独立运行。

有了上述概念，3.3 节将分析之前创建的项目的结构和项目中配置文件的配置内容具体有什么含义。

3.3 创建一个 ACE Java 项目

视频讲解

在对 HarmonyOS 应用程序有了一个初步认知之后，我们使用 DevEco Studio 来创建一个项目，把项目运行起来，先从整体上来了解一下 HarmonyOS 项目的整体结构及开发工具的基本使用。

环境和工具配置好后，就可以创建一个项目了。先创建一个项目，从整体上了解一下 HarmonyOS 应用的整体框架。我们在一个布局里放置一个文本框，用于显示一个数字，再添加一个按钮，每次单击按钮让文本框中的数字加 1。通过这样一个小程序简单演示工具的使用和项目的基本框架。

3.3.1 新建 ACE Java 项目

选择 File→New→New Project 进行项目创建，会弹出如图 3-5 所示的窗口。

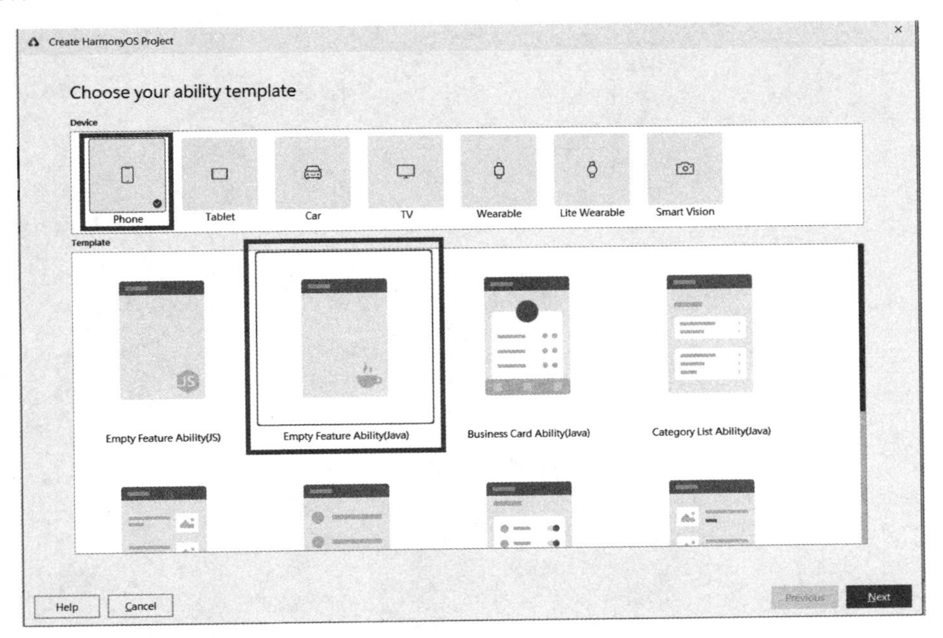

图 3-5　新建 ACE Java 项目

创建项目窗口可分为两块，其中 Device 表示目前支持的设备。设备列表中从左到右依次为手机、平板、车机、电视机、智慧屏、穿戴设备、轻型穿戴设备等。

上边是新建项目时供选择的模板，因为笔者使用 Java 开发，所以选择第二个(Empty Feature Ability Java)，单击 Next 按钮进入下个页面，如图 3-6 所示。

配置完项目名、包名、使用的 SDK 版本及项目的保存路径后，单击 Finish 按钮即可，创建完成后项目会自动构建。构建成功后项目整体结构如图 3-7 所示。

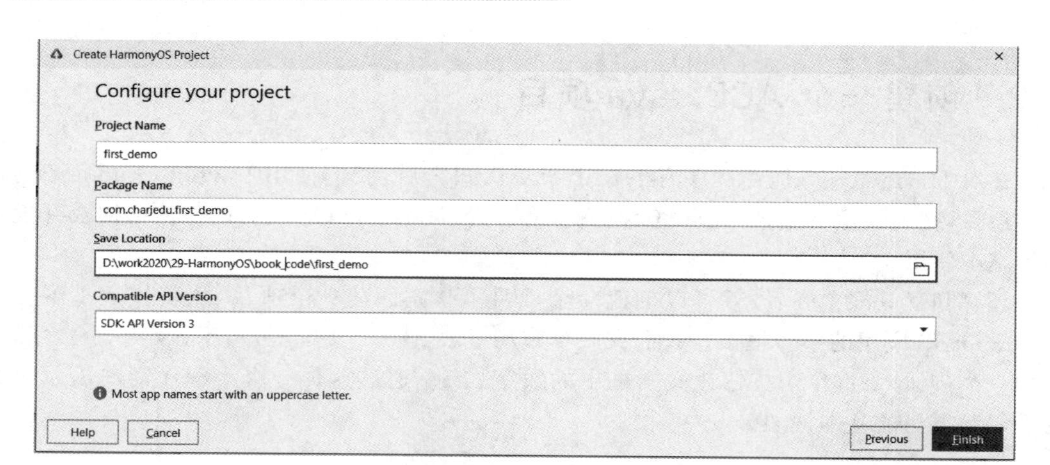

图 3-6　选择项目模板

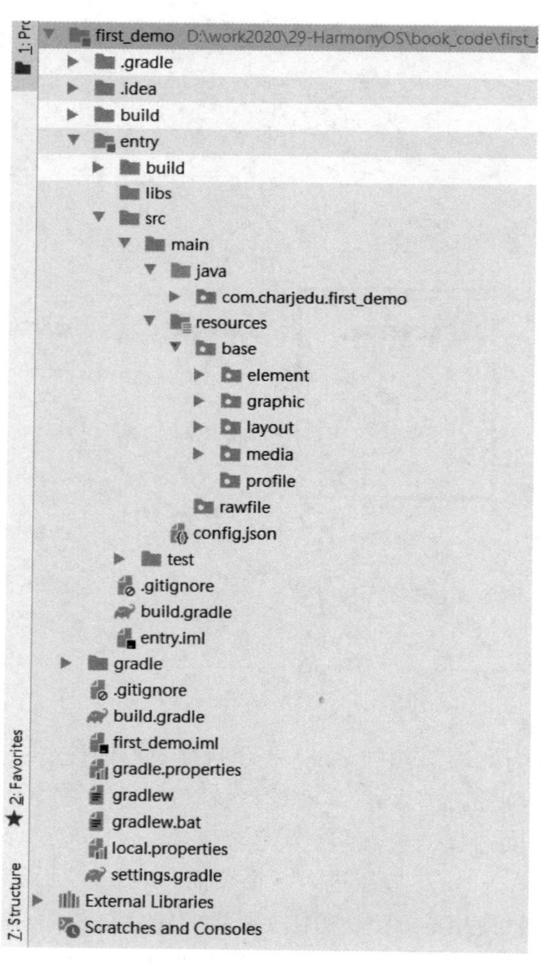

图 3-7　ACE Java 项目目录结构

先完成我们的功能,后续再了解目录及其作用。首先打开 resources/base/layout 目录中自动生成的布局文件,然后单击工具右侧的预览,如图 3-8 所示。

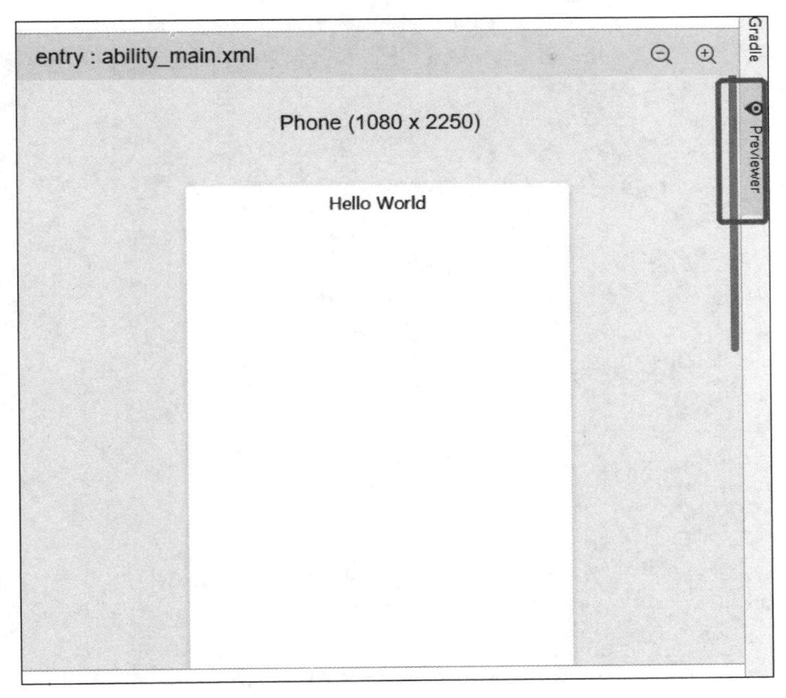

图 3-8　ACE Java 项目预览

这是工具提供的一个预览器,当写 xml 布局的时候可以近乎实时地看到我们所写代码的效果。

3.3.2　编写界面布局

在 ability_main.xml 文件中,添加一个 Text 组件和一个 Button 组件。这里为 Button 组件添加了一个背景文件,通过 Button 属性 background_element 引入。

background_element 的值是 $graphic:background_ability_main。$graphic 表示引用 graphic 文件夹,background_ability_main 是 graphic 文件夹下的 xml 文件,如代码示例 3-1 所示。

代码示例 3-1　编写界面布局:first_demo/entry/layout/ability_main.xml

```
<?xml version = "1.0" encoding = "utf - 8"?>
< DirectionalLayout
    xmlns:ohos = "http://schemas.huawei.com/res/ohos"
    ohos:height = "match_parent"
```

```
    ohos:width = "match_parent"
    ohos:orientation = "vertical">

< Text
        ohos:id = " $ + id:text"
        ohos:height = "match_content"
        ohos:width = "match_content"
        ohos:layout_alignment = "horizontal_center"
        ohos:text = "1"
        ohos:text_size = "60fp"
        />
< Button
        ohos:height = "match_content"
        ohos:width = "match_content"
        ohos:text_size = "18fp"
        ohos:id = " $ + id:btn_add"
        ohos:layout_alignment = "horizontal_center"
        ohos:background_element = " $ graphic:background_ability_main"
        ohos:text_color = "white"
        ohos:padding = "15vp"
        ohos:text = "单击加 1"/>
</DirectionalLayout >
```

此外还需要修改 graphic 目录下 background_ability_main. xml 文件，将背景设置为 10 像素的圆角，将填充色设置为 # 007CFD，如代码示例 3-2 所示。

代码示例 3-2　设置背景：first_demo/entry/layout/background_ability_main. xml

```
<?xml version = "1.0" encoding = "utf - 8"?>
< shape xmlns:ohos = "http://schemas.huawei.com/res/ohos"
       ohos:shape = "rectangle">
< corners
        ohos:radius = "10"/>
< solid
        ohos:color = " # 007CFD"/>
</shape >
```

此时在预览器中看到的效果如图 3-9 所示。

3.3.3　编写界面逻辑代码

接下来实现“单击加 1”功能，打开由项目自动创建的 MainAbilitySlice 文件，如代码示例 3-3 所示。这里通过 Button 事件，每单击一次，设置“count++；”。

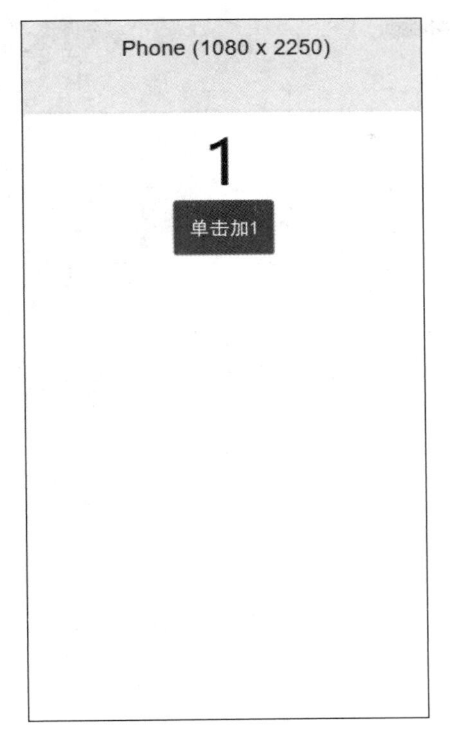

图 3-9 预览效果图

代码示例 3-3 设置背景：first_demo/entry/java/MainAbilitySlice. java

```
public class MainAbilitySlice extends AbilitySlice {
   private int count = 1;
@Override
 public void onStart(Intent intent) {
        super.onStart(intent);
        super.setUIContent(ResourceTable.Layout_ability_main);
        Text text = (Text) findComponentById(ResourceTable.Id_text);
        Button button = (Button) findComponentById(ResourceTable.Id_btn_add);
        button.setClickedListener(new Component.ClickedListener() {
            @Override
            public void onClick(Component component) {
                count++;
                text.setText(count + "");
            }
        });
    }

    @Override
```

```
    public void onActive() {
        super.onActive();
    }

    @Override
    public void onForeground(Intent intent) {
        super.onForeground(intent);
    }
}
```

3.3.4 通过模拟器预览效果

到此为止，功能已经开发完毕，单击工具上方菜单栏的 Tools 下边的 HVD Manager，弹出华为账号授权界面，如图 3-10 所示。

图 3-10 华为账号授权

这里需要登录已注册的华为账号并且授权，单击"允许"按钮，跳转到网页授权，授权完成后，会弹出下面的界面，如图 3-11 所示。

这里可以选择对应的远程设备，因为我们创建项目时选的是手机类型的项目，因此这里选手机 P40，单击后边的蓝色箭头，之后在工具右侧的预览区会变成所选择的手机 P40。

选择手机 P40 模拟器，如图 3-12 所示，等待模拟器启动完成后，就可以单击工具右上方的按钮启动项目了。

单击工具右上方的按钮启动项目，如图 3-13 所示。

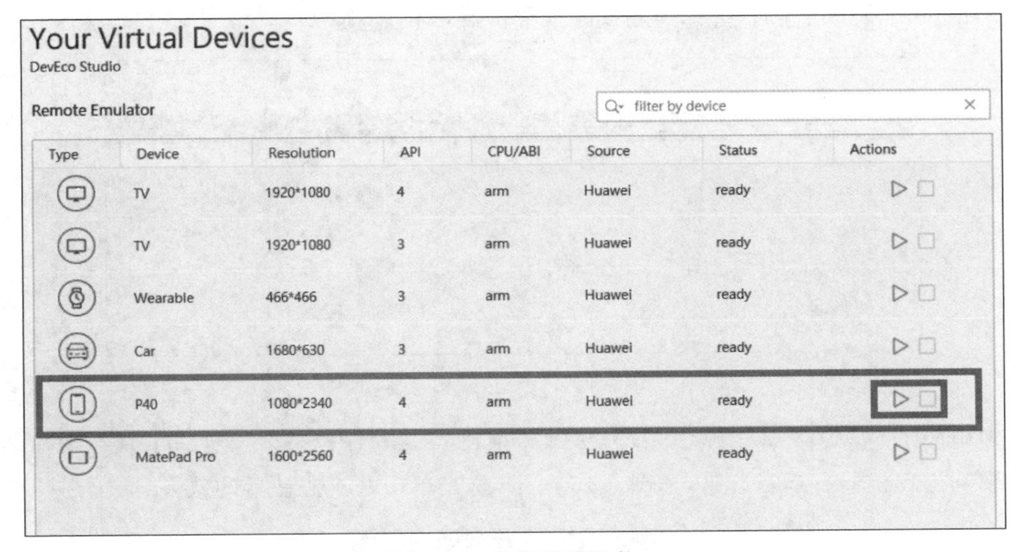

图 3-11 选择远程设备

图 3-12 P40 虚拟机预览效果

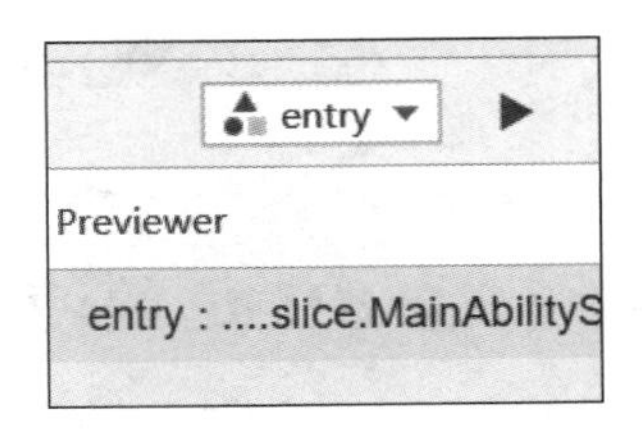

图 3-13 启动项目

单击启动编译完成后,选择要连接的远程机器,如图 3-14 所示。

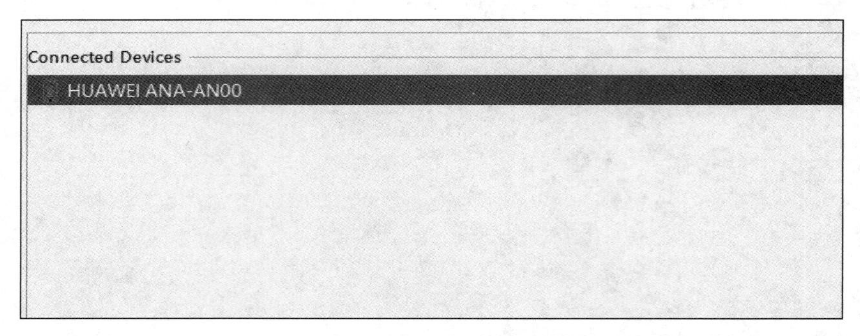

图 3-14　选择远程模拟器

图中的 HUAWEI ANA-AN00 就是刚才选择的 P40 远程机器。之后工具会把你的项目安装到远程机器并运行起来,如图 3-15 所示,这是单击两次按钮后的效果。

图 3-15　P40 预览项目效果

3.3.5　日志 HiLog 的使用

在编写程序的过程中,需要通过打印一条条日志来掌握程序运行的状态,下面就来讲解鸿蒙操作系统中的 HiLog 日志工具的具体使用方法。

使用 HiLog 前必须在 HiLog 的一个辅助类 HiLogLabel 中定义日志类型、服务域和标记。一般把它定义为常量并放在类的最上面，代码如下：

```
static final HiLogLabel label = new HiLogLabel(HiLog.LOG_APP, 0x00201, "MY_TAG");
```

上面的代码有 3 个参数：

(1) HiLog.LOG_APP(日志类型)，我们的应用一般取一个常量值：HiLog.LOG_APP，表示是第三方应用。

(2) 0x00201(服务域)，十六进制整数形式，取值范围是 0x0 ～0xFFFFF。一般情况下，我们建议把这个十六进制数分成两组，前面 3 个数表示应用中的模块编号，后面两个数表示模块中类的编号。

(3) "MY_TAG"(一个字符串常量)，它标识方法调用的类或服务行为。一般情况下写类的名字，可用这个标记对日志进行过滤。

1. 日志的级别

和其他日志一样，HiLog 也分为几个日志级别，越往下信息级别越高。

- debug：调试信息；
- info：普通信息；
- warn：警告信息；
- error：错误信息；
- fatal：致命错误信息。

2. 高级应用(private 和 public 修饰符)

```
String URL = "www.baidu.com";
int errno = 0;
HiLog.warn(label, "Failed to visit %{private}s, reason: %{public}d.", URL, errno);
```

private 所处的位置%{private}s，其中 s 是要输出的内容，输出时不显示内容。

public 所处的位置%{public}d，其中 d 是要输出的内容，输出时正确显示。

3.4 ACE Java 项目目录结构

本节介绍鸿蒙 ACE Java 应用开发的项目目录结构及其作用，以及配置文件的基础配置信息。

3.4.1 项目整体结构

通过 DevEco Studio 创建一个 ACE Java 项目后，在 Project 左边栏中预览项目结构如图 3-16 所示。

首先有一个 entry 目录，一个应用是由一个或多个 HAP 包所组成的，HAP 包又可以分

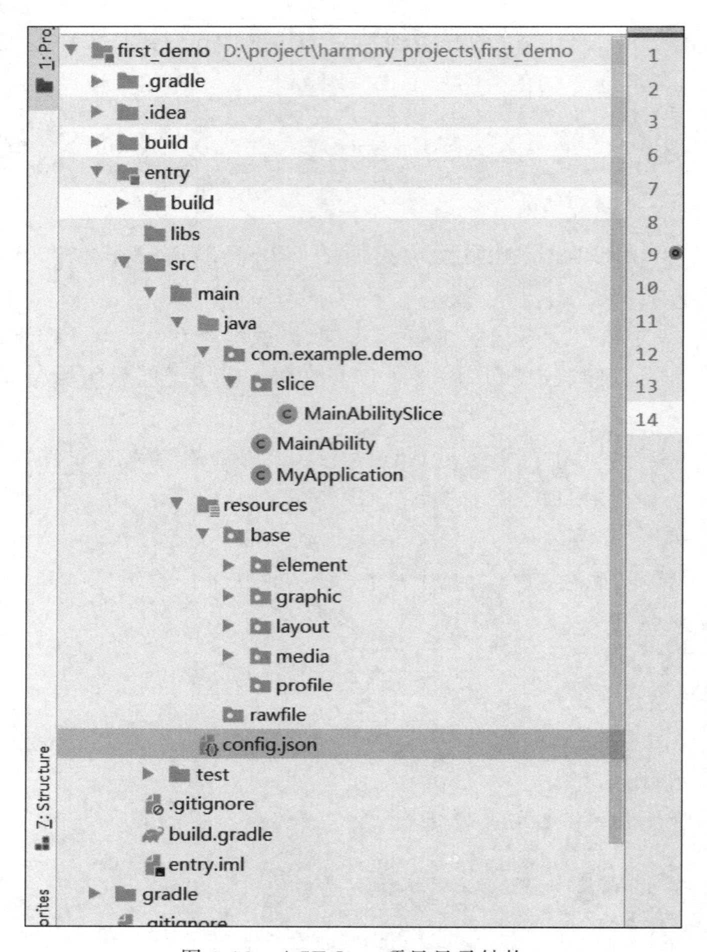

图 3-16 ACE Java 项目目录结构

为 entry 类型和 feature 类型，每个 HAP 包由代码、资源、第三方库及应用配置文件组成。所以代码中的 entry 目录其实就是一个应用的 HAP 包，它是一个 entry 类型的 HAP 包。接着我们来看这些资源、代码等分布在 entry 包的位置。

(1) 在 src/main/java 下以包名命名的文件夹内分布着 Java 代码。这里的代码可以用来创建布局、动态调整布局及为交互提供支撑服务。

(2) 和 java 文件夹同级的 resources 目录下分布着应用资源，该目录的 base 目录下，按资源用途又分为多个文件夹资源。

(3) element：表示元素资源，该文件夹下主要存放着 json 格式的文件，主要用来表示字符串、颜色值、布尔值等，可以在其他地方被引用。

(4) graphic：表示可绘制资源。用 xml 文件来表示，例如项目中设置的圆角按钮、按钮颜色等都是通过引用这里的资源统一管理的。

(5) layout：表示布局资源，用 xml 文件来表示，例如页面的布局资源存放在这里。

(6) media：表示媒体资源，包括图片、声频、视频等非文本格式的文件。

除了上述的几类资源，还有其他类型的资源，因为项目暂时还用不到，先不做考虑。

resources 目录存储的内容如图 3-17 所示。

和 main 目录平级的 test 目录是测试目录，可以用于对自己写的功能进行单元测试，确保代码的正确性。

和 src 平级的 libs 目录用来存储或引用第三方包，例如 jar 包、so 包等。

和 entry 目录平级的 build 目录则用来存放最终编译后的包，也就是 HAP 包，编译后该包的内容如图 3-18 所示。HAP 包中包含了项目中用到的图片、布局、代码和各种资源。

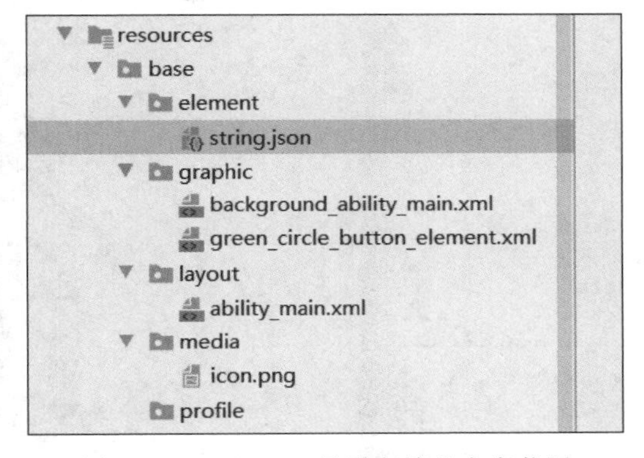

图 3-17　resources 目录存储的内容截图

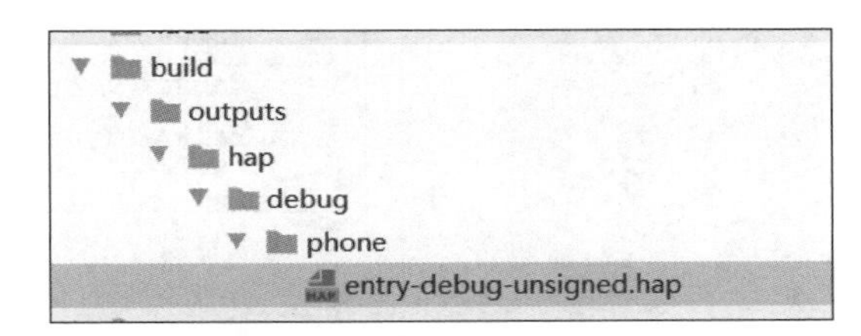

图 3-18　HAP 包

3.4.2　项目的配置文件

每个 HAP 包下都包含了该 HAP 包的配置信息，这个配置文件位于 entry/src/main/目录下，由工具帮我们生成，并被命名为 config.json，HarmonyOS 应用配置采用 json 格式的形式。下面我们来看一下这个配置文件中的内容，并简要介绍配置的作用。该配置文件中，主要有 3 个模块，如图 3-19 所示。

```
{
  "app": {"bundleName": "com.example.demo"...},
  "deviceConfig": {},
  "module": {"name": ".MyApplication"...}
}
```

图 3-19　项目的配置文件

(1) app：表示应用的全局配置信息。同一个应用的不同 HAP 包的 app 配置必须保持一致。

（2）deviceConfig：表示应用在具体设备上的配置信息。

（3）module：表示 HAP 包的配置信息。该标签下的配置只对当前 HAP 包生效。

配置文件采用 json 格式，其中的属性不分先后顺序，每个属性只允许出现一次。

下面具体看一下项目中出现的配置项都有哪些，以及它们的作用。

1. app 下的属性

app 的属性如图 3-20 所示，包含应用的包名、开发应用的厂商、版本等信息。

```
"app": {
  "bundleName": "com.example.demo",
  "vendor": "example",
  "version": {
    "code": 1,
    "name": "1.0"
  },
  "apiVersion": {
    "compatible": 3,
    "target": 4,
    "releaseType": "Beta1"
  }
```

图 3-20 app 属性配置

bundleName：表示应用的包名，用于标识应用的唯一性。通常采用反转的域名。

vendor：表示开发应用的厂商。

version：code 表示内部版本号，用于系统管理版本使用，对用户不可见，name 表示应用的版本号，用于向用户呈现。

apiVersion：包含 3 个选项。

（1）compatible：表示应用运行需要的 API 最低版本。

（2）target：表示应用运行需要的 API 目标版本。

（3）releaseType：表示应用运行需要的 API 目标版本的类型，取值为 CanaryN、BetaN 或者 Release，其中，N 代表大于零的整数。

Canary：受限发布的版本。

Beta：公开发布的 Beta 版本。

Release：公开发布的正式版本。

deviceConfig：表示应用在具体设备上的配置信息，这里暂时没有用到。

2. module 下的配置

```
"module": {
"package": "com.example.demo",//表示 HAP 的包结构名称，在应用内应保证唯一性.采用反向域名
                              //格式
```

```
"name": ".MyApplication",      //表示 HAP 的类名,采用反向域名方式表示,因为我们指定了
                               //package,所以可以直接以.加类名的形式指定
"deviceType": [                //表示允许 Ability 运行的设备类型.phone 表示手机
"phone"
    ],
"distro": {                    //表示 HAP 发布的具体描述
"deliveryWithInstall": true,   //表示当前 HAP 是否支持随应用安装,true 为支持随应用安装
"moduleName": "entry",         //表示当前 HAP 的名称.
"moduleType": "entry"          //表示当前 HAP 的类型,包括两种类型:entry 和 feature
    },
"abilities": [                 //表示当前模块内的所有 Ability
      {
"skills": [                    //表示 Ability 能够接收的 Intent 的特征
          {
"entities": [
"entity.system.home"           //表示能够接收的 Intent 的 Ability 的类别(如视频、桌面应用等),
                               //可以包含一个或多个 entity
            ],
"actions": [
"action.system.home"           //表示能够接收的 Intent 的 action 值,可以包含一个或多
                               //个 action
            ]
          }
        ],
"orientation": "unspecified",//表示该 Ability 的显示模式,这里表示由系统自动判断方向
"name": "com.example.demo.MainAbility",  //表示 Ability 名称.取值可采用反向域名方式表示,
                               //由包名和类名组成,也可以采用.开头的形式表示
"icon": "$media:icon",         //表示 Ability 图标资源文件的索引,$media 表示引用 media 目
                               //录下的 icon 资源
"description": "$string:mainability_description",  //表示对 Ability 的描述
"label": "first_demo",         //表示 Ability 对用户显示的名称,也就是应用安装后用户设备显
                               //示的名称
"type": "page",                //表示 Ability 的 Type 类型,可以为 page、service 或 data
"launchType": "standard"       //表示 Ability 的启动模式,支持 standard 和 singleton 两种模
                               //式, standard 表示可以有多个实例,single 则表示只有一个实例
      }
    ]
  }
```

3. 设置 Ability 配置不同的主题

可以为每个 Ability 配置不同的主题,在 abilities 数组中的每个 ability 的配置后面可以添加一个 metaData 项,在 value 中可以设置不同的主题。例如添加下图中的主题后,页面就不会显示头部横条了,如图 3-21 所示。

也可以使用不同的主题,可以选择的部分主题如图 3-22 所示。

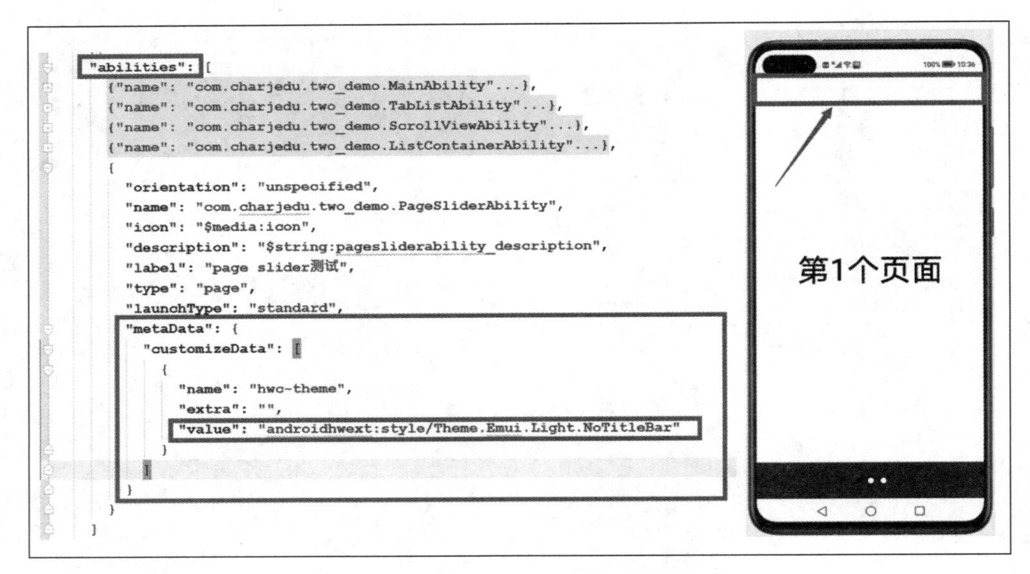

图 3-21　为每个 Ability 配置不同的主题

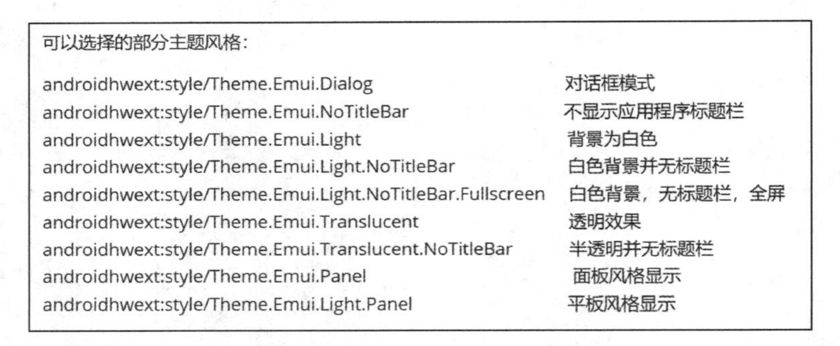

可以选择的部分主题风格:

androidhwext:style/Theme.Emui.Dialog	对话框模式
androidhwext:style/Theme.Emui.NoTitleBar	不显示应用程序标题栏
androidhwext:style/Theme.Emui.Light	背景为白色
androidhwext:style/Theme.Emui.Light.NoTitleBar	白色背景并无标题栏
androidhwext:style/Theme.Emui.Light.NoTitleBar.Fullscreen	白色背景，无标题栏，全屏
androidhwext:style/Theme.Emui.Translucent	透明效果
androidhwext:style/Theme.Emui.Translucent.NoTitleBar	半透明并无标题栏
androidhwext:style/Theme.Emui.Panel	面板风格显示
androidhwext:style/Theme.Emui.Light.Panel	平板风格显示

图 3-22　可以选择的部分主题

3.4.3　资源文件的使用方式

应用的资源文件(字符串、图片、声频等)统一存放于 resources 目录下，便于开发者使用和维护。resources 目录包括两大类目录，一类为 base 目录与限定词目录，另一类为 rawfile 目录。

base 目录下面可以创建资源组目录，包括 element(元素资源)、media(媒体资源，包括图片、声频、视频等非文本格式的文件)、animation(动画资源)、layout、graphic(图形)、profile(表示其他类型文件，以原始文件形式保存)。目录中的资源文件除 profile 目录中的文件外会被编译成二进制文件，并赋予资源文件 ID，可以用 ResourceTable 引用，如表 3-1 所示。

表 3-1 element(元素资源)详细表

element	表示元素资源,以下每一类数据都采用相应的 JSON 文件来表征。 boolean:布尔型 color:颜色 float:浮点型 intarray:整型数组 integer:整型 pattern:样式 plural:复数形式 strarray:字符串数组 string:字符串	element 目录中的文件名称建议与下面的文件名保持一致。每个文件中只能包含同一类型的数据。 boolean.json color.json float.json intarray.json integer.json pattern.json plural.json strarray.json string.json

限定词目录需要开发者自行创建。目录名称由一个或多个表征应用场景或设备特征的限定词组合而成,包括语言、文字、国家或地区、横竖屏、设备类型和屏幕密度等 6 个维度,限定词之间通过下画线“_”或者中画线“-”连接。开发者在创建限定词目录时,需要掌握限定词目录的命名要求及与限定词目录与设备状态的匹配规则。

限定词目录的命名要求如表 3-2 所示。

表 3-2 每类限定词的取值范围

限定词类型	含义与取值说明
语言	表示设备使用的语言类型,由 2 个小写字母组成。例如:zh 表示中文,en 表示英语 详细取值范围,参见 ISO 639-1(ISO 制定的语言编码标准)
文字	表示设备使用的文字类型,由 1 个大写字母(首字母)和 3 个小写字母组成。例如:Hans 表示简体中文,Hant 表示繁体中文 详细取值范围,参见 ISO 15924(ISO 制定的文字编码标准)
国家或地区	表示用户所在的国家或地区,由 2～3 个大写字母或者 3 个数字组成,例如:CN 表示中国,GB 表示英国 详细取值范围,参见 ISO 3166-1(ISO 制定的国家或地区编码标准)
横竖屏	表示设备的屏幕方向,取值如下: vertical:竖屏 horizontal:横屏
设备类型	表示设备的类型,取值如下: car:车机 tv:智慧屏 wearable:智能穿戴

续表

限定词类型	含义与取值说明
屏幕密度	表示设备的屏幕密度(单位为 dpi),取值如下: sdpi:表示小规模的屏幕密度(Small-scale Dots Per Inch),适用于 dpi 取值为(0,120]的设备 mdpi:表示中规模的屏幕密度(Medium-scale Dots Per Inch),适用于 dpi 取值为(120,160]的设备 ldpi:表示大规模的屏幕密度(Large-scale Dots Per Inch),适用于 dpi 取值为(160,240]的设备 xldpi:表示特大规模的屏幕密度(Extra Large-scale Dots Per Inch),适用于 dpi 取值为(240,320]的设备 xxldpi:表示超大规模的屏幕密度(Extra Extra Large-scale Dots Per Inch),适用于 dpi 取值为(320,480]的设备 xxxldpi:表示超特大规模的屏幕密度(Extra Extra Extra Large-scale Dots Per Inch),适用于 dpi 取值为(480,640]的设备

(1) 限定词的组合顺序:语言_文字_国家或地区-横竖屏-设备类型-屏幕密度。开发者可以根据应用的使用场景和设备特征,选择其中的一类或几类限定词组成目录名称。

(2) 限定词的连接方式:语言、文字、国家或地区之间采用下画线"_"连接,除此之外的其他限定词之间均采用中画线"-"连接。例如:zh_Hant_CN、zh_CN-car-ldpi。

(3) 限定词的取值范围:每类限定词的取值必须符合表 3-2 中的条件,否则,将无法匹配目录中的资源文件。

rawfile 目录支持创建多层子目录,目录名称可以自定义,文件夹内可以自由放置各类资源文件。rawfile 目录的文件不会根据设备状态匹配不同的资源。目录中的资源文件会被直接打包进应用,不经过编译,不可以用 ResourceTable 引用。

资源文件的使用方法如下。

(1) base 目录与限定词目录中的资源文件:通过指定资源类型(type)和资源名称(name)来引用。

通过 Java 引用资源文件:

- 普通资源引用:ResourceTable. type_name;
- 系统资源引用:ohos. global. systemres. ResourceTable. type_name;
- 引用 string. json 文件中类型为 String、名称为 app_name 的资源;
- 引用 color. json 文件中类型为 Color、名称为 red 的资源。

示例代码如下:

```
ResourceManager resourceManager = getResourceManager();
resourceManager.getElement(ResourceTable.String_app_name).getString();
resourceManager.getElement(ResourceTable.Color_red).getColor();
```

xml 文件引用资源文件：

- 资源文件：$type:name；
- 系统资源：$ohos:type:name；
- 引用 string.json 文件中类型为 String、名称为 app_name 的资源，示例代码如下：

```
<?xml version = "1.0" encoding = "utf - 8"?>
< DirectionalLayout xmlns:ohos = "http://schemas.huawei.com/res/ohos"
    ohos:width = "match_parent"
    ohos:height = "match_parent"
    ohos:orientation = "vertical">
< Text ohos:text = " $ string:app_name"/>
</DirectionalLayout >
```

(2) rawfile 目录中的资源文件：通过指定文件路径和文件名称来引用。

在 Java 文件中，引用一个路径为 resources/rawfile/、名称为 example.js 的资源文件，示例代码如下：

```
Resource resource  =  null;
String rawfileURL  =  "resources/rawfile/example.js"
try {
    resource  =  getResourceManager().getRawFileEntry(rawfileURL).openRawFile();
    InputStreamReader inputStreamReader  =  new InputStreamReader(resource,"utf - 8");
    BufferedReader bufferedReader  =  new BufferedReader(inputStreamReader);
    String lineTxt  =  "";
    while ((lineTxt  =  bufferedReader.readLine()) != null){
        System.out.println(lineTxt);
    }
}catch (Exception e){}
```

3.5　ACE Java UI 布局

在 ACE Java UI 框架中，提供了两种编写布局的方式：代码方式创建布局和 XML 方式声明 UI 布局。

这两种方式创建出的布局没有本质差别，在 XML 中声明布局，加载后同样可在代码中对该布局进行修改。

XML 声明布局的方式更加简便直观。每个 Component 和 ComponentContainer 对象中的大部分属性支持在 XML 中进行设置，它们都有各自的 XML 属性列表。某些属性仅适用于特定的组件，例如：只有 Text 组件支持 text_color 属性，但如果不支持该属性的组件添加了该属性，则该属性会被忽略。具有继承关系的组件子类将继承父类的属性列表，Component 作为组件的基类，拥有各个组件常用的属性，例如 ID、布局参数等。

编程创建布局的方式比较麻烦，需要在 AbilitySlice 中分别创建组件和布局，并将它们进行组织关联。

3.5.1 通过 XML 的方式创建布局

XML 声明布局的方式更加简便直观。项目默认使用 XML 的方式创建页面布局。

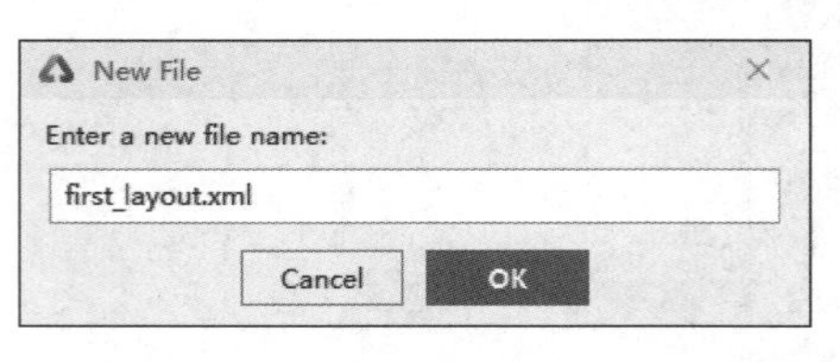

图 3-23 通过 DevEco Studio 创建 XML 布局

1. 创建 XML 布局

在 DevEco Studio 的 Project 窗口，打开 entry→src→main→resources→base，右击 layout 文件夹，选择 New→File，将文件命名为 first_layout. xml，如图 3-23 所示。

打开新创建的 first_layout. xml 布局文件，修改其中的内容，对布局和组件的属性和层级进行描述。如代码示例 3-4 所示。

代码示例 3-4 编写界面布局：first_demo/entry/layout/first_layout. xml

```
<?xml version = "1.0" encoding = "utf - 8"?>
< DirectionalLayout
    xmlns:ohos = "http://schemas.huawei.com/res/ohos"
    ohos:width = "match_parent"
    ohos:height = "match_parent"
    ohos:orientation = "vertical"
    ohos:padding = "32">
< Text
        ohos:id = " $ + id:text"
        ohos:width = "match_content"
        ohos:height = "match_content"
        ohos:layout_alignment = "horizontal_center"
        ohos:text = "My name is Text. "
        ohos:text_size = "25fp"/>
< Button
        ohos:id = " $ + id:button"
        ohos:margin = "50"
        ohos:width = "match_content"
        ohos:height = "match_content"
        ohos:layout_alignment = "horizontal_center"
        ohos:text = "My name is Button. "
        ohos:text_size = "50"/>
</DirectionalLayout >
```

在上面的 XML 布局中，向组件添加的宽和高分为有固定的单位 vp、match_parent 和 match_content。

- 具体的数值：10(以像素为单位)、10vp(以屏幕相对像素为单位)；
- match_parent：表示组件大小将扩展为父组件允许的最大值，它将占据父组件方向上的剩余大小；

- match_content：表示组件大小与它的内容占据的大小范围相适应。

2. 加载 XML 布局

在代码中需要加载 XML 布局，并添加为根布局或作为其他布局的子 Component。如代码示例 3-5 所示。

代码示例 3-5 编写界面布局：first_demo/entry/java/ExampleAbilitySlice. java

```
public class ExampleAbilitySlice extends AbilitySlice {
    @Override
    public void onStart(Intent intent) {
        super.onStart(intent);
        //加载 XML 布局作为根布局
        super.setUIContent(ResourceTable.Layout_first_layout);
        Button button = (Button) findComponentById(ResourceTable.Id_button);
        if (button != null) {
            //设置组件的属性
            ShapeElement background = new ShapeElement();
            background.setRgbColor(new RgbColor(0, 125, 255));
            background.setCornerRadius(25);
            button.setBackground(background);

            button.setClickedListener(new Component.ClickedListener() {
                @Override
                //在组件中增加对单击事件的检测
                public void onClick(Component component) {
                    //此处添加当按钮被单击时需要执行的操作
                }
            });
        }
    }
}
```

3.5.2 通过编码的方式创建布局

代码创建布局需要在 AbilitySlice 中分别创建组件和布局，并将它们进行组织关联。

1. 声明布局

创建一个方向布局(DirectionalLayout)组件，代码如下：

```
DirectionalLayout directionalLayout = new DirectionalLayout(getContext());
```

设置方向布局大小，代码如下：

```
directionalLayout.setWidth(ComponentContainer.LayoutConfig.MATCH_PARENT);
directionalLayout.setHeight(ComponentContainer.LayoutConfig.MATCH_PARENT);
```

设置方向布局属性，代码如下：

```
directionalLayout.setOrientation(Component.VERTICAL);
```

将组件添加到方向布局中(视布局需要对组件所设置的布局属性进行约束)，button 为已在 3.3.2 节中定义的 Button 组件，代码如下：

```
directionalLayout.addComponent(button);
```

将布局添加到组件树中，代码如下：

```
setUIContent(directionalLayout);
```

2. 创建组件

上面创建好了一个布局，接下来需要将一些组件放在布局中，例如声明一个 Button 组件，代码如下：

```
Button button = new Button(getContext());
```

设置 Botton 组件大小，代码如下：

```
button.setWidth(ComponentContainer.LayoutConfig.MATCH_CONTENT);
button.setHeight(ComponentContainer.LayoutConfig.MATCH_CONTENT);
```

设置 Botton 组件属性，代码如下：

```
button.setText("My name is Button.");
button.setTextSize(50);
```

3. 编码布局案例

下面通过编程的方式创建一个简单的布局页面，页面显示效果如图 3-24 所示。

根据以上步骤创建组件和布局后的界面显示效果如图 3-24 所示。如代码示例 3-6 所示，为组件设置了一个按键回调，在按键被按下后，应用会执行自定义的操作。

代码示例 3-6　页面逻辑：first_demo/entry/java/ExampleAbilitySlice.java

```
public class ExampleAbilitySlice extends AbilitySlice {
    @Override
    public void onStart(Intent intent) {
        super.onStart(intent);
        //声明布局
        DirectionalLayout directionalLayout = new DirectionalLayout(getContext());
        //设置布局大小
```

图 3-24 通过编程的方式创建布局

```
        directionalLayout.setWidth(ComponentContainer.LayoutConfig.MATCH_PARENT);
        directionalLayout.setHeight(ComponentContainer.LayoutConfig.MATCH_PARENT);
        //设置布局属性
        directionalLayout.setOrientation(Component.VERTICAL);
        directionalLayout.setPadding(32, 32, 32, 32);

        Text text = new Text(getContext());
        text.setText("编码创建视图布局");
        text.setTextSize(80);
        text.setId(100);
        //为组件添加对应布局的布局属性
        DirectionalLayout.LayoutConfig layoutConfig = new DirectionalLayout.LayoutConfig
(ComponentContainer.LayoutConfig.MATCH_CONTENT, ComponentContainer.LayoutConfig.MATCH_CONTENT);
        layoutConfig.alignment = LayoutAlignment.HORIZONTAL_CENTER;
        text.setLayoutConfig(layoutConfig);

        //将 Text 添加到布局中
```

```
        directionalLayout.addComponent(text);
        //类似地添加一个 Button
        Button button = new Button(getContext());
        layoutConfig.setMargins(0, 50, 0, 0);
        button.setLayoutConfig(layoutConfig);
        button.setText("更新信息");
        button.setTextSize(50);
        ShapeElement background = new ShapeElement();
        background.setRgbColor(new RgbColor(0, 125, 255));
        background.setCornerRadius(25);
        button.setBackground(background);
        button.setPadding(10, 10, 10, 10);
button.setTextColor(Color.WHITE);
        button.setClickedListener(new Component.ClickedListener() {
            @Override
            //在组件中增加对单击事件的检测
            public void onClick(Component component) {
                //此处添加当按钮被单击时需要执行的操作
            }
        });
        directionalLayout.addComponent(button);
        //将布局作为根布局添加到视图树中
        super.setUIContent(directionalLayout);
    }
}
```

在代码示例中，可以看到设置组件大小的方法有两种：

- 通过 setWidth/setHeight 直接设置宽和高；
- 通过 setLayoutConfig 方法设置布局属性来设定宽和高。

这两种方法的区别是后者还可以增加更多的布局属性设置，例如使用 alignment 设置水平居中的约束。另外，这两种方法设置的宽和高以最后设置的数值作为最终结果。它们的取值一致，可以是以下取值：

- 具体以像素为单位的数值，如 button.setTextSize(50)；
- MATCH_PARENT：表示组件大小将扩展为父组件允许的最大值，它将占据父组件方向上的剩余大小；
- MATCH_CONTENT：表示组件大小与它的内容所占据的大小范围相适应。

3.5.3 鸿蒙常见布局方式

在 Java UI 框架中，具体的布局类通常以 XXLayout 命名，如 DirectionalLayout、

DependentLayout 等，如图 3-25 所示。完整的用户界面是一个布局，用户界面中的一部分也可以是一个布局。鸿蒙的常见布局有定向布局、依赖布局、位置布局、堆叠布局、自适应布局。

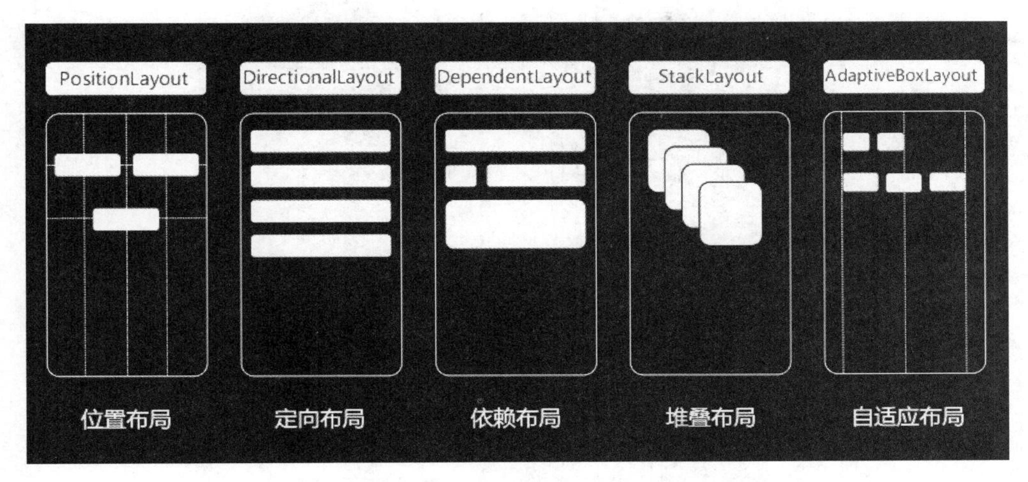

图 3-25　鸿蒙常见的布局分类

Java UI 框架提供了标准布局功能的容器，都继承自 ComponentContainer。有些组件继承自 StackLayout，但命名没有以 Layout 结尾，如 ScrollView、PageSlider 等。ListContainer 组件同样继承自 ComponentContainer。

1. 定向布局 DirectionalLayout

DirectionalLayout 布局类似于 Android 中的 LinearLayout 布局。

DirectionalLayout 的 orientation 属性的取值有两个：horizontal 和 vertical，代表横向和纵向布局。

DirectionalLayout 用于将一组组件（Component）按照水平或者垂直方向排布，能够方便地对齐布局内的组件。

定向布局有几个重要的属性：排列属性、对齐方式属性、权重属性，接下来分别看一看这些属性的用法。

1）排列属性

DirectionalLayout 布局分为两种模式：vertical 垂直排列子元素，horizontal 水平排列子元素，如图 3-26 所示。

如果垂直排列的子元素 height 的总和超过了父元素，则会被截取。如果水平排列的子元素 width 的总和超过了父元素，则会被截取。

在下面的案例中，首先在一个 DirectionalLayout 布局组件内，放置 3 个 DirectionalLayout 子组件。外部 DirectionalLayout 布局组件通过将 orientation 属性值设置为 vertical 或者通过 horizontal 来控制子组件的排列方向，如代码示例 3-7 所示。

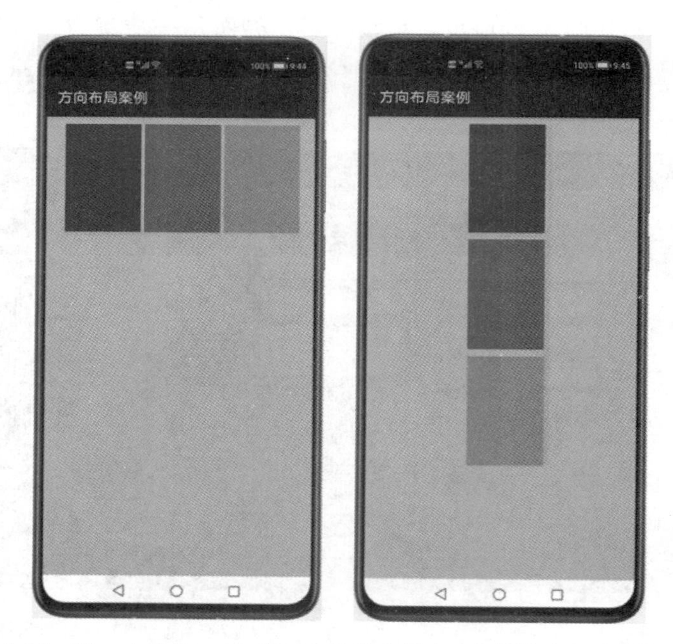

图 3-26　DirectionalLayout 布局分为两种模式：水平模式和垂直模式

代码示例 3-7　DirectionalLayout：first_demo/layout/ability_second.xml

```
<?xml version = "1.0" encoding = "utf - 8"?>
< DirectionalLayout
    xmlns:ohos = "http://schemas.huawei.com/res/ohos"
    ohos:height = "match_parent"
    ohos:width = "match_parent"
    ohos:background_element = " # ccc"
    ohos:alignment = "horizontal_center"
    ohos:orientation = "vertical">


< DirectionalLayout
        ohos:height = "150vp"
        ohos:width = "200vp"
        ohos:top_margin = "10vp"
        ohos:background_element = " # 00f"
        ohos:left_margin = "5vp"
        ohos:orientation = "vertical">
</DirectionalLayout >

< DirectionalLayout
        ohos:height = "150vp"
        ohos:width = "200vp"
```

```
        ohos:left_margin = "5vp"
        ohos:top_margin = "10vp"
        ohos:background_element = "red"
        ohos:orientation = "vertical">
</DirectionalLayout>

<DirectionalLayout
        ohos:height = "150vp"
        ohos:width = "200vp"
        ohos:left_margin = "5vp"
        ohos:top_margin = "10vp"
        ohos:background_element = "green"
        ohos:orientation = "vertical">
</DirectionalLayout>

</DirectionalLayout>
```

在上面的代码中子组件的宽和高没有超出屏幕的宽和高，但是如果垂直排列的子元素height的总和超过了父元素，则会被截取。如果水平排列的子元素width的总和超过了父元素，则会被截取。我们来看下面的例子，如图3-27所示。因为水平和垂直方向上的宽和高超出屏幕的宽和高，所以超出部分被截取。

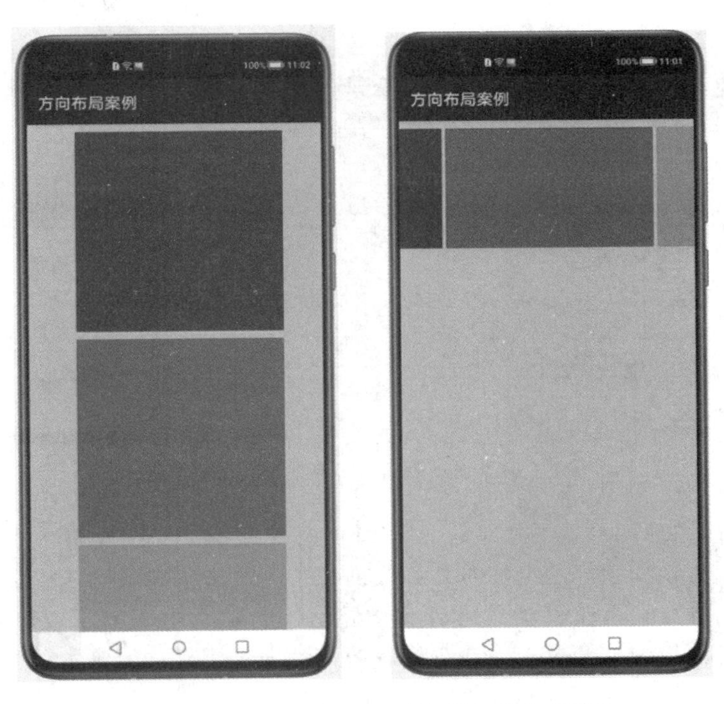

图3-27 DirectionalLayout布局超出部分被截取

2）对齐方式属性

对齐方式属性列表，如表 3-3 所示。

表 3-3 对齐方式属性列表

参　　数	作　　用	可搭配排列方式
left	左对齐	垂直排列
top	顶部对齐	水平排列
right	右对齐	垂直排列
bottom	底部对齐	水平排列
horizontal_center	水平方向居中	垂直排列
vertical_center	垂直方向居中	水平排列
center	垂直与水平方向都居中	水平/垂直排列

将布局的方向设置为 horizontal 后，分别将内部的布局组件的对齐方式属性设置为 top、vertical_center、bottom，显示效果如图 3-28 所示。

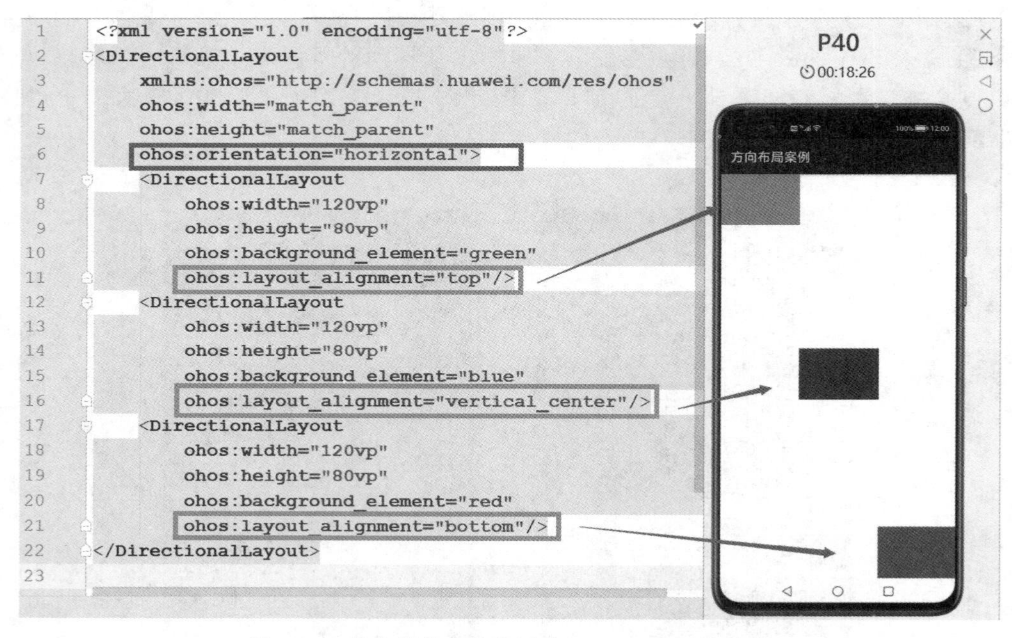

图 3-28 将布局的方向设置为 horizontal 后的效果

这种设置方式的问题是，内部的 3 个组件的宽度是累加的，第一个组件的宽度为 120vp，中间的组件使用顶部的组件的宽度为起点进行累加，所以到了第 3 个组件是以中间的组件的右边为开始显示在底部，如果组件的宽度超出屏幕，将只显示剩余部分的内容。

将布局的方向设置为 vertical 后，分别将内部的布局组件的对齐方式属性设置为 left、horizontal_center、right，显示效果如图 3-29 所示。

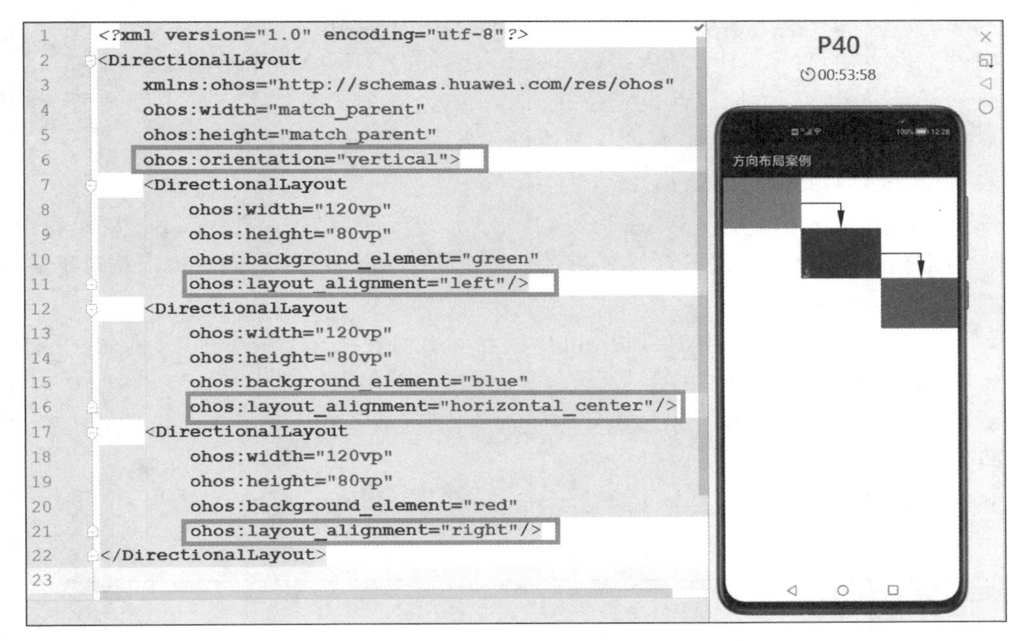

图 3-29　将布局的方向设置为 vertical 后的效果

3）权重属性

权重(weight)就是按比例来分配组件占用父组件的大小，如图 3-30 所示。

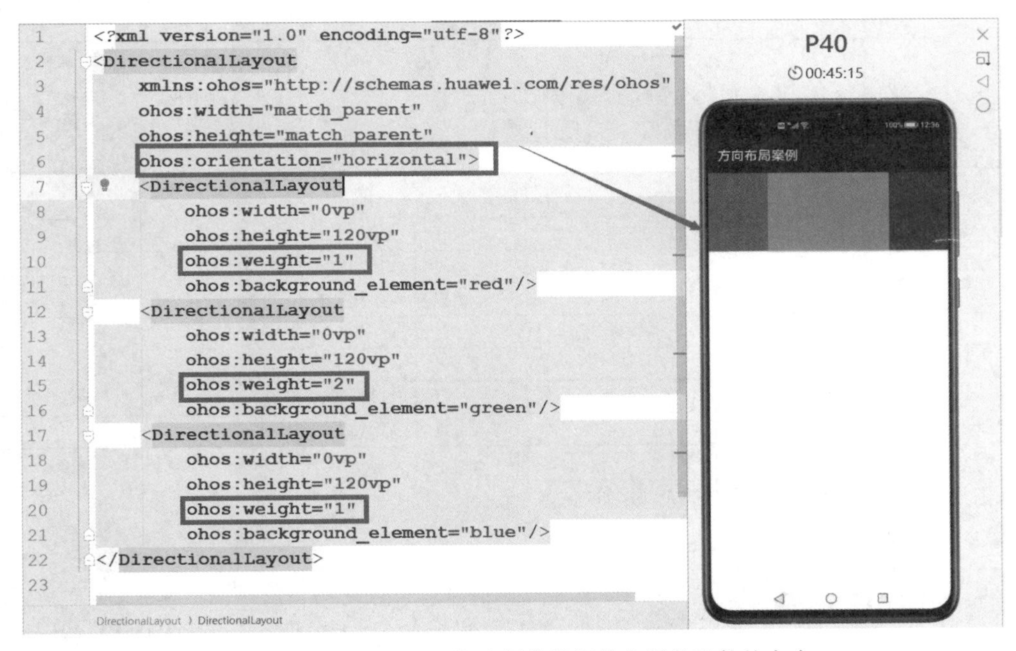

图 3-30　权重(weight)按比例分配组件占用父组件的大小

父布局可分配宽度＝父布局宽度-所有子组件 width 之和。

组件宽度＝(组件 weight/所有组件 weight 之和)×父布局可分配宽度。

将外部布局组件的方向属性设置为 horizontal，内部的子组件不需要设置宽度，这里设置为 0vp，每个子组件通过 weight 的比例值设置所占父容器的宽度比。

2. 依赖布局 DependentLayout

DependentLayout 是 Java UI 系统里的一种常见布局。与 DirectionalLayout 相比，拥有更多的排布方式，每个组件可以指定相对于其他同级元素的位置，或者指定相对于父组件的位置。

- DependentLayout 相当于 Android 中的相对布局 RelativeLayout；
- DependentLayout 与 DirectionalLayout 相比有更多的排列方式；
- 每个组件可以指定相对于其他同级元素的位置，或者指定相对于父组件的位置。

1）排列方式

相对于同级组件的位置布局，如表 3-4 所示。

表 3-4　相对于同级组件的位置布局

位置布局	描述
above	处于同级组件的上侧
below	处于同级组件的下侧
start_of	处于同级组件的起始侧
end_of	处于同级组件的结束侧
left_of	处于同级组件的左侧
right_of	处于同级组件的右侧

相对于父组件的位置布局，如表 3-5 所示。

表 3-5　相对于父组件的位置布局

位置布局	描述
align_parent_left	处于父组件的左侧
align_parent_right	处于父组件的右侧
align_parent_start	处于父组件的起始侧
align_parent_end	处于父组件的结束侧
align_parent_top	处于父组件的上侧
align_parent_bottom	处于父组件的下侧
center_in_parent	处于父组件的中间

相对于同级组件的位置布局，如图 3-31 所示。

2）相对父控件布局

第 1 个 Text 组件被设置在相对于父容器组件的右边 align_parent_right＝"true"；第 2 个 Text 组件被设置在相对于父容器的底部中间位置：这里设置了 align_parent_bottom＝"true"和 center_in_parent＝"true"；第 3 个 Text 组件被设置在父容器的正中间：center_in

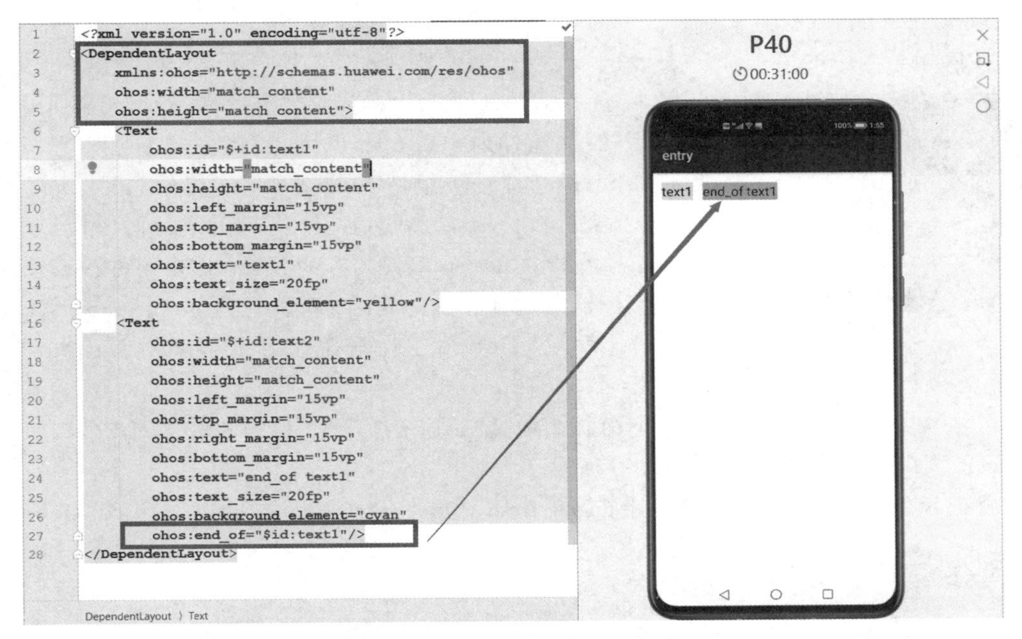

图 3-31　相对于同级组件的位置布局

_parent＝"true"；第 4 个 Text 组件被设置在父容器的左上方：align_parent_left＝"true"和 align_parent_top＝"true"，如图 3-32 所示。

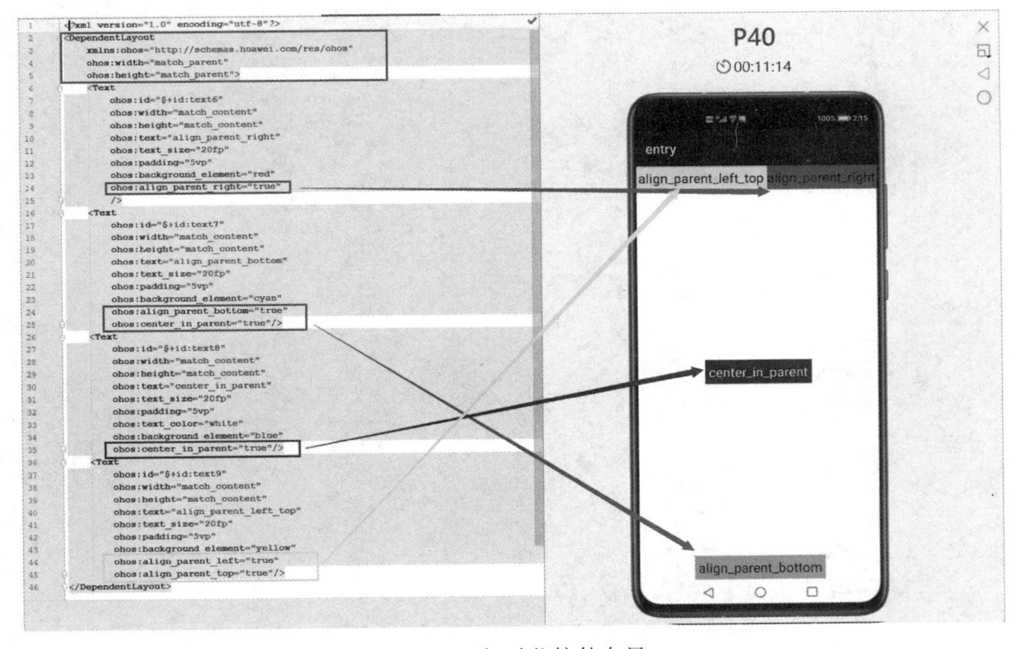

图 3-32　相对父控件布局

3）复杂界面布局

在依赖布局中的第 1 个 Text 组件，无须任何设置，第 2 个 Text 组件内容为 Catalog，只需相对于 Text1 组件进行布局：ohos:below＝"$id:text1"；第 3 个 Text 组件内容为 Content，这个组件同样相对于 Text1 组件布局，位于 Text1 组件下方及 Text2 组件右边，所以需要设置：ohos:below＝"$id:text1"和 ohos:end_of＝"$id:text2"；第 2 个 Button 组件内容为 Next，这个组件被设置在相对于父容器的底部和右边位置，即设置为 align_parent_end＝"true"；第 1 个 Button 组件只需相对于第 2 个 Button 布局：left_of＝"$id:button2"。

图 3-33　复杂界面布局

下面我们通过 XML 的方式实现图 3-33 所示效果，如代码示例 3-8 所示。

代码示例 3-8　依赖布局的复杂布局：first_demo/entry/layout/ability_third.xml

```
<?xml version = "1.0" encoding = "utf - 8"?>
< DependentLayout
    xmlns:ohos = "http://schemas.huawei.com/res/ohos"
    ohos:width = "match_parent"
    ohos:height = "match_content">
< Text
        ohos:id = "$ + id:text1"
        ohos:width = "match_parent"
        ohos:height = "match_content"
        ohos:text_size = "25fp"
        ohos:top_margin = "15vp"
        ohos:left_margin = "15vp"
        ohos:right_margin = "15vp"
        ohos:background_element = "#ccc"
        ohos:text = "Title"
        ohos:text_weight = "1000"
        ohos:text_alignment = "horizontal_center"
        />
< Text
        ohos:id = "$ + id:text2"
        ohos:width = "match_content"
        ohos:height = "120vp"
        ohos:text_size = "10fp"
        ohos:background_element = "#ccc"
        ohos:text = "Catalog"
        ohos:top_margin = "15vp"
        ohos:left_margin = "15vp"
        ohos:right_margin = "15vp"
```

```
            ohos:bottom_margin = "15vp"
            ohos:align_parent_left = "true"
            ohos:text_alignment = "center"
            ohos:multiple_lines = "true"
            ohos:below = " $ id:text1"
            ohos:text_font = "serif"/>
    < Text
            ohos:id = " $ + id:text3"
            ohos:width = "match_parent"
            ohos:height = "120vp"
            ohos:text_size = "25fp"
            ohos:background_element = " # ccc"
            ohos:text = "Content"
            ohos:top_margin = "15vp"
            ohos:right_margin = "15vp"
            ohos:bottom_margin = "15vp"
            ohos:text_alignment = "center"
            ohos:below = " $ id:text1"
            ohos:end_of = " $ id:text2"
            ohos:text_font = "serif"/>
    < Button
            ohos:id = " $ + id:button1"
            ohos:width = "70vp"
            ohos:height = "match_content"
            ohos:text_size = "15fp"
            ohos:background_element = " # ccc"
            ohos:text = "Previous"
            ohos:right_margin = "15vp"
            ohos:bottom_margin = "15vp"
            ohos:below = " $ id:text3"
            ohos:left_of = " $ id:button2"
            ohos:italic = "false"
            ohos:text_weight = "5"
            ohos:text_font = "serif"/>
    < Button
            ohos:id = " $ + id:button2"
            ohos:width = "70vp"
            ohos:height = "match_content"
            ohos:text_size = "15fp"
            ohos:background_element = " # ccc"
            ohos:text = "Next"
            ohos:right_margin = "15vp"
            ohos:bottom_margin = "15vp"
            ohos:align_parent_end = "true"
            ohos:below = " $ id:text3"
            ohos:italic = "false"
```

```
        ohos:text_weight = "5"
        ohos:text_font = "serif"/>
</DependentLayout>
```

3. 表格布局 TableLayout

TableLayout 使用表格的方式划分子组件。TableLayout 类似于 Android 中的 TableLayout，以列表的方式展示内容，使用 TableLayout 时，通过设置行数和列数来控制表格内组件的行和列，如图 3-34 所示。

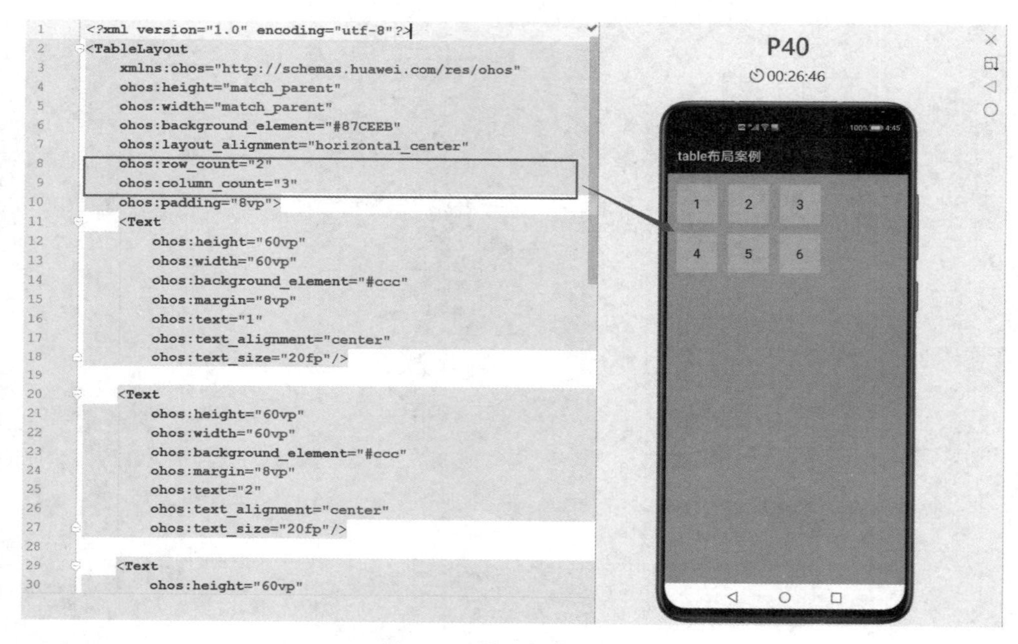

图 3-34　设置行数和列数

下面实现如图 3-34 所示的效果，将 row_count 和 column_count 设置为两行三列的布局，如代码示例 3-9 所示。

代码示例 3-9　表格布局：first_demo/entry/layout/ability_forth.xml

```
<?xml version = "1.0" encoding = "utf - 8"?>
< TableLayout
    xmlns:ohos = "http://schemas.huawei.com/res/ohos"
    ohos:height = "match_parent"
    ohos:width = "match_parent"
    ohos:background_element = "#87CEEB"
    ohos:layout_alignment = "horizontal_center"
    ohos:row_count = "2"
    ohos:column_count = "3"
```

```
    ohos:padding = "8vp">
<Text
        ohos:height = "60vp"
        ohos:width = "60vp"
        ohos:background_element = "#ccc"
        ohos:margin = "8vp"
        ohos:text = "1"
        ohos:text_alignment = "center"
        ohos:text_size = "20fp"/>

<Text
        ohos:height = "60vp"
        ohos:width = "60vp"
        ohos:background_element = "#ccc"
        ohos:margin = "8vp"
        ohos:text = "2"
        ohos:text_alignment = "center"
        ohos:text_size = "20fp"/>

<Text
        ohos:height = "60vp"
        ohos:width = "60vp"
        ohos:background_element = "#ccc"
        ohos:margin = "8vp"
        ohos:text = "3"
        ohos:text_alignment = "center"
        ohos:text_size = "20fp"/>

<Text
        ohos:height = "60vp"
        ohos:width = "60vp"
        ohos:background_element = "#ccc"
        ohos:margin = "8vp"
        ohos:text = "4"
        ohos:text_alignment = "center"
        ohos:text_size = "20fp"/>
<Text
        ohos:height = "60vp"
        ohos:width = "60vp"
        ohos:background_element = "#ccc"
        ohos:margin = "8vp"
        ohos:text = "5"
        ohos:text_alignment = "center"
        ohos:text_size = "20fp"/>

<Text
        ohos:height = "60vp"
```

```
        ohos:width = "60vp"
        ohos:background_element = "#ccc"
        ohos:margin = "8vp"
        ohos:text = "6"
        ohos:text_alignment = "center"
        ohos:text_size = "20fp"/>
</TableLayout>
```

可以通过下面的属性设置对齐方式，如表 3-6 所示。

表 3-6　设置 Table 布局的对齐方式

属　性　值	效　　果
align_edges	表示子组件边界对齐，默认对齐方式
align_contents	表示子组件边距对齐

align_edges 表示子组件边界对齐，此方式为默认对齐方式。

这里将第一个 Text 组件的 margin 设置为"18vp"，将其他的 Text 组件设置为"8vp"。如图 3-35 所示，第二行的组件默认使用屏幕的边界对齐方式。

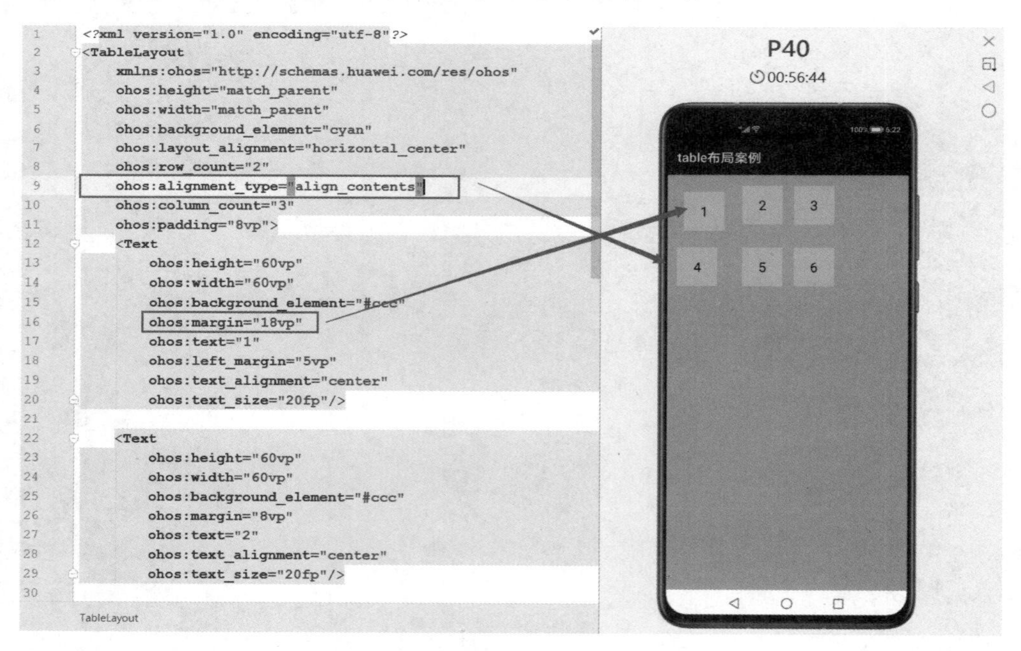

图 3-35　设置行数和列数

align_contents 表示子组件边距对齐。

上面的布局，如果把 alignment_type 设置为 align_contents 的值，则第二行的组件将按第一行的组件对齐，如图 3-36 所示。

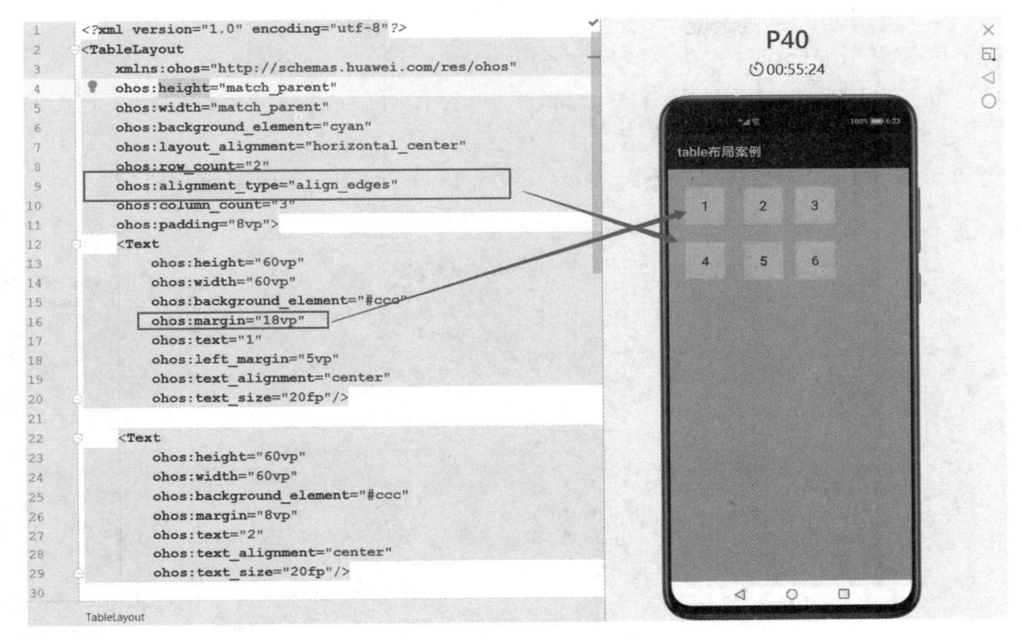

图 3-36 设置行数和列数

4. 位置布局 PositionLayout

PositionLayout 为绝对布局，该布局指定了子组件在其中的具体位置（x/y 坐标）。由于需要指定子组件的 x/y 精确坐标，其布局的灵活性相对较差，在没有绝对定位的情况下相比其他类型的布局更加难以维护，因此不建议使用。

我们通过 PositionLayout 布局实现上面的 UI 图，如图 3-37 所示。

图 3-37 PositionLayout 绝对布局

(1) 以 PositionLayout 为父组件,宽和高占满整个屏幕。

(2) 接下来先进行组件占位,两个 TextField 和一个 Button 组件。

(3) 根据 UI 图的比例,让组件左上角(x/y)为起始点,相对父组件的距离进行移动。如代码示例 3-10 所示。

代码示例 3-10　位置布局:first_demo/entry/layout/ability_six.xml

```
<PositionLayout
    xmlns:ohos="http://schemas.huawei.com/res/ohos"
    ohos:height="match_parent"
    ohos:width="match_parent"
    ohos:background_element="$media:background_login">

<TextField
        ohos:id="$+id:username"
        ohos:height="match_content"
        ohos:width="200vp"
        ohos:text="请输入用户名..."
        ohos:background_element="#EAEDED"
        ohos:text_color="#CCCCCC"
        ohos:text_size="18fp"
        ohos:padding="10vp"/></PositionLayout>
```

我们发现运行后文本框置于整个布局的左上角,并没有到达我们想要的位置,这是因为没有给其绝对位置的定位。我们在 MainAbilitySlice 中根据 ID 找到我们的文本框,然后将其绝对位置 x/y 轴距离进行定位。

```
TextField username = (TextField)
findComponentById(ResourceTable.Id_username);
username.setContentPosition(1110, 380);
```

另外两个组件和第一个文本组件一样,找到绝对位置即可实现,如代码示例 3-11 所示。

代码示例 3-11　位置布局案例:first_demo/entry/layout/ability_six.xml

```
<PositionLayout
    xmlns:ohos="http://schemas.huawei.com/res/ohos"
    ohos:background_element="#76bef7"
    ohos:height="match_parent"
    ohos:width="match_parent">

<Image
        ohos:id="$+id:logo"
        ohos:image_src="$media:log.png"
        ohos:layout_alignment="right|vertical_center"
        ohos:height="match_content"
        ohos:width="match_content">
</Image>
```

```
<Text
        ohos:id = "$ + id:title"
        ohos:text_color = "#000"
        ohos:text = "鸿蒙 TV 管理系统"
        ohos:text_size = "30fp"
        ohos:height = "match_content"
        ohos:width = "match_content">

</Text>

<TextField
        ohos:id = "$ + id:username"
        ohos:height = "match_content"
        ohos:width = "200vp"
        ohos:text = "请输入用户名..."
        ohos:background_element = "#EAEDED"
        ohos:text_color = "#CCCCCC"
        ohos:text_size = "18fp"
        ohos:padding = "10vp"/>

<TextField
        ohos:id = "$ + id:password"
        ohos:height = "match_content"
        ohos:width = "200vp"
        ohos:text = "请输入密码..."
        ohos:background_element = "#EAEDED"
        ohos:text_color = "#CCCCCC"
        ohos:text_size = "18fp"
        ohos:padding = "10vp"/>

<Button
        ohos:id = "$ + id:login_btn"
        ohos:height = "match_content"
        ohos:width = "200vp"
        ohos:text = "登录"
        ohos:text_size = "20fp"
        ohos:padding = "10vp"
        ohos:text_color = "#FFFFFF"
        ohos:background_element = "#0EAB8D"/>
</PositionLayout>
```

通过编码的方式指定每个组件的位置坐标，通过 setContentPosition 方法设置 x 轴和 y 轴坐标，如代码示例 3-12 所示。

代码示例 3-12　位置布局控制组件：first_demo/entry/java/SixAbility. java

```
public void onStart(Intent intent) {
    super.onStart(intent);
    super.setUIContent(ResourceTable.Layout_ability_six);
    Text title = (Text) findComponentById(ResourceTable.Id_title);
    title.setContentPosition(1110, 280);
    TextField username = (TextField) findComponentById(ResourceTable.Id_username);
    username.setContentPosition(1110, 380);
    TextField password = (TextField) findComponentById(ResourceTable.Id_password);
    password.setContentPosition(1110, 522);
    Button loginBtn = (Button) findComponentById(ResourceTable.Id_login_btn);
    loginBtn.setContentPosition(1110, 675);

    Image logo = (Image) findComponentById(ResourceTable.Id_logo);
    logo.setContentPosition(320, 320);
}
```

至此，完成了绝对布局中组件的绝对定位，但不建议在项目中使用该布局，因为如果设备分辨率发生变化，则整个布局会出现错乱。

5. 堆叠布局 StackLayout

StackLayout 直接在屏幕上开辟出一块空白的区域，添加到这个布局中的视图都以层叠的方式显示，而它会把这些视图默认放到这块区域的左上角，第一个添加到布局中的视图显示在最底层，最后一个被放在最顶层，上一层的视图会覆盖下一层的视图，如图 3-38 所示。

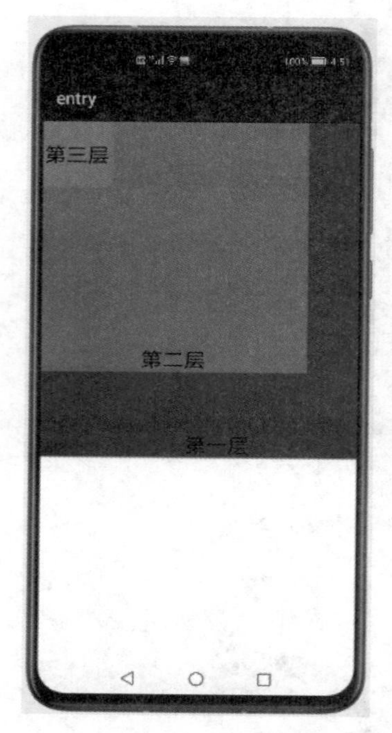

图 3-38　StackLayout 布局

StackLayout 布局有以下特征：

- StackLayout 相当于 Android 中的帧布局 FrameLayout；
- StackLayout 直接在屏幕上开辟出一块空白的区域，添加到这个布局中的视图都以层叠的方式显示；
- 第一个添加到布局中的视图显示在最底层，最后一个被放在最顶层；
- 上一层的视图会覆盖下一层的视图。

使用默认布局添加组件，StackLayout 中组件的布局默认在区域的左上角，并且以后创建的组件会堆叠在上层。

具体的代码实现，如代码示例 3-13 所示。

代码示例 3-13　堆叠布局：first_demo/entry/layout/ability_seven. xml

```
<?xml version = "1.0" encoding = "utf - 8"?>
< StackLayout
```

```
    xmlns:ohos = "http://schemas.huawei.com/res/ohos"
    ohos:id = "$ + id:stack_layout"
    ohos:height = "match_parent"
    ohos:width = "match_parent">

  < Text
        ohos:id = "$ + id:text_blue"
        ohos:text_alignment = "bottom|horizontal_center"
        ohos:text_size = "24fp"
        ohos:text = "第一层"
        ohos:height = "400vp"
        ohos:width = "400vp"
        ohos:background_element = "#3F56EA" />

  < Text
        ohos:id = "$ + id:text_light_purple"
        ohos:text_alignment = "bottom|horizontal_center"
        ohos:text_size = "24fp"
        ohos:text = "第二层"
        ohos:height = "300vp"
        ohos:width = "300vp"
        ohos:background_element = "#00AAEE" />

  < Text
        ohos:id = "$ + id:text_orange"
        ohos:text_alignment = "center"
        ohos:text_size = "24fp"
        ohos:text = "第三层"
        ohos:height = "80vp"
        ohos:width = "80vp"
        ohos:background_element = "#00BFC9" />
  </StackLayout >
```

使用相对位置添加组件，如代码示例 3-14 所示。使用 layout_alignment 属性可以指定组件在 StackLayout 中的相对位置，如将 Button 组件添加至 StackLayout 的右面，如图 3-39 所示。

图 3-39 使用 layout_alignment 属性

代码示例 3-14　使用相对位置添加组件

```
<?xml version = "1.0" encoding = "utf - 8"?>
< StackLayout
    xmlns:ohos = "http://schemas.huawei.com/res/ohos"
    ohos:id = " $ + id:stack_layout"
    ohos:height = "match_parent"
    ohos:width = "match_parent">
< Button
        ohos:id = " $ + id:button"
        ohos:height = "40vp"
        ohos:width = "80vp"
        ohos:layout_alignment = "right"
        ohos:background_element = " ＃3399FF"/>
</StackLayout >
```

将子视图从后面移到前面显示，如代码示例 3-15 所示，单击 moveChildToFront 方法让底部组件移动到最前面。

代码示例 3-15　将子视图从后面移到前面显示

```
ComponentContainer stackLayout = (ComponentContainer) findComponentById(ResourceTable.Id_
stack_layout);
Text textFirst = (Text) findComponentById(ResourceTable.Id_text_blue);
textFirst.setClickedListener(new Component.ClickedListener() {
    @Override
    public void onClick(Component component) {
        stackLayout.moveChildToFront(component);
    }
});
```

6. 自适应布局 AdaptiveBoxLayout

AdaptiveBoxLayout 意为自适应盒模式布局，是将整个 UI 划分为相同宽度，但高度有可能不同的行和列的盒子，也可以理解为将整个 UI 划分为多块。其中盒子的宽度取决于布局的宽度和每行中盒子的数量，这个需要在布局策略中指定。子组件的排列只有在前一行被填满后才会开始在新一行中占位。

自适应布局 AdaptiveBoxLayout 的特点如下。

(1) 布局中盒子(块)的宽度取决于布局的宽度和每行中盒子的数量，这些属性需要在布局策略中指定。

(2) 每个盒子(块)的高度由其子组件的高度决定。

(3) 子组件的排列只有在前一行被填满后才会开始在新一行中占位。

(4) 每个盒子(块)包含一个子组件。

(5) 每一行的高度由该行最高盒子的高度决定。

(6) 布局的宽度只能为 MATCH_PARENT 或固定值。

(7) 可以为布局中的组件设置长度、宽度和对齐方式。

如图 3-40 所示的相册中的图片会随着屏幕的变化自动选择适当的方式显示图片，如在大屏 TV 上方向布局使用水平方向，在小屏手机上方向布局使用垂直方向。

图 3-40　通过 AdaptiveBoxLayout 实现的自适应相册

我们来看一看如何实现图 3-40 的布局效果，如代码示例 3-16 所示。

这里创建一个 AdaptiveBoxLayout 布局组件，在 AdaptiveBoxLayout 组件中定义了两个 DirectionalLayout 组件，这两个组件会随着屏幕大小的改变自动选择以水平还是以垂直方式显示。

代码示例 3-16　自适应布局：first_demo/entry/layout/ability_eight. xml

```
< AdaptiveBoxLayout
        ohos:id = " $  + id:title_bar"
        ohos:height = "match_content"
        ohos:width = "match_parent"
        ohos:top_margin = "30vp">
```

```
<DirectionalLayout
            ohos:height = "match_content"
            ohos:width = "match_parent"
            ohos:margin = "10vp"
            ohos:orientation = "vertical">

<DirectionalLayout
                ohos:height = "match_content"
                ohos:width = "match_parent"
                ohos:orientation = "horizontal">

<Text
                    ohos:height = "match_content"
                    ohos:width = "match_content"
                    ohos:layout_alignment = "left|vertical_center"
                    ohos:multiple_lines = "false"
                    ohos:text = "推荐"
                    ohos:text_color = "#FF000000"
                    ohos:text_size = "20vp"
                    ohos:weight = "1"/>

<Text
                    ohos:height = "match_content"
                    ohos:width = "match_content"
                    ohos:layout_alignment = "right|vertical_center"
                    ohos:multiple_lines = "false"
                    ohos:text = "共 51 张"
                    ohos:text_color = "#FF787878"
                    ohos:text_size = "15vp"/>

<Image
                    ohos:height = "match_content"
                    ohos:width = "match_content"
                    ohos:image_src = "$media:icon_arrow.png"
                    ohos:layout_alignment = "right|vertical_center"/>
</DirectionalLayout>

<Image
                ohos:id = "$+id:pic1"
                ohos:height = "match_content"
                ohos:width = "match_parent"
                ohos:image_src = "$media:pic1.png"
                ohos:layout_alignment = "center"
                ohos:top_margin = "20vp"/>
</DirectionalLayout>
```

```
<DirectionalLayout
            ohos:height = "match_content"
            ohos:width = "match_parent"
            ohos:margin = "10vp"
            ohos:orientation = "vertical">

<DirectionalLayout
                ohos:height = "match_content"
                ohos:width = "match_parent"
                ohos:orientation = "horizontal">

<Text
                    ohos:height = "match_content"
                    ohos:width = "match_content"
                    ohos:layout_alignment = "left|vertical_center"
                    ohos:multiple_lines = "false"
                    ohos:text = "照片"
                    ohos:text_color = "#FF000000"
                    ohos:text_size = "20vp"
                    ohos:weight = "1"/>

<Text
                    ohos:height = "match_content"
                    ohos:width = "match_content"
                    ohos:layout_alignment = "right|vertical_center"
                    ohos:multiple_lines = "false"
                    ohos:text = "共 32 张"
                    ohos:text_color = "#FF787878"
                    ohos:text_size = "15vp"/>

<Image
                    ohos:height = "match_content"
                    ohos:width = "match_content"
                    ohos:image_src = "$media:icon_arrow.png"
                    ohos:layout_alignment = "right|vertical_center"/>
</DirectionalLayout>

<Image
                ohos:id = "$+id:pic2"
                ohos:height = "match_content"
                ohos:width = "match_parent"
                ohos:image_src = "$media:pic2.png"
                ohos:layout_alignment = "center"
                ohos:top_margin = "20vp"/>
</DirectionalLayout>
</AdaptiveBoxLayout>
```

实现完整效果如代码示例 3-17 所示。

代码示例 3-17　自适应布局：first_demo/entry/layout/ability_eight.xml

```
<?xml version = "1.0" encoding = "utf - 8"?>
< DirectionalLayout
    xmlns:ohos = "http://schemas.huawei.com/res/ohos"
    ohos:id = " $ + id:play_music_root"
    ohos:height = "match_parent"
    ohos:width = "match_parent"
    ohos:left_padding = "24vp"
    ohos:orientation = "vertical"
    ohos:right_padding = "24vp">

< Text
        ohos:height = "match_content"
        ohos:width = "match_parent"
        ohos:layout_alignment = "left"
        ohos:multiple_lines = "false"
        ohos:text = "图库"
        ohos:text_color = " # FF000000"
        ohos:text_size = "30vp"
        ohos:top_margin = "20vp"/>

< DirectionalLayout
        ohos:height = "match_content"
        ohos:width = "match_parent"
        ohos:orientation = "horizontal">

< Text
            ohos:height = "match_content"
            ohos:width = "match_content"
            ohos:layout_alignment = "left|vertical_center"
            ohos:multiple_lines = "false"
            ohos:text = "用户可自定义新的图库分组"
            ohos:text_color = " # FF787878"
            ohos:text_size = "15vp"
            ohos:weight = "1"/>

< Image
            ohos:id = " $ + id:remote"
            ohos:height = "match_content"
            ohos:width = "match_content"
            ohos:image_src = " $ media:icon_display_off.png"
            ohos:layout_alignment = "right|vertical_center"
            ohos:right_margin = "10vp"/>
```

```
<Image
        ohos:height = "match_content"
        ohos:width = "match_content"
        ohos:image_src = "$media:icon_more.png"
        ohos:layout_alignment = "right|vertical_center"/>
</DirectionalLayout>

<AdaptiveBoxLayout
    ohos:id = "$+id:title_bar"
    ohos:height = "match_content"
    ohos:width = "match_parent"
    ohos:top_margin = "30vp">

<DirectionalLayout
        ohos:height = "match_content"
        ohos:width = "match_parent"
        ohos:margin = "10vp"
        ohos:orientation = "vertical">

<DirectionalLayout
            ohos:height = "match_content"
            ohos:width = "match_parent"
            ohos:orientation = "horizontal">

<Text
                ohos:height = "match_content"
                ohos:width = "match_content"
                ohos:layout_alignment = "left|vertical_center"
                ohos:multiple_lines = "false"
                ohos:text = "推荐"
                ohos:text_color = "#FF000000"
                ohos:text_size = "20vp"
                ohos:weight = "1"/>

<Text
                ohos:height = "match_content"
                ohos:width = "match_content"
                ohos:layout_alignment = "right|vertical_center"
                ohos:multiple_lines = "false"
                ohos:text = "共 51 张"
                ohos:text_color = "#FF787878"
                ohos:text_size = "15vp"/>

<Image
```

```
                ohos:height = "match_content"
                ohos:width = "match_content"
                ohos:image_src = " $ media:icon_arrow.png"
                ohos:layout_alignment = "right|vertical_center"/>
</DirectionalLayout>

<Image
            ohos:id = " $ + id:pic1"
            ohos:height = "match_content"
            ohos:width = "match_parent"
            ohos:image_src = " $ media:pic1.png"
            ohos:layout_alignment = "center"
            ohos:top_margin = "20vp"/>
</DirectionalLayout>

<DirectionalLayout
        ohos:height = "match_content"
        ohos:width = "match_parent"
        ohos:margin = "10vp"
        ohos:orientation = "vertical">

<DirectionalLayout
            ohos:height = "match_content"
            ohos:width = "match_parent"
            ohos:orientation = "horizontal">

<Text
                ohos:height = "match_content"
                ohos:width = "match_content"
                ohos:layout_alignment = "left|vertical_center"
                ohos:multiple_lines = "false"
                ohos:text = "照片"
                ohos:text_color = "#FF000000"
                ohos:text_size = "20vp"
                ohos:weight = "1"/>

<Text
                ohos:height = "match_content"
                ohos:width = "match_content"
                ohos:layout_alignment = "right|vertical_center"
                ohos:multiple_lines = "false"
                ohos:text = "共 32 张"
                ohos:text_color = "#FF787878"
                ohos:text_size = "15vp"/>
```

```
    < Image
                    ohos:height = "match_content"
                    ohos:width = "match_content"
                    ohos:image_src = " $ media:icon_arrow.png"
                    ohos:layout_alignment = "right|vertical_center"/>
    </DirectionalLayout >

    < Image
                ohos:id = " $ + id:pic2"
                ohos:height = "match_content"
                ohos:width = "match_parent"
                ohos:image_src = " $ media:pic2.png"
                ohos:layout_alignment = "center"
                ohos:top_margin = "20vp"/>
    </DirectionalLayout >
    </AdaptiveBoxLayout >
    </DirectionalLayout >
```

3.6 ACE Java UI 基础组件

鸿蒙 ACE Java UI 框架为开发人员提供了开发中所需要的基础组件,可以通过组装这些基础组件来完成页面开发,本节学习鸿蒙 ACE Java UI 的基础组件的详细用法。

3.6.1 组件与组件容器

在学习基础组件之前,首先需要了解鸿蒙的组件分类及组件与组件容器的关系。根据组件(Component)的功能,可以将组件分为布局类、显示类、交互类三类,如图 3-41 所示。

组件类别	组件名称	功能描述
布局类	PositionLayout、DirectionalLayout、StackLayout、DependentLayout、TableLayout、AdaptiveBoxLayout	提供了不同布局规范的组件容器，例如以单一方向排列的DirectionalLayout、以相对位置排列的DependentLayout、以确切位置排列的PositionLayout等
显示类	Text、Image、Clock、TickTimer、ProgressBar	提供了单纯的内容显示，例如用于文本显示的Text，用于图像显示的Image等
交互类	TextField、Button、Checkbox、RadioButton/RadioContainer、Switch、ToggleButton、Slider、Rating、ScrollView、TabList、ListContainer、PageSlider、PageFlipper、PageSliderIndicator、Picker、TimePicker、DatePicker、SurfaceProvider、ComponentProvider	提供了具体场景下与用户交互响应的功能，例如Button提供了单击响应功能，Slider提供了进度选择功能等

图 3-41 组件分类

组件的定义:用户界面元素统称为组件,组件根据一定的层级结构进行组合形成布局。组件在未被添加到布局中时,既无法显示也无法交互,因此一个用户界面至少包含一个布局。

布局的定义：在 UI 框架中，具体的布局类通常以 XXLayout 命名，完整的用户界面是一个布局，用户界面中的一部分也可以是一个布局。布局中包含多个 Component 与 ComponentContainer 对象，如图 3-42 所示。

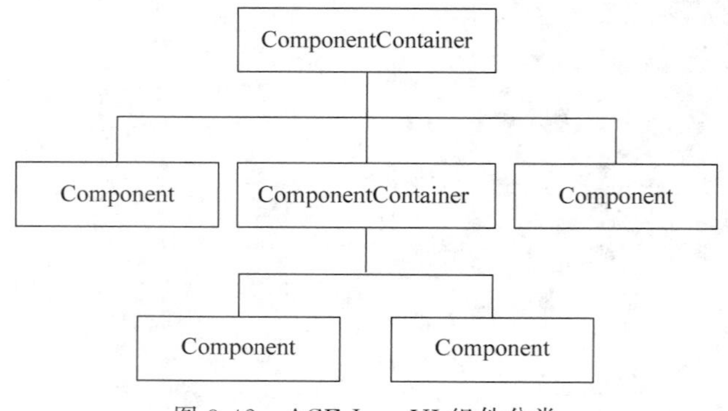

图 3-42　ACE Java UI 组件分类

Component：提供内容显示，是界面中所有组件的基类，可以给 Component 设置事件处理回调来创建一个可交互的组件。

Java UI 框架提供了一些常用的界面元素，也可称为组件，组件一般直接继承 Component 或它的子类，如 Text、Image 等。

ComponentContainer：作为容器容纳 Component 或 ComponentContainer 对象，并对它们进行布局。

Java UI 框架提供了一些标准布局功能的容器，它们继承自 ComponentContainer，一般以 Layout 结尾，如 DirectionalLayout、DependentLayout 等。

每种布局都根据自身特点提供 LayoutConfig 供子 Component 设定布局属性和参数，通过指定布局属性可以对子 Component 在布局中的显示效果进行约束。如 width、height 是最基本的布局属性，它们指定了组件的大小。

LayoutConfig 主要分为两种：DirectionalLayout 和 DependentLayout，如图 3-43 所示。

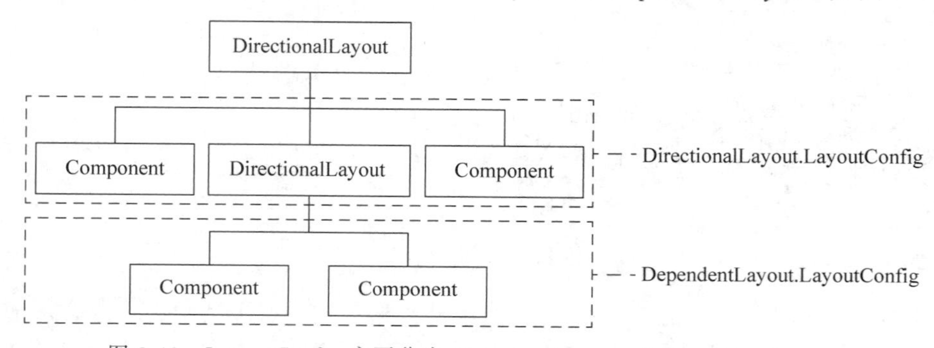

图 3-43　LayoutConfig 主要分为 DirectionalLayout 和 DependentLayout

代码示例 3-18 中通过 DirectionalLayout. LayoutConfig 创建一个布局配置对象，可以把这个创建好的配置对象设置给指定的组件，如代码中的 text. setLayoutConfig(layoutConfig)。

代码示例 3-18 为组件添加对应布局的布局属性

```
Text text = new Text(getContext());
 text.setText("编码创建视图布局");
 text.setTextSize(80);
 text.setId(100);
 //为组件添加对应布局的布局属性
DirectionalLayout. LayoutConfig layoutConfig = new DirectionalLayout. LayoutConfig
(ComponentContainer. LayoutConfig. MATCH _ CONTENT, ComponentContainer. LayoutConfig. MATCH _
CONTENT);
layoutConfig.alignment = LayoutAlignment.HORIZONTAL_CENTER;
text.setLayoutConfig(layoutConfig);
```

3.6.2 文本组件 Text

Text 是用来显示字符串的组件，在界面上显示为一块文本区域。Text 作为一个基本组件，有很多扩展，常见的有按钮组件 Button 和文本编辑组件 TextField。

Text 组件继承自 Component 类，如图 3-44 所示。

```
public class Text extends Component {
    public static final int AUTO_CURSOR_POSITION = -1;
    public static final int AUTO_SCROLLING_FOREVER = -1;

    public Text(Context context) {
        super((Context)null);
        throw new RuntimeException("Stub!");
    }
```

图 3-44 Text 组件继承自 Component 类

Text 组件支持根据文本长度自动调整文本的字体大小和换行。Text 组件可以通过 background_element 属性设置背景。常用的背景如文本背景、按钮背景，可以采用 XML 格式放置在 graphic 目录下。

在 graphic 目录下创建 background_text. xml 文件，如代码示例 3-19 所示。

在 Project 窗口，打开 entry→src→main→resources→base，右击 graphic 文件夹，选择 New→File，将文件命名为 background_text. xml，在 background_text. xml 文件中定义文本的背景。

代码示例 3-19 定义文本的背景

```
<?xml version = "1.0" encoding = "utf - 8"?>
< shape xmlns:ohos = "http://schemas.huawei.com/res/ohos"
```

```
    ohos:shape = "rectangle">
<cornersohos:radius = "20"/>
<solid ohos:color = "#878787"/>
<stroke ohos:width = "10" ohos:color = "red" />
</shape>
```

solid 标签：设置矩形的填充颜色。

stroke 标签：设置矩形边框的宽度、每段虚线的长度和两段虚线之间的颜色，以及矩形的颜色。

corners：设置形状的圆角。

可以通过 text_alignment 设置文本对齐方式，如设置为 horizontal_center|bottom。这里通过“|”设置了两个值，文字底部居中对齐，如图 3-45 和代码示例 3-20 所示。

图 3-45　设置文本对齐方式的效果

代码示例 3-20　text_alignment 设置文本对齐方式

```
<Text
    ohos:id = "$ + id:text"
    ohos:width = "300vp"
    ohos:height = "100vp"
    ohos:text = "Text"
    ohos:text_size = "28fp"
    ohos:text_color = "#0000FF"
    ohos:italic = "true"
    ohos:text_weight = "700"
    ohos:text_font = "serif"
    ohos:left_margin = "15vp"
    ohos:bottom_margin = "15vp"
    ohos:right_padding = "15vp"
    ohos:left_padding = "15vp"
    ohos:text_alignment = "horizontal_center|bottom"
    ohos:background_element = "$graphic:background_text"/>
```

如果 Text 组件的文字太长，则可以通过 multiple_lines 设置文本换行和最大显示行数，效果如图 3-46 所示，如代码示例 3-21 所示。

代码示例 3-21 设置文本换行和最大显示行数

```
<Text
    ohos:id = "$ + id:text"
    ohos:width = "75vp"
    ohos:height = "match_content"
    ohos:text = "TextText"
    ohos:text_size = "28fp"
    ohos:text_color = "#0000FF"
    ohos:italic = "true"
    ohos:text_weight = "700"
    ohos:text_font = "serif"
    ohos:multiple_lines = "true"
    ohos:max_text_lines = "2"
    ohos:background_element = "$graphic:background_text"/>
```

Text 对象支持根据文本长度自动调整文本的字体大小和换行，如图 3-47 所示。

图 3-46 设置文本换行和最大显示行数的效果

图 3-47 自动调节字体大小

设置自动换行、最大显示行数和自动调节字体大小，如代码示例 3-22 所示。

代码示例 3-22 设置自动换行、最大显示行数和自动调节字体大小

```
<Text
    ohos:id = "$ + id:text"
    ohos:width = "90vp"
    ohos:height = "match_content"
    ohos:min_height = "30vp"
    ohos:text = "T"
    ohos:text_color = "#0000FF"
    ohos:italic = "true"
    ohos:text_weight = "700"
    ohos:text_font = "serif"
    ohos:multiple_lines = "true"
    ohos:max_text_lines = "1"
    ohos:auto_font_size = "true"
    ohos:right_padding = "8vp"
    ohos:left_padding = "8vp"
    ohos:background_element = "$graphic:background_text"/>
```

通过 setAutoFontSizeRule 设置自动调整规则，3 个入参分别是最小的字体大小、最大的字体大小、每次调整文本字体大小的步长，如代码示例 3-23 所示。

代码示例 3-23　通过 setAutoFontSizeRule 设置自动调整规则

```
Text text = (Text) findComponentById(ResourceTable.Id_text);
//设置自动调整规则
text.setAutoFontSizeRule(30, 100, 1);
//设置单击一次增多一个字母"T"
text.setClickedListener(new Component.ClickedListener() {
    @Override
    public void onClick(Component component) {
        text.setText(text.getText() + "T");
    }
});
```

当文本过长时，可以设置跑马灯效果，如图 3-48 所示，实现文本滚动显示，其前提是文本换行被关闭且最大显示行数为 1，默认情况下即可满足前提要求。

图 3-48　自动调节字体大小

下面通过 XML 方式实现布局，同时需要通过代码进行跑马灯设置，如代码示例 3-24 所示。

代码示例 3-24 跑马灯文本效果 Text

```
<Text
    ohos:id = "$ + id:text"
    ohos:width = "75vp"
    ohos:height = "match_content"
    ohos:text = "TextText"
    ohos:text_size = "28fp"
    ohos:text_color = "#0000FF"
    ohos:italic = "true"
    ohos:text_weight = "700"
    ohos:text_font = "serif"
    ohos:background_element = "$graphic:background_text"/>
```

如果需要文字也有走马灯的效果，还需要在代码中将 setTruncationMode 属性值设置为 Text. TruncationMode. AUTO_SCROLLING，同时需要通过 startAutoScrolling 启动跑马灯效果，如代码示例 3-25 所示。

代码示例 3-25 setTruncationMode 设置跑马灯效果

```
public void onStart(Intent intent) {
    super.onStart(intent);
    super.setUIContent(ResourceTable.Layout_ability_main);

    Text text = (Text) findComponentById(ResourceTable.Id_text);
    //跑马灯效果
    text.setTruncationMode(Text.TruncationMode.AUTO_SCROLLING);
    //始终处于自动滚动状态
    text.setAutoScrollingCount(Text.AUTO_SCROLLING_FOREVER);
    //启动跑马灯效果
    text.startAutoScrolling();
}
```

3.6.3 按钮组件 Button

Button 是一种常见的组件，单击可以触发相应的操作，通常由文本或图标组成，也可以由图标和文本共同组成。

Button 组件继承自 Text 组件，是一种特殊的 Text 组件，如图 3-49 所示。

按照按钮的形状，按钮可以分为普通按钮、椭圆按钮、胶囊按钮、圆形按钮等。

按钮的形状可通过 Button 组件的 background_element 属性进行设置。background_element 可以直接被设置为颜色值，也可以通过在 graphic 文件夹中定义一个 xml 文件，该 xml 文件用来定义组件的形状，如图 3-50 所示。

可以在 graphic 文件夹中创建一个定义组件形状的 xml 文件，如代码示例 3-26 所示。

```java
public class Button extends Text {
    public Button(Context context) {
        super((Context)null);
        throw new RuntimeException("Stub!");
    }

    public Button(Context context, AttrSet attrSet) {
        super((Context)null);
        throw new RuntimeException("Stub!");
    }

    public Button(Context context, AttrSet attrSet, String styleName) {
        super((Context)null);
        throw new RuntimeException("Stub!");
    }
}
```

图 3-49 Button 组件继承自 Text 组件，是一种特殊的 Text

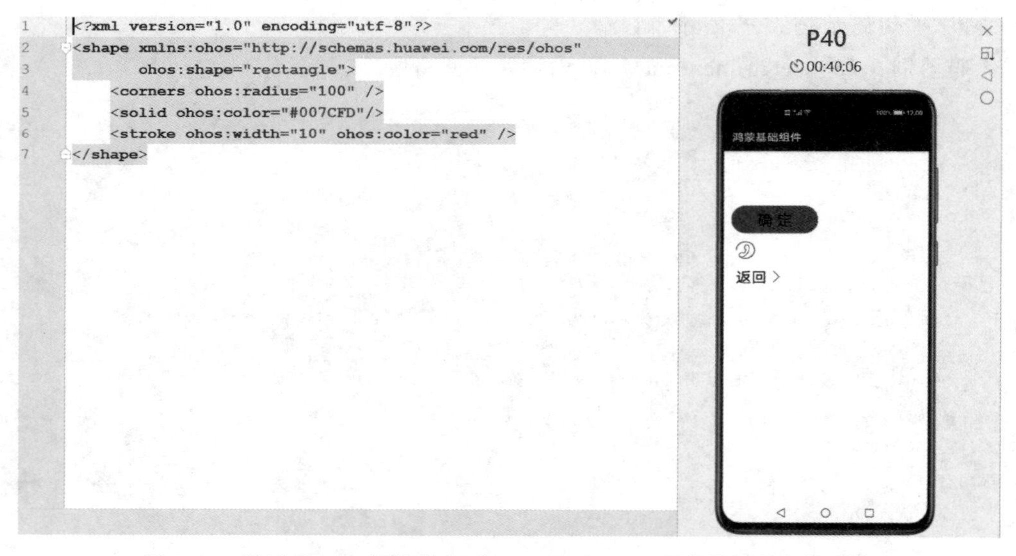

图 3-50 通过 Button 组件的 background_element 属性设置按钮的形状

代码示例 3-26 创建 Button 形状的 xml 文件

```xml
<?xml version = "1.0" encoding = "utf-8"?>
<shape xmlns:ohos = "http://schemas.huawei.com/res/ohos"
        ohos:shape = "rectangle">
<corners ohos:radius = "100" />
<solid ohos:color = "#007CFD"/>
<stroke ohos:width = "10" ohos:color = "red" />
</shape>
```

上面的 xml 文件通过 shape 标签定义了一个矩形，矩形的填充颜色为＃007CFD。shape 的默认形状是 rectangle，此外还有 oval（椭圆）、line（线）、ring（圆环）等形状。

注意： 如果将 shape 设置为 oval 或者 line、ring，则不能使用 corners 设置圆角，但是可以设置 solid 和 stroke。

普通按钮和其他按钮的区别在于不需要设置任何形状，只需设置文本和背景颜色即可。

background_element 属性指定自定义按钮的形状文件名：color_button_element. xml。通过 $graphic 指定 xml 文件名称，如代码示例 3-27 所示。

代码示例 3-27　自定义按钮的形状

```
<Button
    ohos:width="150vp"
    ohos:height="50vp"
    ohos:text_size="27fp"
    ohos:text="button"
    ohos:background_element="$graphic:color_button_element"
    ohos:left_margin="15vp"
    ohos:bottom_margin="15vp"
    ohos:right_padding="8vp"
    ohos:left_padding="8vp"
/>
```

Button 组件可以设置图片，如果需要为 Button 组件设置图片，则可以通过 svg 文件转换成 xml 文件，再通过 ohos:element_left、element_right、element_top、element_button 分别表示图片的居于文字的左、右、上、下 4 个位置。

Button 组件上的图片，可以通过一些开源的字体图标网站下载 svg 文件，然后通过 DevEco Studio 导出为 xml 文件，如图 3-51 所示。

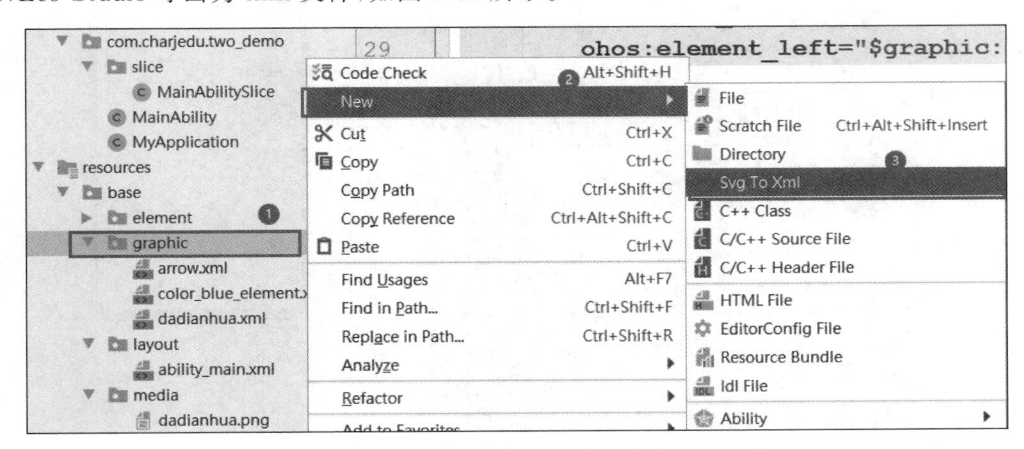

图 3-51　把外部的图片 svg 文件转换成 xml 文件

选择需要导入的 svg 文件，DevEco Studio 会自动把 svg 文件导出为 xml 文件，并保存到 graphic 目录下，如图 3-52 所示。

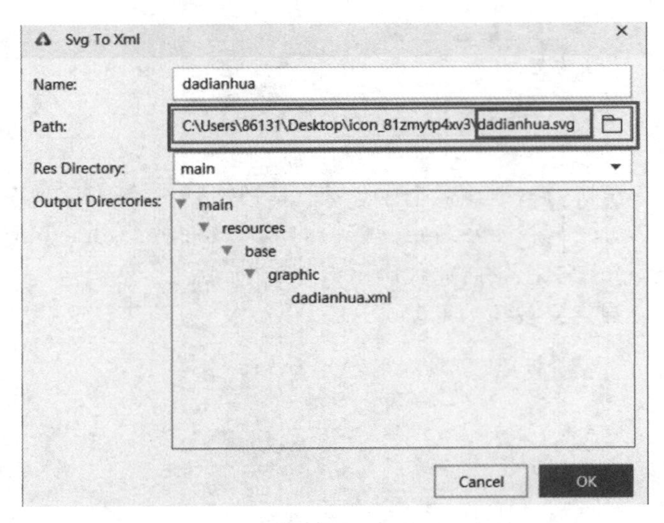

图 3-52 选择需要导入的 svg 文件

通过 element_right 引用导入的 svg 文件名，如果 Button 组件不设置 text 文字，则只显示图片。如果需要文字和图片一起显示，则可以通过 element_right|left|top|bottom 设置图片相对于文字的位置，如图 3-53 所示。

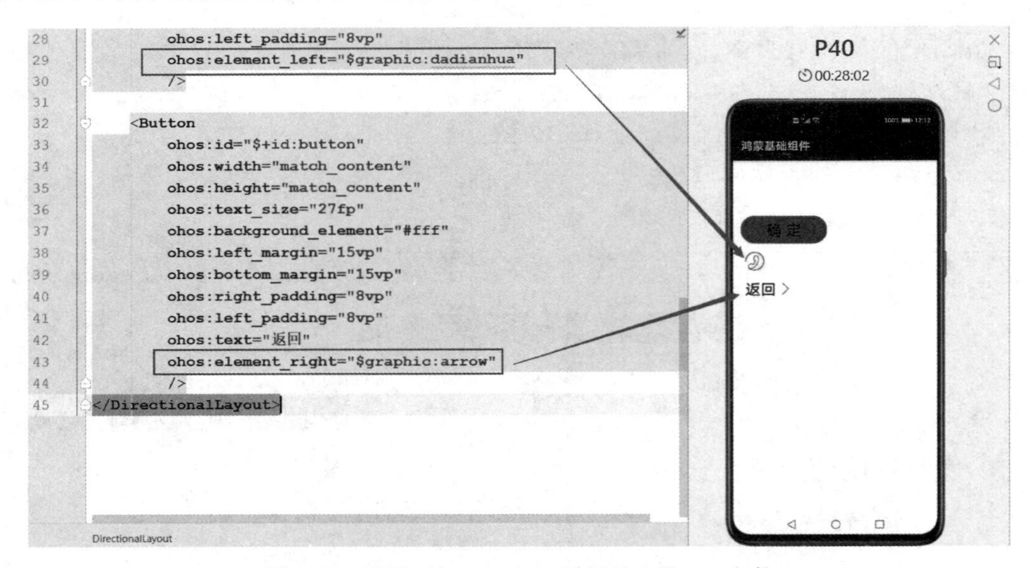

图 3-53 通过 element_right 引用导入的 svg 文件

3.6.4　文本输入框组件 TextField

TextField 文本输入框组件用来接收用户的输入，例如在用户的登录或者注册页面都会使用文本输入框组件。TextField 文本输入框组件继承自 Text 组件，如图 3-54 所示。

```
public class TextField extends Text {
    public TextField(Context context) {
        super((Context)null);
        throw new RuntimeException("Stub!");
    }

    public TextField(Context context, AttrSet attrSet) {
        super((Context)null);
        throw new RuntimeException("Stub!");
    }

    public TextField(Context context, AttrSet attrSet, String styleName) {
        super((Context)null);
        throw new RuntimeException("Stub!");
    }

    protected void setCursorChangedListener() { throw new RuntimeException("Stub!"); }
```

图 3-54　TextField 文本输入框组件继承自 Text 组件

文本输入框组件的背景默认为白色，可以通过 background_element 设置背景，hint 属性可以设置输入内容的提示信息，element_cursor_bubble 属性可以自定义光标提示气泡，如图 3-55 所示。

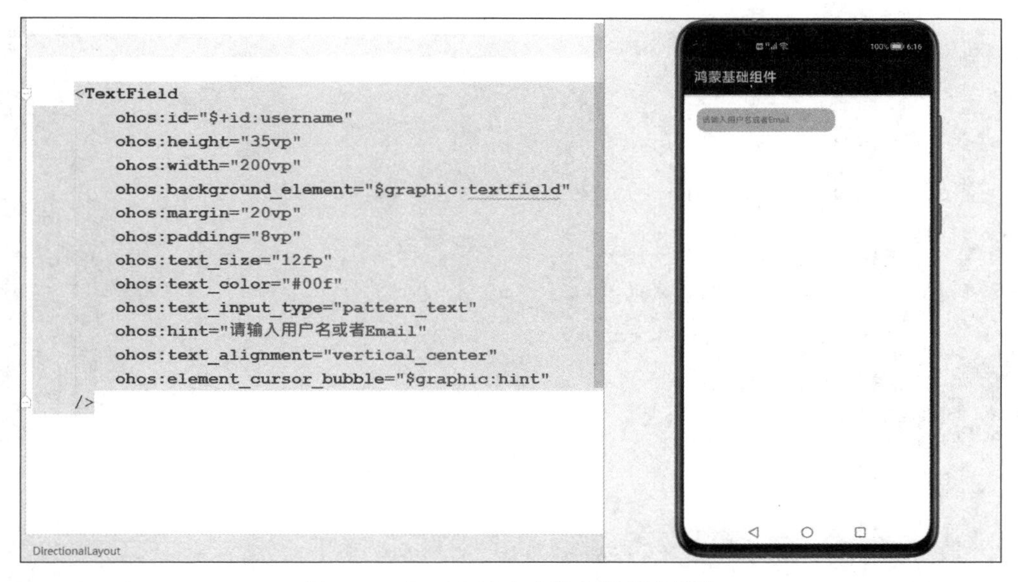

图 3-55　TextField 文本输入框基础用法

TextField 通过 text_input_type 属性设置不同类型的输入值：

- text_input_type="pattern_password"：密码类型；
- text_input_type="pattern_text"：文本类型；
- text_input_type="pattern_number"：数字类型；
- text_input_type="pattern_null"：空类型。

TextField 文本输入框组件的基本用法，如代码示例 3-28 所示。

代码示例 3-28　TextField 基本用法

```
<TextField
    ohos:id = "$ + id:username"
    ohos:height = "35vp"
    ohos:width = "200vp"
    ohos:element_cursor_bubble = "$graphic:ele_cursor_bubble"
    ohos:hint = "请输入密码"
    ohos:margin = "20vp"
    ohos:padding = "8vp"
    ohos:text_alignment = "vertical_center"
    ohos:text_color = "#00f"
    ohos:text_input_type = "pattern_password"
    ohos:text_size = "12fp"
    />
```

设置光标自定义气泡 Bubble，代码如下：

```
<TextField
    ...
    ohos:element_cursor_bubble = "$graphic:ele_cursor_bubble" />
```

ele_cursor_bubble.xml 用来自定义光标的气泡效果，代码如下：

```
<?xml version = "1.0" encoding = "UTF - 8" ?>
<shape xmlns:ohos = "http://schemas.huawei.com/res/ohos" ohos:shape = "rectangle">
<corners ohos:radius = "40"/>
<solid ohos:color = "#6699FF"/>
<stroke ohos:color = "#0066FF" ohos:width = "10"/>
</shape>
```

设置提示文字，代码如下：

```
<TextField
    ohos:hint = "Enter phone number or email"
    ohos:text_alignment = "vertical_center"/>
```

设置 TextField 的内边距，代码如下：

```
<TextField
    ohos:left_padding = "24vp"
    ohos:right_padding = "24vp"
    ohos:top_padding = "8vp"
    ohos:bottom_padding = "8vp"/>
```

设置 TextField 的多行显示，代码如下：

```
<TextField
    ohos:multiple_lines = "true"/>
```

通过 TextField 的 Enable 属性控制文本框是否可用，当设置成 false 时，无法在文本框输入，代码如下：

```
TextField textField = (TextField)
findComponentById(ResourceTable.Id_text_field);
textField.setEnabled(false);
```

响应焦点变化，代码如下：

```
textField.setFocusChangedListener(((component, isFocused) -> {
    if (isFocused) {
        //获取焦点
    } else {
        //失去焦点
    }
}));
```

下面是一个登录页面案例，如图 3-56 所示。本案例有 3 个文本输入框，第 1 个文本输入框用来填写用户手机号或者邮箱信息，第 2 个文本输入框用来填写密码，第 3 个文本输入框用来填写短信验证码。

单击登录按钮，在第 1 个文本框上会显示验证错误信息，验证错误信息覆盖在第 1 个文本框之上。第 2 个文本输入框需要对密码进行保护，将文本输入框类型修改为密码类型。

第 1 个文本输入框上面覆盖一个隐藏的文本框，我们需要采用堆叠布局方式，StatckLayout 可以实现堆叠显示效果。第 3 个输入框右边的“获取验证码”的文字效果同样需要使用堆叠布局实现。

第 1 个文本输入框的堆叠布局效果的实现，如代码示例 3-29 所示。

图 3-56 登录页面案例效果图

代码示例 3-29 第 1 个文本输入框添加叠加验证

```
<StackLayout
        ohos:height = "match_content"
        ohos:width = "match_parent"
        ohos:layout_alignment = "center"
        ohos:top_margin = "60vp">

<TextField
            ohos:id = "$ + id:name_textField"
            ohos:height = "match_content"
            ohos:width = "300vp"
            ohos:background_element = "$graphic:textfield"
            ohos:hint = "请输入邮箱或者手机号码"
            ohos:layout_alignment = "center"
            ohos:padding = "10vp"
            ohos:text_alignment = "center_vertical"
            ohos:text_size = "18fp"/>

<Text
            ohos:id = "$ + id:error_tip_text"
            ohos:height = "match_content"
```

```
                ohos:width = "300vp"
                ohos:layout_alignment = "right"
                ohos:padding = "10vp"
        ohos:text = "账号或者密码错误"
                ohos:text_color = "red"
                ohos:text_size = "18fp"
                ohos:visibility = "hide"/>
</StackLayout>
```

第3个"短信验证码"输入框的堆叠布局实现，如代码示例3-30所示，输入内容有可能超出堆叠在上面的"获取验证码"文本，可以通过控制底部输入文本框TextField的ohos: right_padding="100vp"来限制输入内容不会超过"获取验证码"文本框。

代码示例3-30 "短信验证码"输入框实现

```
<StackLayout
        ohos:height = "match_content"
        ohos:width = "300vp"
        ohos:top_margin = "40vp"
        ohos:layout_alignment = "center"
>

<TextField
                ohos:height = "match_content"
                ohos:width = "match_parent"
                ohos:background_element = "$graphic:textfield"
                ohos:hint = "短信验证码"
                ohos:top_padding = "10vp"
                ohos:bottom_padding = "10vp"
                ohos:left_padding = "10vp"
                ohos:right_padding = "100vp"
                ohos:text_alignment = "vertical_center"
                ohos:text_size = "18fp"
                />

<Text
                ohos:height = "match_content"
                ohos:width = "match_content"
                ohos:layout_alignment = "right"
                ohos:text = "获取验证码"
                ohos:padding = "10vp"
                ohos:text_color = "blue"
                ohos:text_alignment = "vertical_center"
                ohos:text_size = "18fp"
                />
</StackLayout>
```

当单击"登录"按钮时，在第 1 个输入框上面会显示错误信息提示，如代码示例 3-31 所示。

代码示例 3-31 单击按钮显示错误信息提示

```
//当单击登录时，改变相应组件的样式
Button button = (Button) findComponentById(ResourceTable.Id_ensure_button);
button.setClickedListener((component -> {
    //显示错误提示的 Text
    Text text = (Text) findComponentById(ResourceTable.Id_error_tip_text);
    text.setVisibility(Component.VISIBLE);

    //显示 TextField 错误状态下的样式
    ShapeElement errorElement = new ShapeElement(this, ResourceTable.Graphic_background_
text_field_error);
    TextField textField = (TextField) findComponentById(ResourceTable.Id_name_textField);
    textField.setBackground(errorElement);

    //TextField 失去焦点
    textField.clearFocus();
}));
```

下面，整体看一下登录布局页 XML 布局，如代码示例 3-32 所示。

代码示例 3-32 登录布局页 XML 布局：chapter03/demo6/textfeild_demo.xml

```
<?xml version = "1.0" encoding = "utf - 8"?>
<DirectionalLayout
    xmlns:ohos = "http://schemas.huawei.com/res/ohos"
    ohos:height = "match_parent"
    ohos:width = "match_parent"
    ohos:orientation = "vertical">


<StackLayout
        ohos:height = "match_content"
        ohos:width = "match_parent"
        ohos:layout_alignment = "center"
        ohos:top_margin = "60vp">

<TextField
            ohos:id = " $ + id:name_textField"
            ohos:height = "match_content"
            ohos:width = "300vp"
            ohos:background_element = " $ graphic:textfield"
            ohos:hint = "请输入邮箱或者手机号码"
            ohos:layout_alignment = "center"
```

```
                ohos:padding = "10vp"
                ohos:text_alignment = "center_vertical"
                ohos:text_size = "18fp"/>


    < Text
                ohos:id = " $ + id:error_tip_text"
                ohos:height = "match_content"
                ohos:width = "300vp"
                ohos:layout_alignment = "right"
                ohos:padding = "10vp"
                ohos:text = "账号或者密码错误"
                ohos:text_color = "red"
                ohos:text_size = "18fp"
                ohos:visibility = "hide"/>
    </StackLayout >


    < TextField
            ohos:id = " $ + id:password_text_field"
            ohos:height = "match_content"
            ohos:width = "300vp"
            ohos:background_element = " $ graphic:textfield"
            ohos:hint = "请输入密码"
            ohos:layout_alignment = "center"
            ohos:multiple_lines = "false"
            ohos:padding = "10vp"
            ohos:text_alignment = "center_vertical"
            ohos:text_size = "18fp"
            ohos:top_margin = "40vp"/>

    < StackLayout
            ohos:height = "match_content"
            ohos:width = "300vp"
            ohos:top_margin = "40vp"
            ohos:layout_alignment = "center"
    >

    < TextField
                ohos:height = "match_content"
                ohos:width = "match_parent"
                ohos:background_element = " $ graphic:textfield"
                ohos:hint = "短信验证码"
                ohos:top_padding = "10vp"
                ohos:bottom_padding = "10vp"
                ohos:left_padding = "10vp"
```

```
                ohos:right_padding = "100vp"
                ohos:text_alignment = "vertical_center"
                ohos:text_size = "18fp"
                />

    <Text
                ohos:height = "match_content"
                ohos:width = "match_content"
                ohos:layout_alignment = "right"
                ohos:text = "获取验证码"
                ohos:padding = "10vp"
                ohos:text_color = "blue"
                ohos:text_alignment = "vertical_center"
                ohos:text_size = "18fp"
                />
    </StackLayout>


    <Button
            ohos:id = "$ + id:ensure_button"
            ohos:height = "35vp"
            ohos:width = "120vp"
            ohos:background_element = "$ graphic:textfield"
            ohos:layout_alignment = "horizontal_center"
            ohos:text = "登 录"
            ohos:text_size = "20fp"
            ohos:top_margin = "40vp"/>

    </DirectionalLayout>
```

3.6.5 图片组件 Image

本地的图片一般存放在 resource 目录下的 media 文件夹中，可以使用图片 Image 组件显示这些本地的图片，可以对图片设置缩放和裁剪，如图 3-57 所示。

创建 Image 的方式有两种：可以在 XML 中创建 Image，也可以在代码中创建 Image。

在 XML 中创建 Image，image_src 属性用于设置图片的位置，这里通过 $ media:plant 指定 media 目录下的 plant.png 图片，代码如下：

```
<Image
    ohos:id = "$ + id:image"
    ohos:width = "match_content"
    ohos:height = "match_content"
    ohos:layout_alignment = "center"
    ohos:image_src = "$ media:plant"/>
```

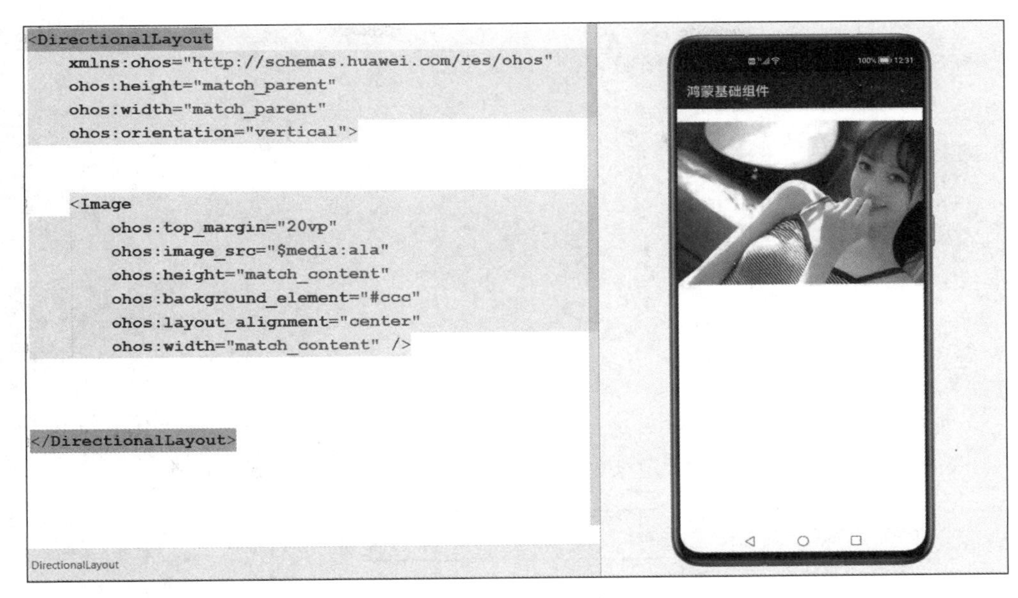

图 3-57　在 XML 中创建 Image

上面的代码可以通过代码的方式创建 Image,代码如下:

```
Image image = new Image(getContext());
image.setPixelMap(ResourceTable.Media_plant);
```

当图片尺寸与 Image 尺寸不同时,可以根据不同的缩放方式对图片进行缩放,如设置 Image 的宽和高为 200vp。ohos:scale_mode="zoom_center":表示按比例缩小居中显示,其他缩放方式取值如表 3-7 所示。

表 3-7　Scale_mode 缩放方式

缩放方式	值
按比例将原图扩大(缩小)到 Image 的宽度,居中显示	zoom_center
按比例将原图扩大(缩小)到 Image 的宽度,显示在 Image 的上部	zoom_start
按比例将原图扩大(缩小)到 Image 的宽度,显示在 Image 的下部	zoom_end
不按比例将图片扩大/缩小到 Image 的大小并显示	stretch
保持原图大小,显示在 Image 的中心。当原图的尺寸大于 Image 的尺寸时,超过部分进行裁剪处理	center
按比例将原图缩小到 Image 的宽度,将图片的内容完整居中显示	inside
按比例将原图扩大(缩小)到 Image 的宽度和高度中较大的值。如当设置的高度值较大时,在垂直方向上完整显示,而在水平方向上将超出 Image 宽度的部分进行裁剪处理	clip_center

加载和显示网络图片步骤如下。

（1）在 config.json 中添加网络权限，代码如下：

```
{
"module": {
"reqPermissions": [
        {
"name": "ohos.permission.INTERNET"
        }
      ]
  }
}
```

（2）配置网络明文访问白名单，这里将域名 blog.51itcto.com 设置为白名单，如图 3-58 所示。cleartextPermitted 表示自定义的网域范围内是否允许明文流量传输。

```
{
  "app": {"bundleName": "com.charjedu.migration_demo"...},
  "deviceConfig": {
    "default": {
      "network": {
        "usesCleartext": true,
        "securityConfig": {
          "domainSettings": {
            "cleartextPermitted": true,
            "domains": [
              {
                "subDomains": true,
                "name": "blog.51itcto.com"
              }
            ]
          }
        }
      }
    }
  },
  "module": {"name": ".MyApplication"...}
}
```

图 3-58 配置网络明文访问白名单

（3）在 XML 布局文件中添加 Image 组件，然后通过 ID 设置网络的图片，代码如下：

```
<Image
    ohos:top_margin = "10vp"
    ohos:background_element = "#ccc"
    ohos:id = "$ + id:photo"
    ohos:height = "match_content"
    ohos:width = "match_content"
/>
```

(4) 定义一个加载图片的方法,代码如下:

```
public void loadNetImageURL() {
    String URLImage = "";
    HttpURLConnection connection = null;

    try {
        URL URL = new URL(URLImage);
        URLConnection URLConnection = URL.openConnection();
        if (URLConnection instanceof HttpURLConnection) {
            connection = (HttpURLConnection) URLConnection;
        }
        if (connection != null) {
            connection.connect();
            //之后可进行 URL 的其他操作
            //得到服务器返回的流对象
            InputStream inputStream = URLConnection.getInputStream();
            ImageSource imageSource = ImageSource.create(inputStream, new ImageSource.
SourceOptions());
            ImageSource.DecodingOptions decodingOptions = new ImageSource.DecodingOptions();
            decodingOptions.desiredPixelFormat = PixelFormat.ARGB_8888;
            //普通解码叠加旋转、缩放、裁剪
            PixelMap pixelMap = imageSource.createPixelmap(decodingOptions);
            //普通解码
            getUITaskDispatcher().syncDispatch(() -> {
                new ToastDialog(MainAbilitySlice.this)
                        .setText("333")
                        .show();
                img.setPixelMap(pixelMap);
                img.release();
            });
        }

    } catch (Exception e) {
        System.out.println("-------------------------------");
        System.out.println(e.fillInStackTrace());
        System.out.println("-------------------------------");
    }
}
```

(5) 在 onStart 方法中,需要开启新的线程以便处理图片的加载,代码如下:

```
img = (Image)findComponentById(ResourceTable.Id_photo);

new Thread(new Runnable(){
```

```
        @Override
        public void run() {
            loadNetImageURL();
        }
    }).start();
```

3.6.6 TabList 和 Tab 组件

TabList 可以实现多个页签栏的切换，Tab 为某个页签。子页签通常放在内容区上方，用于展示不同的分类。页签名称应该简洁明了，清晰描述分类的内容，如图 3-59 所示。

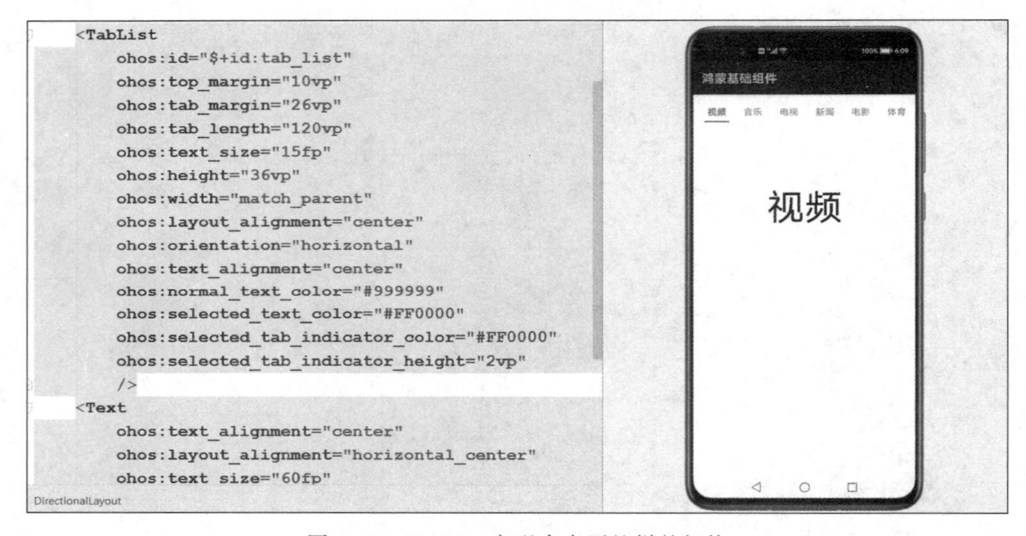

图 3-59 TabList 实现多个页签栏的切换

首先，在 XML 中创建 TabList 组件，如代码示例 3-33 所示。

代码示例 3-33 在 XML 中创建 TabList 组件

```
<TabList
    ohos:id = "$+id:tab_list"
    ohos:top_margin = "10vp"
    ohos:tab_margin = "24vp"
    ohos:tab_length = "140vp"
    ohos:text_size = "20fp"
    ohos:height = "36vp"
    ohos:width = "match_parent"
    ohos:layout_alignment = "center"
    ohos:orientation = "horizontal"
    ohos:text_alignment = "center"/>
```

设置默认状态和选中状态的字体颜色和 indicator 的颜色，如代码示例 3-34 所示。

代码示例 3-34 设置默认状态和选中状态的字体颜色

```
<TabList
    ohos:normal_text_color="#999999"
    ohos:selected_text_color="#FFFFFF"
    ohos:selected_tab_indicator_color="#FFFFFF"
    ohos:selected_tab_indicator_height="2vp"/>
```

通过代码向 TabList 中添加 Tab 子组件，代码如下：

```
TabList tabList = (TabList) findComponentById(ResourceTable.Id_tab_list);
TabList.Tab tab = tabList.new Tab(getContext());
tab.setText("Image");
tabList.addTab(tab);
```

在代码中设置 Tab 的布局，代码如下：

```
tabList.setTabLength(60); //设置 Tab 的宽度
tabList.setTabMargin(26); //设置两个 Tab 之间的间距
```

将 FixedMode 默认值设置为 false，在该模式下 TabList 的总宽度是各 Tab 宽度的总和，若固定了 TabList 的宽度，当超出可视区域时，则可以通过滑动 TabList 来显示。如果设置为 true，则 TabList 的总宽度将与可视区域相同，各个 Tab 的宽度也会根据 TabList 的宽度而进行平均分配，该模式适用于 Tab 较少的情况，代码如下：

```
tabList.setFixedMode(true);
```

TabList 常用接口如表 3-8 所示。

表 3-8 TabList 常用接口

方法	说明
getSelectedTab	返回选中的 Tab
getSelectedTabIndex	返回选中的 Tab 的位置索引
getTabCount	获取 Tab 的个数
getTabAt	获取某个 Tab
removeTab	移除某个位置的 Tab
setOrientation	设置横或竖方向

接下来，实现如图 3-59 所示页面的显示效果，TabList 组件在方向布局中排列在最上面，将上边距设置为 10vp，TabList 中子组件排列方式 orientaion 为水平方向排列。如代码示例 3-35 所示。

代码示例 3-35　实现如图 3-59 所示的页面显示效果

```
<?xml version = "1.0" encoding = "utf - 8"?>
< DirectionalLayout
    xmlns:ohos = "http://schemas.huawei.com/res/ohos"
    ohos:height = "match_parent"
    ohos:width = "match_parent"
    ohos:orientation = "vertical">

< TabList
        ohos:id = " $ + id:tab_list"
        ohos:top_margin = "10vp"
        ohos:tab_margin = "26vp"
        ohos:tab_length = "120vp"
        ohos:text_size = "15fp"
        ohos:height = "36vp"
        ohos:width = "match_parent"
        ohos:layout_alignment = "center"
        ohos:orientation = "horizontal"
        ohos:text_alignment = "center"
        ohos:normal_text_color = " # 999999"
        ohos:selected_text_color = " # FF0000"
        ohos:selected_tab_indicator_color = " # FF0000"
        ohos:selected_tab_indicator_height = "2vp"
        />
< Text
        ohos:text_alignment = "center"
        ohos:layout_alignment = "horizontal_center"
        ohos:text_size = "60fp"
        ohos:top_margin = "100vp"
        ohos:id = " $ + id:content"
        ohos:height = "match_content"
        ohos:width = "match_content"/>
</DirectionalLayout >
```

有了上面的 TabList 布局 xml 文件，首先根据数据创建 TabList 中的 Tab 组件，定义一个 Tab 组件的数组，代码如下：

```
private String[] str = {"相册","视频","音乐","电视","新闻","电影","体育","旅游"};
```

通过循环 Tab 组件的数组，为 TabList 添加 Tab 组件，代码如下：

```
for (int i = 0; i < str.length; i++) {
      TabList.Tab tab = tabList.new Tab(getContext());
      tab.setText(str[i]);
```

```
tabList.addTab(tab);
 }
```

监听 TabList 的选中事件，在某个 Tab 组件被选中的时候，会触发 onSelected 事件，并通过事件的参数返回选中的 Tab 组件信息，代码如下：

```
tabList.addTabSelectedListener(new TabList.TabSelectedListener() {
     @Override
     public void onSelected(TabList.Tab tab) {
          //当某个 Tab 从未选中状态变为选中状态时的回调
          content.setText(tab.getText());
     }
     @Override
     public void onUnselected(TabList.Tab tab) {
          //当某个 Tab 从选中状态变为未选中状态时的回调
     }
     @Override
     public void onReselected(TabList.Tab tab) {
          //当某个 Tab 已处于选中状态,再次被单击时的回调
     }
 });
```

完整的 TabList 实现代码，如代码示例 3-36 所示。

代码示例 3-36　TabList 代码实现

```
public class TabListAbilitySlice extends AbilitySlice {
    private  String[] str = {"相册","视频","音乐","电视","新闻","电影","体育","旅游"};
    Text content;
    @Override
    public void onStart(Intent intent) {
        super.onStart(intent);
        super.setUIContent(ResourceTable.Layout_ability_main);
        content = (Text)findComponentById(ResourceTable.Id_content);
        initview();
    }

    private void initview() {
        TabList tabList = (TabList) findComponentById(ResourceTable.Id_tab_list);
        tabList.setFixedMode(true);
        if(tabList!= null){
            for (int i = 0; i < str.length; i++) {
                TabList.Tab tab = tabList.new Tab(getContext());
                tab.setText(str[i]);

                tabList.addTab(tab);
```

```
            }

            tabList.selectTabAt(1);
            content.setText(tabList.getSelectedTab().getText());

            tabList.setTabLength(120); //设置 Tab 的宽度
            tabList.setTabMargin(26);  //设置两个 Tab 之间的间距

// tabList.setFixedMode(true);
            tabList.addTabSelectedListener(new TabList.TabSelectedListener() {
                @Override
                public void onSelected(TabList.Tab tab) {
                    //当某个 Tab 从未选中状态变为选中状态时的回调
                    content.setText(tab.getText());
                }
                @Override
                public void onUnselected(TabList.Tab tab) {
                    //当某个 Tab 从选中状态变为未选中状态时的回调
                }
                @Override
                public void onReselected(TabList.Tab tab) {
                    //当某个 Tab 已处于选中状态,再次被单击时的回调
                }
            });
        }
    }

}
```

3.6.7 Picker 组件

Picker 组件提供了滑动选择器，允许用户从预定义范围中进行选择，如图 3-60 所示。在 XML 中创建 Picker 组件，代码如下：

```
<Picker
    ohos:id="$+id:test_picker"
    ohos:height="match_content"
    ohos:width="300vp"
    ohos:background_element="#E1FFFF"
    ohos:layout_alignment="horizontal_center"
    ohos:normal_text_size="16fp"
    ohos:selected_text_size="16fp"/>
```

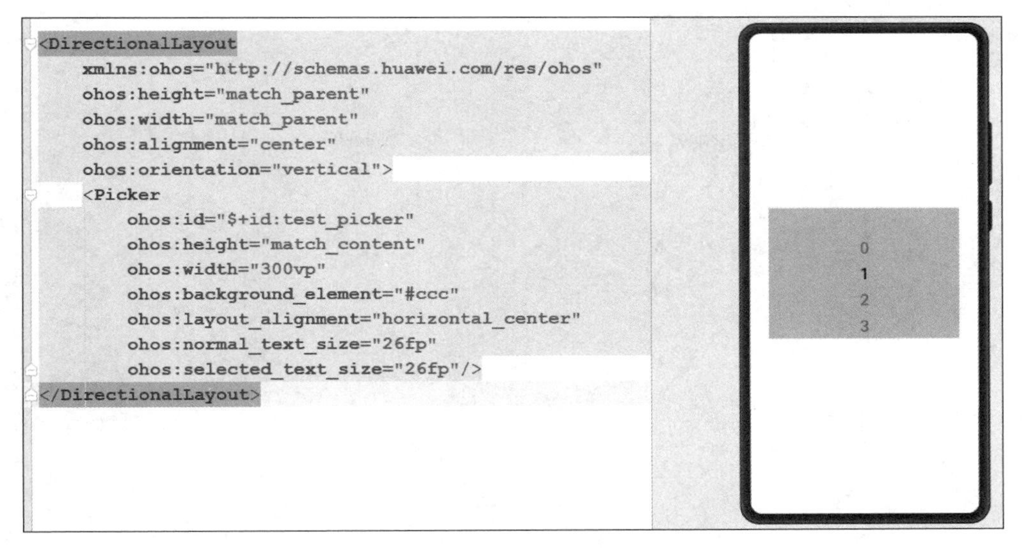

图 3-60　Picker 组件

设置 Picker 组件的取值范围，代码如下：

```
Picker picker = (Picker) findComponentById(ResourceTable.Id_test_picker);
picker.setMinValue(0);          //设置选择器中的最小值
picker.setMaxValue(6);          //设置选择器中的最大值
```

响应选择器变化，代码如下：

```
picker.setValueChangedListener((picker1, oldVal, newVal) -> {
    //oldVal:上一次选择的值; newVal: 最新选择的值
});
```

通过 Picker 组件的 setFormatter(Formatter formatter)方法，如图 3-61 所示，用户可以将 Picker 选项中显示的字符串修改为特定的格式，代码如下：

```
picker.setFormatter(i -> {
            String value;
            switch (i) {
                case 0:
                    value = "星期一";
                    break;
                case 1:
                    value = "星期二";
                    break;
                case 2:
                    value = "星期三";
```

```
                break;
            case 3:
                value = "星期四";
                break;
            case 4:
                value = "星期五";
                break;
            case 5:
                value = "星期六";
                break;
            case 6:
                value = "星期日";
                break;
            default:
                value = "" + i;
        }
        return value;
    });
```

图 3-61　Picker 组件修改格式后的选择器

对于不直接显示数字的组件，该方法可以设置字符串与数字一一对应。字符串数组长度必须等于取值范围值的总数。用户在使用时需要注意，该方法会覆盖 picker.setFormatter(Formatter formatter)方法。

Java 代码中，通过 setDisplayData 方法添加一个字符串数组，代码如下：

```
picker.setDisplayedData(new String[]{"星期一", "星期二", "星期三", "星期四", "星期五", "星期六", "星期日"});
```

下面两个 Picker 组件属于时间选择器：DatePicker 和 TimePicker。

1. DatePicker 主要供用户选择日期

在 XML 中创建 DatePicker，显示效果如图 3-62 所示，代码如下：

```
<DatePicker
    ohos:id = "$ + id:date_pick"
    ohos:height = "match_content"
    ohos:width = "300vp">
</DatePicker>
```

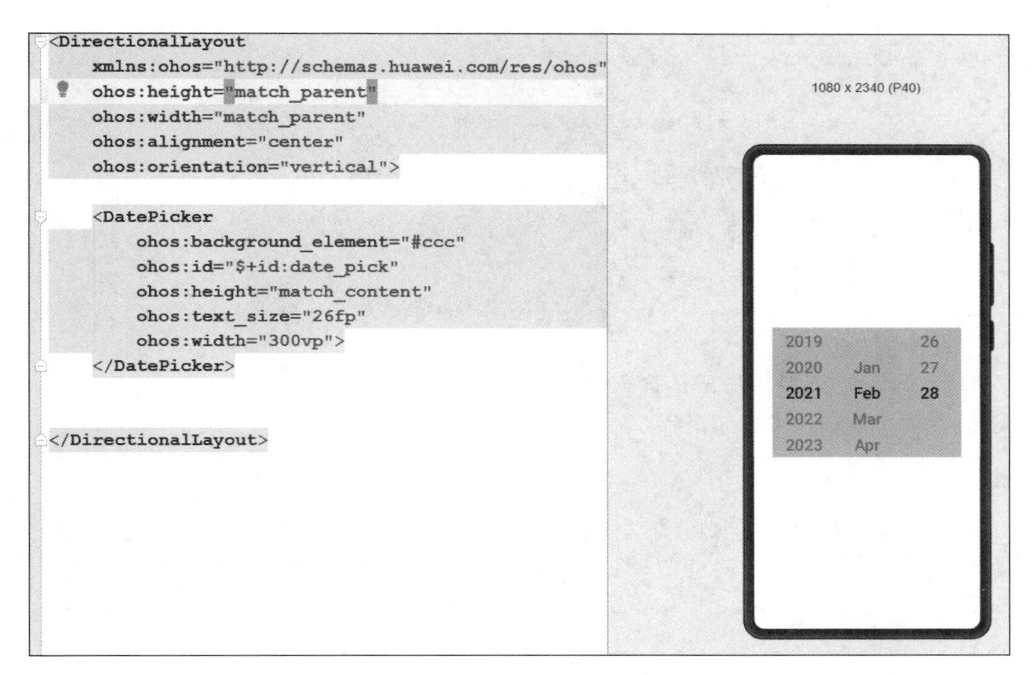

图 3-62　创建默认的 DatePicker

获取当前日期，年/月/日，DatePicker 默认选择当前日期，代码如下：

```
//获取 DatePicker 实例
DatePicker datePicker = (DatePicker)
findComponentById(ResourceTable.Id_date_pick);
int day = datePicker.getDayOfMonth();
int month = datePicker.getMonth();
int year = datePicker.getYear();
```

响应日期改变事件，在 XML 中添加 Text 用于显示所选择的日期，代码如下：

```
<Text
    ohos:id="$+id:text_date"
    ohos:height="match_content"
    ohos:width="match_parent"
    ohos:hint="date"
    ohos:margin="8vp"
    ohos:padding="4vp"
    ohos:text_size="14fp">
</Text>
```

在 Java 代码中响应日期改变事件，代码如下：

```
Text selectedDate = (Text) findComponentById(ResourceTable.Id_text_date);
datePicker.setValueChangedListener(
```

```
        new DatePicker.ValueChangedListener() {
            @Override
            public void onValueChanged(DatePicker datePicker, int year, int monthOfYear, int
dayOfMonth) {
                selectedDate.setText(String.format("%02d/%02d/%4d", dayOfMonth,
monthOfYear, year));
            }
        }
);
```

2. TimePicker 主要供用户选择时间

创建 TimePicker,代码如下:

```
<TimePicker
    ohos:id="$+id:time_picker"
    ohos:height="match_content"
    ohos:width="match_parent" />
```

通过下面的代码可以设置 TimePicker 的时间,效果如图 3-63 所示。

```
timePicker.setHour(15);
timePicker.setMinute(06);
timePicker.setSecond(25);
```

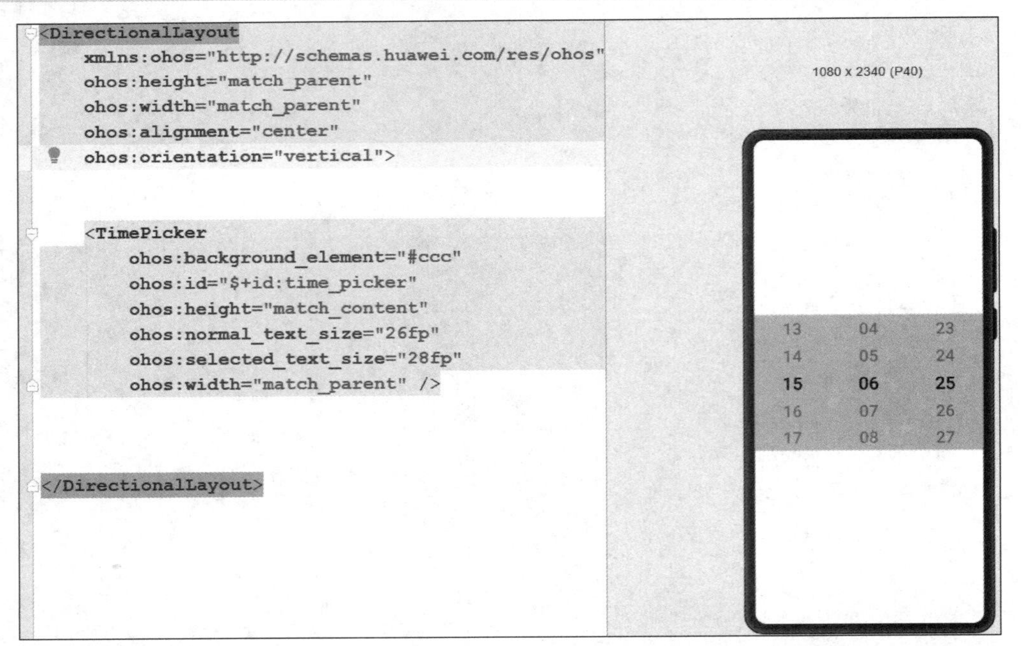

图 3-63　创建一个默认的 TimePicker 的效果

获取时间，代码如下：

```
TimePicker timePicker = (TimePicker)
findComponentById(ResourceTable.Id_time_picker);
int hour = timePicker.getHour();
int minute = timePicker.getMinute();
int second = timePicker.getSecond();
```

响应时间改变事件，代码如下：

```
timePicker.setTimeChangedListener(new TimePicker.TimeChangedListener() {
    @Override
    public void onTimeChanged(TimePicker timePicker, int hour, int minute, int second) {
        ...
    }
});
```

3.6.8　复选框组件 CheckBox

CheckBox 可以实现选中和取消选中功能，如图 3-64 所示。

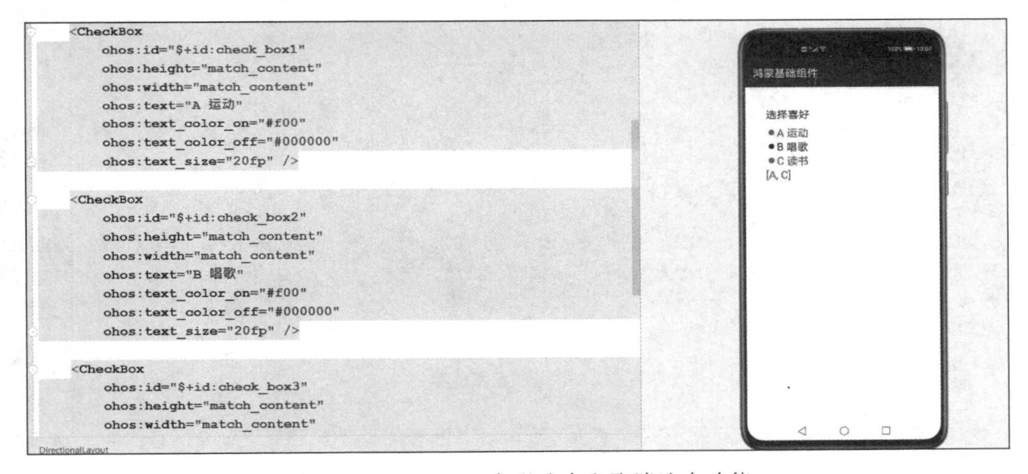

图 3-64　CheckBox 实现选中和取消选中功能

在 xml 中设置 CheckBox 的背景，如图 3-65 所示，代码如下：

```
< CheckBox
    ohos:check_element = " $ graphic:background_checkbox_check" />
```

graphic 目录下 xml 文件（例：background_checkbox_check.xml）的示例代码如下：

```
<?xml version = "1.0" encoding = "UTF - 8" ?>
< shape xmlns:ohos = "http://schemas.huawei.com/res/ohos"
```

```
        ohos:shape = "oval">
    < solid ohos:color = " # 00FFCC"/>
</shape >
```

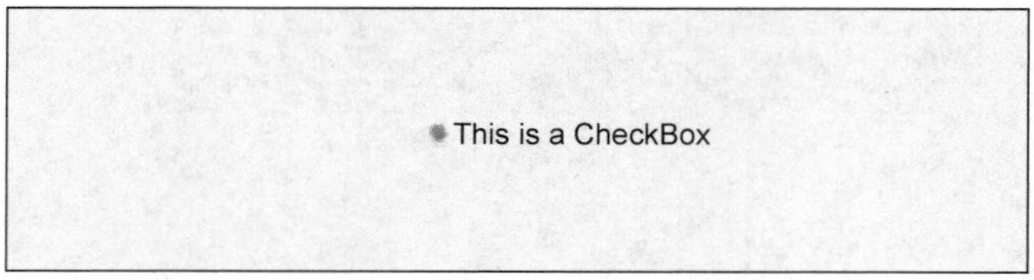

图 3-65　使用 xml 设置 CheckBox 背景的效果

使用 Java 代码设置 CheckBox 在选中与取消选中时的背景，如图 3-66 所示，如代码示例 3-37 所示。

代码示例 3-37　设置 CheckBox 在选中与取消选中时的背景

```
ShapeElement elementButtonOn = new ShapeElement();
elementButtonOn.setRgbColor(RgbPalette.RED);
elementButtonOn.setShape(ShapeElement.RECTANGLE);
elementButtonOn.setCornerRadius(0.0f);

ShapeElement elementButtonOff = new ShapeElement();
elementButtonOff.setRgbColor(RgbPalette.BLACK);
elementButtonOff.setShape(ShapeElement.RECTANGLE);
elementButtonOff.setCornerRadius(0.0f);

StateElement checkElement = new StateElement();
checkElement.addState(new int[]{ComponentState.COMPONENT_STATE_CHECKED}, elementButtonOn);
checkElement.addState(new int[]{ComponentState.COMPONENT_STATE_EMPTY}, elementButtonOff);
CheckBox checkbox = (CheckBox)
findComponentById(ResourceTable.Id_check_box);
checkBox.setButtonElement(checkElement);
```

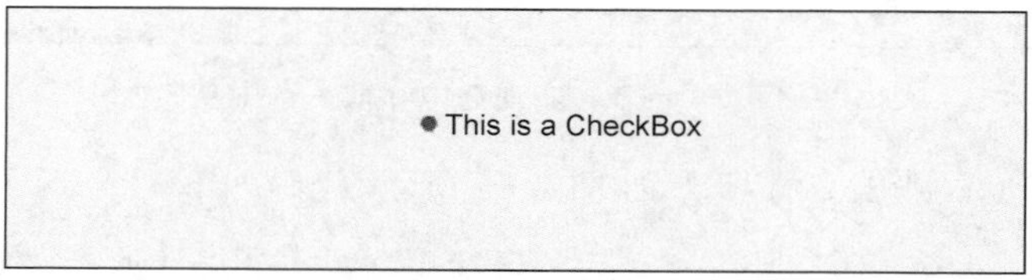

图 3-66　使用 Java 代码设置 CheckBox 背景的效果

设置 CheckBox 的文字在选中和取消选中时的颜色，如图 3-67 所示，代码如下：

```
< CheckBox
    ...
```

```
    ohos:text_color_on = "#00AAEE"
    ohos:text_color_off = "#000000" />
```


图 3-67 设置 CheckBox 文字颜色的效果

设置 CheckBox 的选中状态，代码如下：

```
checkbox.setChecked(true);
```

设置不同状态之间的切换：如果当前为选中状态，则将变为未选中；如果当前是未选中状态，则将变为选中状态，代码如下：

```
checkbox.toggle();
```

设置响应 CheckBox 状态变更的事件，代码如下：

```
//state 表示是否被选中
checkbox.setCheckedStateChangedListener((component, state) -> {
    //状态改变的逻辑
});
```

实现如图 3-64 所示的页面显示效果，如代码示例 3-38 所示。

代码示例 3-38　实现如图 3-64 所示的页面显示效果

```
<?xml version = "1.0" encoding = "utf-8"?>
<DirectionalLayout
    xmlns:ohos = "http://schemas.huawei.com/res/ohos"
    ohos:height = "match_parent"
    ohos:width = "match_parent"
    ohos:orientation = "vertical"
    ohos:left_padding = "40vp"
    ohos:top_padding = "40vp">

<Text
        ohos:height = "match_content"
        ohos:width = "match_content"
        ohos:text_size = "20vp"
        ohos:bottom_margin = "10vp"
        ohos:text_alignment = "horizontal_center"
```

```
            ohos:text = "选择喜好"/>

    < CheckBox
            ohos:id = "$ + id:check_box1"
            ohos:height = "match_content"
            ohos:width = "match_content"
            ohos:text = "A 运动"
            ohos:text_color_on = "#f00"
            ohos:text_color_off = "#000000"
            ohos:text_size = "20fp" />

    < CheckBox
            ohos:id = "$ + id:check_box2"
            ohos:height = "match_content"
            ohos:width = "match_content"
            ohos:text = "B 唱歌"
            ohos:text_color_on = "#f00"
            ohos:text_color_off = "#000000"
            ohos:text_size = "20fp" />

    < CheckBox
            ohos:id = "$ + id:check_box3"
            ohos:height = "match_content"
            ohos:width = "match_content"
            ohos:text = "C 读书"
            ohos:text_color_on = "#f00"
            ohos:text_color_off = "#000000"
            ohos:text_size = "20fp" />

    < Text
            ohos:id = "$ + id:result"
            ohos:height = "match_parent"
            ohos:width = "match_content"
            ohos:text_color = "red"
            ohos:text_size = "20vp"
            ohos:text_alignment = "horizontal_center"
            ohos:text = "[]"
            />
</DirectionalLayout >
```

上面，创建好了页面的布局，首先需要定义一个用来保存选中结果的集合，代码如下：

```
private Set < String > selectedSet  =  new HashSet <>();
```

为每个 CheckBox 绑定 setCheckedStateChangedListener 事件监听器，当监听选中时，把选择的编号添加到 Set 集合中，如代码示例 3-39 所示。

代码示例 3-39　setCheckedStateChangedListener 事件监听器

```
public class MainAbilitySlice extends AbilitySlice {

    //保存最终选中的结果
    private Set<String> selectedSet = new HashSet<>();
    @Override
    public void onStart(Intent intent) {
        super.onStart(intent);
        super.setUIContent(ResourceTable.Layout_ability_main);
        initCheckBox();
    }

    //初始化 CheckBox
    private void initCheckBox() {
        CheckBox checkbox1 = (CheckBox) findComponentById(ResourceTable.Id_check_box1);
        checkbox1.setButtonElement(elementButtonInit());
        checkbox1.setCheckedStateChangedListener((component, state) -> {
            if (state) {
                selectedSet.add("A");
            } else {
                selectedSet.remove("A");
            }
            showAnswer();
        });

        CheckBox checkbox2 = (CheckBox) findComponentById(ResourceTable.Id_check_box2);
        checkbox2.setButtonElement(elementButtonInit());
        checkbox2.setCheckedStateChangedListener((component, state) -> {
            if (state) {
                selectedSet.add("B");
            } else {
                selectedSet.remove("B");
            }
            showAnswer();
        });

        CheckBox checkbox3 = (CheckBox) findComponentById(ResourceTable.Id_check_box3);
        checkbox3.setButtonElement(elementButtonInit());
        checkbox3.setCheckedStateChangedListener((component, state) -> {
            if (state) {
                selectedSet.add("C");
            } else {
                selectedSet.remove("C");
            }
            showAnswer();
        });
    }
```

```
    //设置 CheckBox 背景
    private StateElement elementButtonInit() {
        ShapeElement elementButtonOn = new ShapeElement();
        elementButtonOn.setRgbColor(RgbPalette.RED);
        elementButtonOn.setShape(ShapeElement.OVAL);

        ShapeElement elementButtonOff = new ShapeElement();
        elementButtonOff.setRgbColor(RgbPalette.BLACK);
        elementButtonOff.setShape(ShapeElement.OVAL);

        StateElement checkElement = new StateElement();
        checkElement.addState(new int[]{ComponentState.COMPONENT_STATE_CHECKED},
elementButtonOn);
        checkElement.addState(new int[]{ComponentState.COMPONENT_STATE_EMPTY},
elementButtonOff);

        return checkElement;
    }

    //显示结果
    private void showAnswer() {
        Text answerText = (Text) findComponentById(ResourceTable.Id_result);
        String answer = selectedSet.toString();
        answerText.setText(answer);
    }
}
```

3.6.9 单选按钮组件 RadioButton

RadioButton 用于多选一操作，需要搭配 RadioContainer 使用，实现单选效果，如图 3-68 所示。

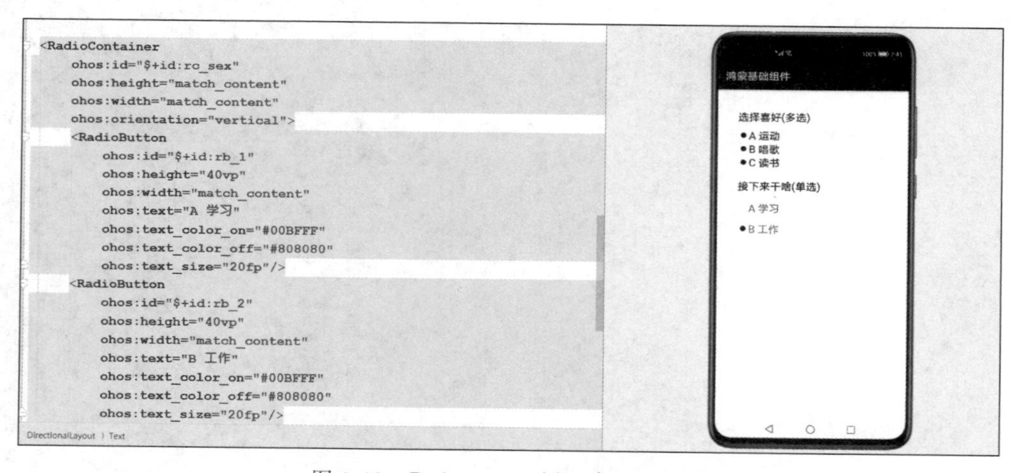

图 3-68 RadioButton 用于多选一操作

RadioContainer 是 RadioButton 的容器，在其包裹下的 RadioButton 保证只有一个被选中，如代码示例 3-40 所示。

代码示例 3-40　RadioContainer 包裹 RadioButton 实现单选

```
<RadioContainer
    ohos:id="$+id:radio_container"
    ohos:height="match_content"
    ohos:width="match_content"
    ohos:orientation="vertical">
<RadioButton
        ohos:id="$+id:rb_1"
        ohos:height="40vp"
        ohos:width="match_content"
        ohos:text="A 学习"
        ohos:text_color_on="#00BFFF"
        ohos:text_color_off="#808080"
        ohos:text_size="20fp"/>
<RadioButton
        ohos:id="$+id:rb_2"
        ohos:height="40vp"
        ohos:width="match_content"
        ohos:text="B 工作"
        ohos:text_color_on="#00BFFF"
        ohos:text_color_off="#808080"
        ohos:text_size="20fp"/>
</RadioContainer>
```

设置响应 RadioContainer 状态改变的事件，如代码示例 3-41 所示。

代码示例 3-41　设置响应 RadioContainer 状态改变的事件

```
RadioContainer container = (RadioContainer) findComponentById(ResourceTable.Id_radio_container);
container.setMarkChangedListener(new RadioContainer.CheckedStateChangedListener() {
    @Override
    public void onCheckedChanged(RadioContainer radioContainer, int index) {
    }
});
```

根据索引值设置指定 RadioButton 为选定状态，代码如下：

```
container.mark(0);
```

清除 RadioContainer 中所有 RadioButton 的选定状态，代码如下：

```
container.cancelMarks();
```

设置 RadioButton 的布局方向：将 orientation 设置为 horizontal，表示横向布局；将 orientation 设置为 vertical，表示纵向布局。默认为纵向布局。

在 xml 中设置，代码如下：

```
<RadioContainer
    ohos:orientation = "horizontal">
</RadioContainer>
```

在 Java 代码中设置，代码如下：

```
container.setOrientation(Component.HORIZONTAL);
```

3.6.10 信息提示框组件 ToastDialog

ToastDialog 是在窗口上方弹出的对话框，是通知操作的简单反馈。ToastDialog 会在一段时间后消失，在此期间，用户还可以操作当前窗口的其他组件。

ToastDialog 继承自 CommonDialog 类，如图 3-69 所示。

```
public class ToastDialog extends CommonDialog {
    public ToastDialog(Context context) {
        super((Context)null);
        throw new RuntimeException("Stub!");
    }
```

图 3-69 ToastDialog 继承自 CommonDialog 类

通过 setAlignment 方法设置提示框显示的位置，代码如下：

```
new ToastDialog(getContext())
    .setText("This is a ToastDialog displayed in the middle")
    .setAlignment(LayoutAlignment.CENTER)
    .show();
```

自定义 ToastDialog 的 Component，通过 setComponent 方法设置自定义的布局组件，如代码示例 3-42 所示。

代码示例 3-42 setComponent 方法设置自定义的布局组件

```
DirectionalLayout toastLayout = (DirectionalLayout) LayoutScatter.getInstance(this)
                        .parse(ResourceTable.Layout_layout_toast, null, false);
new ToastDialog(getContext())
    .setComponent(toastLayout)
    .setSize(DirectionalLayout.LayoutConfig.MATCH_CONTENT, DirectionalLayout.LayoutConfig.
MATCH_CONTENT)
```

```
    .setAlignment(LayoutAlignment.CENTER)
    .show();
```

创建自定义布局文件：layout_toast.xml，如代码示例 3-43 所示。

代码示例 3-43　创建自定义布局文件

```
<?xml version = "1.0" encoding = "utf - 8"?>
< DirectionalLayout
    xmlns:ohos = "http://schemas.huawei.com/res/ohos"
    ohos:height = "match_content"
    ohos:width = "match_content"
    ohos:orientation = "vertical">
< Text
        ohos:id = " $ + id:msg_toast"
        ohos:height = "match_content"
        ohos:width = "match_content"
        ohos:left_padding = "16vp"
        ohos:right_padding = "16vp"
        ohos:top_padding = "4vp"
        ohos:bottom_padding = "4vp"
        ohos:layout_alignment = "center"
        ohos:text_size = "16fp"
        ohos:text = "This is a ToastDialog for the customized component"
        ohos:background_element = " $ graphic:background_toast_element"/>
</DirectionalLayout >
```

在上面的布局文件中定义背景文件：background_toast_element.xml，如代码示例 3-44 所示。

代码示例 3-44　定义背景文件

```
<?xml version = "1.0" encoding = "utf - 8"?>
< shape xmlns:ohos = "http://schemas.huawei.com/res/ohos"
        ohos:shape = "rectangle">
< corners
        ohos:radius = "30vp"/>
< solid
        ohos:color = " # 66808080"/>
</shape >
```

3.6.11　弹框组件 CommonDialog

在 3.6.10 节讲解了信息提示框组件，此组件继承自 CommonDialog，可以基于 CommonDialog 封装更多的弹框。下面，通过 CommonDialog 实现一个确认框，需要用户单击“确定”按钮后将其关闭，如图 3-70 所示。

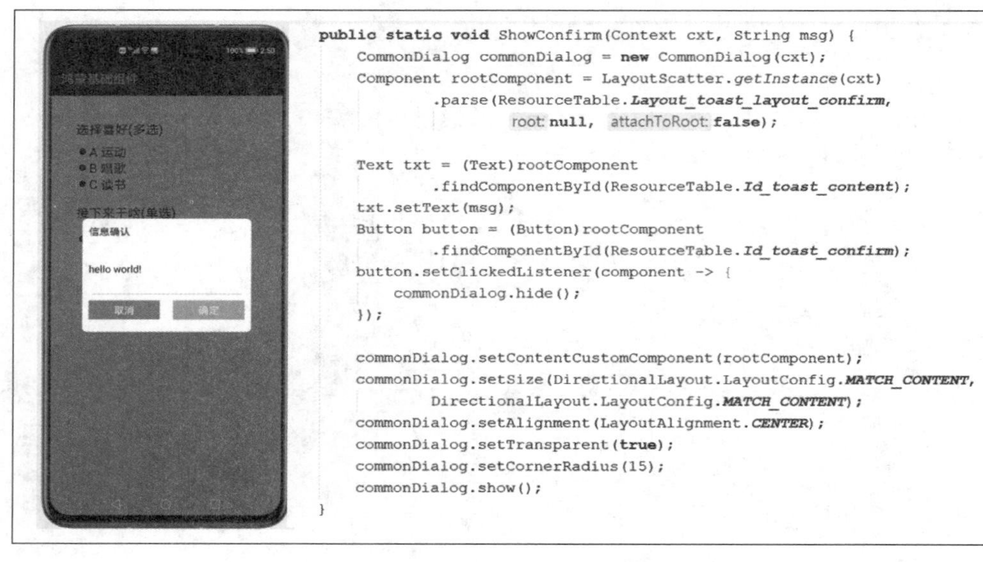

图 3-70 通过 CommonDialog 实现一个确认框

CommonDialog 实现一个确认框，首先需要定义一个静态方法 ShowConfirm，当需要弹框的时候，只需调用该方法，传入需要的数据就可以了，如代码示例 3-45 所示。

代码示例 3-45 定义一个静态方法 ShowConfirm

```
public static void ShowConfirm(Context cxt, String msg) {
//创建一个弹框对象
    CommonDialog commonDialog = new CommonDialog(cxt);
//通过 LayoutScatter 查找布局文件,并返回转换的组件对象
    Component rootComponent = LayoutScatter.getInstance(cxt)
            .parse(ResourceTable.Layout_toast_layout_confirm,
                    null, false);
    //通过组件对象查找布局文件中声明的 Text、Button 组件,并设置 Text 的值
    Text txt = (Text)rootComponent
            .findComponentById(ResourceTable.Id_toast_content);
    txt.setText(msg);
//给布局中的 Button 添加时间监听,当 Button 被单击时,关闭弹框
    Button button = (Button)rootComponent
            .findComponentById(ResourceTable.Id_toast_confirm);
    button.setClickedListener(component ->{
        commonDialog.hide();
    });

//把布局组件添加到弹框中
    commonDialog.setContentCustomComponent(rootComponent);
//弹框属性设置
```

```
    commonDialog.setSize(DirectionalLayout.LayoutConfig.MATCH_CONTENT,
            DirectionalLayout.LayoutConfig.MATCH_CONTENT);
//设置对齐方式
    commonDialog.setAlignment(LayoutAlignment.CENTER);
//设置背景透明
    commonDialog.setTransparent(true);
//设置弹框圆角
    commonDialog.setCornerRadius(15);
//弹出弹框
    commonDialog.show();
}
```

下面是弹框需要装载的布局文件：Toast_layout_confirm.xml，如代码示例 3-46 所示。

代码示例 3-46　定义弹框需要装载的布局文件

```
<?xml version = "1.0" encoding = "utf - 8"?>
< DirectionalLayout
    xmlns:ohos = "http://schemas.huawei.com/res/ohos"
    ohos:height = "match_content"
    ohos:width = "match_parent">

< DirectionalLayout
        ohos:id = " $ + id:toast_container"
        ohos:background_element = " $ graphic:background_white_radius"
        ohos:height = "match_content"
        ohos:width = "match_parent"
        ohos:left_margin = "50vp"
        ohos:right_margin = "50vp"
        ohos:orientation = "vertical">
< Text
            ohos:margin = "10vp"
            ohos:height = "match_content"
            ohos:width = "match_parent"
            ohos:text = "信息确认"
            ohos:text_color = "black"
            ohos:text_alignment = "vertical_center"
            ohos:text_size = "16fp"/>

< Text
            ohos:id = " $ + id:toast_content"
            ohos:height = "match_parent"
            ohos:width = "match_parent"
            ohos:bottom_margin = "30vp"
            ohos:left_margin = "10vp"
            ohos:right_margin = "10vp"
            ohos:text = "自定义 Dialog 内容"
```

```
                ohos:text_alignment = "vertical_center"
                ohos:text_color = "black"
                ohos:text_size = "16fp"
                ohos:top_margin = "30vp"
                ohos:weight = "1"/>

<Component
                ohos:height = "1vp"
                ohos:width = "match_parent"
                ohos:background_element = "#ccc"
                ohos:left_margin = "15vp"
                ohos:right_margin = "15vp"
                ohos:text_alignment = "vertical_center"
                ohos:text_size = "16fp"/>

<DirectionalLayout
                ohos:bottom_margin = "10vp"
                ohos:height = "match_content"
                ohos:width = "match_parent"
                ohos:orientation = "horizontal">

<Button
                    ohos:height = "match_content"
                    ohos:width = "match_parent"
                    ohos:background_element = "gray"
                    ohos:margin = "10vp"
                    ohos:padding = "5vp"
                    ohos:text = "取消"
                    ohos:text_color = "white"
                    ohos:text_size = "16fp"
                    ohos:weight = "1"/>

<Button
                    ohos:id = "$ + id:toast_confirm"
                    ohos:height = "match_content"
                    ohos:width = "match_parent"
                    ohos:background_element = "#fff9a825"
                    ohos:margin = "10vp"
                    ohos:padding = "5vp"
                    ohos:text = "确定"
                    ohos:text_color = "white"
                    ohos:text_size = "16fp"
                    ohos:weight = "1"/>
</DirectionalLayout>
</DirectionalLayout>

</DirectionalLayout>
```

3.6.12　进度条组件 ProgressBar

ProgressBar 用于显示内容或操作的进度。ProgressBar 继承自 Component，如图 3-71 所示。

```
public class ProgressBar extends Component {
    public ProgressBar(Context context) {
        super((Context)null);
        throw new RuntimeException("Stub!");
    }

    public ProgressBar(Context context, AttrSet attrSet) {
        super((Context)null);
        throw new RuntimeException("Stub!");
    }

    public ProgressBar(Context context, AttrSet attrSet, String styleName) {
        super((Context)null);
        throw new RuntimeException("Stub!");
    }
```

图 3-71　ProgressBar 继承自 Component

将 ProgressBar 方向设置为垂直，如图 3-72 和代码示例 3-47 所示。

代码示例 3-47　将 ProgressBar 方向设置为垂直

```
<ProgressBar
    ohos:orientation = "vertical"
    ohos:top_margin = "20vp"
    ohos:height = "150vp"
    ohos:width = "60vp"
    ohos:progress_width = "10vp"
    ohos:max = "100"
    ohos:min = "0"
    ohos:progress = "60"/>
```

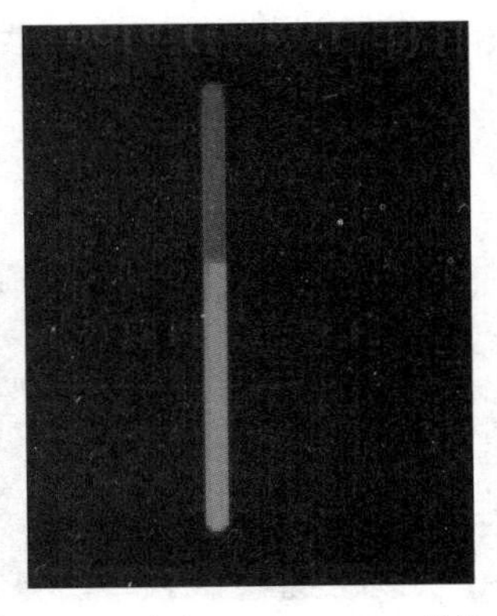

图 3-72　垂直 ProgressBar 的效果

设置当前进度，在 xml 中进行设置，代码如下：

```
<ProgressBar
    ...
ohos:progress = "60"/>
```

或者在 Java 中进行设置，代码如下：

```
progressBar.setProgressValue(60);
```

设置最大值和最小值，如图 3-73 所示。

图 3-73　设置最大值、最小值及进度的效果

在 xml 中进行设置，代码如下：

```
<ProgressBar
    ...
ohos:max = "400"
    ohos:min = "0"/>
```

或者在 Java 中设置，代码如下：

```
progressBar.setMaxValue(400);
progressBar.setMinValue(0);
```

设置 ProgressBar 进度颜色，效果如图 3-74 所示，代码如下：

```
<ProgressBar
    ...
ohos:progress_element = "#FF9900" />
```

图 3-74　设置 ProgressBar 颜色效果

设置 ProgressBar 底色颜色，效果如图 3-75 所示，代码如下：

```
<ProgressBar
    ...
ohos:background_instruct_element = "#FFFFFF" />
```

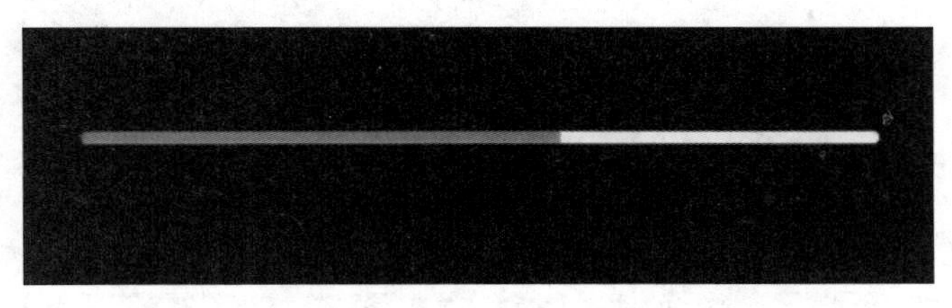

图 3-75 设置底色颜色效果

设置 ProgressBar 分割线，效果如图 3-76 所示。

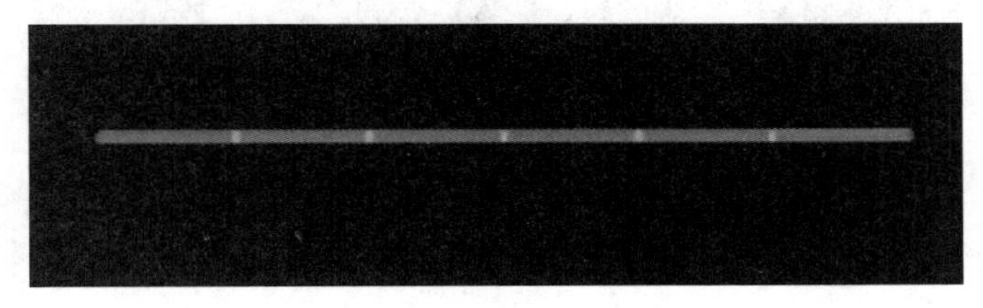

图 3-76 添加分割线效果

在 xml 中配置，代码如下：

```
<ProgressBar
    ...
ohos:divider_lines_enabled = "true"
    ohos:divider_lines_number = "5"/>
```

在 Java 代码中配置，代码如下：

```
progressBar.enableDividerLines(true);
progressBar.setDividerLinesNumber(5);
```

设置 ProgressBar 分割线颜色，效果如图 3-77 所示，代码如下：

```
progressBar.setDividerLineColor(Color.MAGENTA);
```

设置 ProgressBar 提示文字，效果如图 3-78 所示，代码如下：

```
<ProgressBar
    ...
```

```
    ohos:progress_hint_text = "20 % "
     ohos:progress_hint_text_color = " # FFCC99" />
```

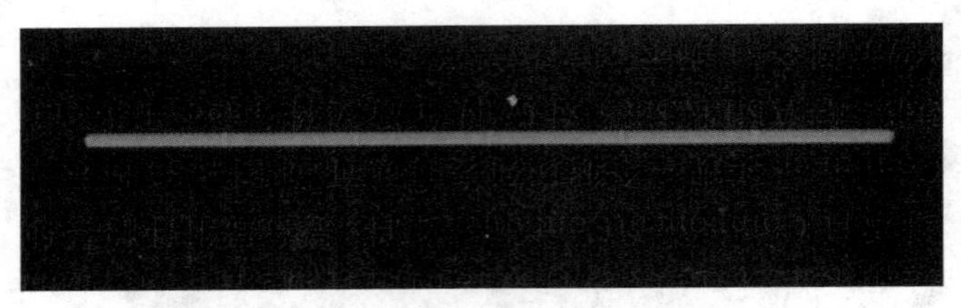

图 3-77　设置分割线颜色效果

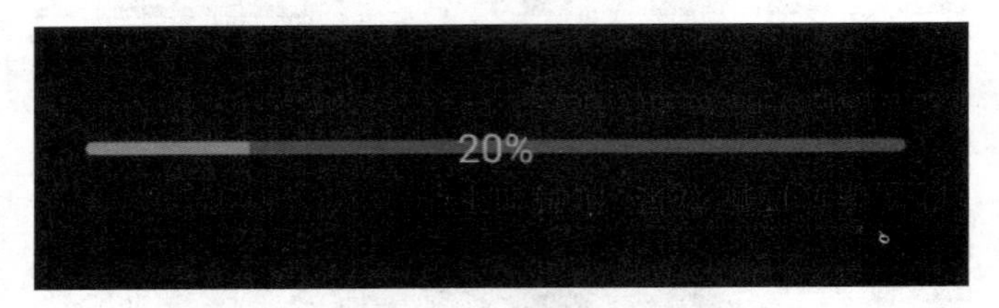

图 3-78　设置提示文字效果

3.6.13　滑块组件 Slider

注意该 Slider 组件与进度条 ProgressBar 组件的区别，ProgressBar 不能拖动，只有显示功能，而 Slider 组件继承自 AbsSlider，除了具有显示功能外还可以拖动，如图 3-79 所示。

```
public class Slider extends AbsSlider {
    protected Slider.ValueChangedListener mValueChangedListener;

    public Slider(Context context) {
        super((Context)null);
        throw new RuntimeException("Stub!");
    }

    public Slider(Context context, AttrSet attrSet) {
        super((Context)null);
        throw new RuntimeException("Stub!");
    }

    public Slider(Context context, AttrSet attrSet, String styleName) {
        super((Context)null);
        throw new RuntimeException("Stub!");
    }
```

图 3-79　Slider 组件继承自 AbsSlider

AbsSlider 组件继承自 ProgressBar 组件，如图 3-80 所示。

布局中设置的 Slider 拖动条，效果如图 3-81 所示。

```java
public abstract class AbsSlider extends ProgressBar {
    public AbsSlider(Context context) {
        super((Context)null);
        throw new RuntimeException("Stub!");
    }

    public AbsSlider(Context context, AttrSet attrSet) {
        super((Context)null);
        throw new RuntimeException("Stub!");
    }

    public AbsSlider(Context context, AttrSet attrSet, String styleName) {
        super((Context)null);
        throw new RuntimeException("Stub!");
    }
```

图 3-80 AbsSlider 组件继承自 ProgressBar 组件

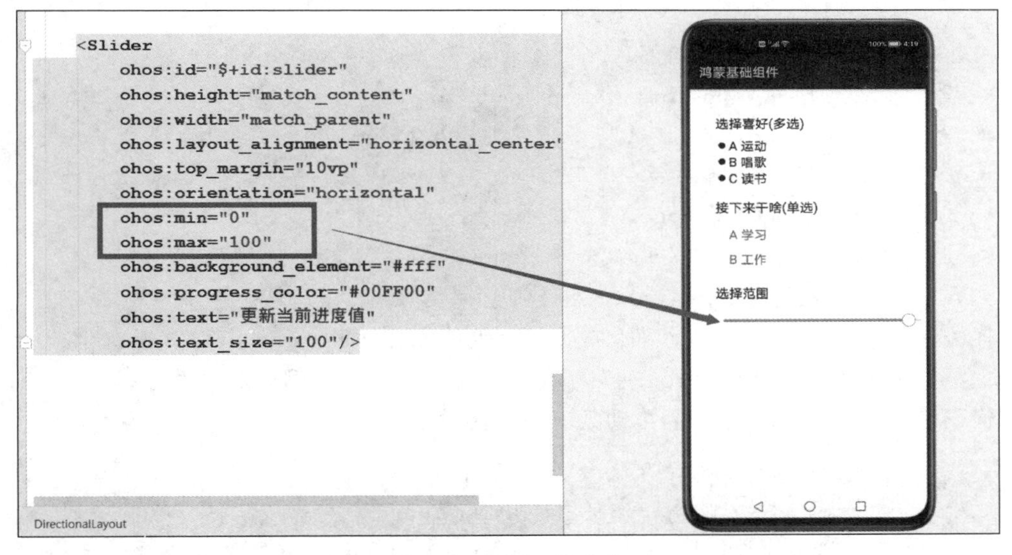

图 3-81 Slider 拖动条

接下来，实现 Slider 拖动条功能，如代码示例 3-48 所示。

代码示例 3-48 布局中设置的 Slider 拖动条

```xml
<?xml version = "1.0" encoding = "utf - 8"?>
< DirectionalLayout
    xmlns:ohos = "http://schemas.huawei.com/res/ohos"
    ohos:height = "match_parent"
    ohos:width = "match_parent"
    ohos:orientation = "vertical">
```

```
<Slider
        ohos:id = "$ + id:button"
        ohos:height = "match_content"
        ohos:width = "match_parent"
        ohos:layout_alignment = "horizontal_center"
        ohos:top_margin = "200"
        ohos:orientation = "horizontal"
        ohos:min = "0"
        ohos:max = "100"
ohos:progress = "66"
        ohos:background_element = "#000000"
        ohos:progress_color = "#00FF00"
        ohos:text_size = "100"/>

</DirectionalLayout>
```

Slider 相关标签属性说明如下。

- 设置拖动条方向：ohos:orientation="horizontal"，水平方向；
- 设置最小值 ：ohos:min="0"；
- 设置最大值 ：ohos:max="100"；
- 设置当前值 ：ohos:progress="66"；
- 设置背景颜色 ：ohos:background_element="#000000"，黑色；
- 设置进度条颜色 ：ohos:progress_color="#00FF00"，绿色。

代码中控制拖动条 Slider 组件，向 Slider 组件添加 ValueChangedListener 监听器，当 Slider 值发生变化时，触发 onProgressUpdated 方法，如代码示例 3-49 所示。

代码示例 3-49　代码中控制拖动条 Slider 组件

```
//获取布局文件中的拖动条 Slider
  Slider slider = (Slider) findComponentById(ResourceTable.Id_slider);
  slider.setValueChangedListener(new Slider.ValueChangedListener() {
//当进度变化时触发
//组件信息,i: 进度值 ;  b: 是否允许用户改变进度值,默认返回值为 true
        @Override
        public void onProgressUpdated(Slider slider, int i, boolean b) {
            new ToastDialog(MainAbilitySlice.this)
                    .setText("进度: " + slider.getProgress())
                    .setAlignment(LayoutAlignment.CENTER)
                    .show();
        }

        @Override
```

```
            public void onTouchStart(Slider slider) {

            }

            @Override
            public void onTouchEnd(Slider slider) {

            }
        });
    }
```

onProgressUpdated：当进度变化时触发，参数说明如下。

- slider：组件信息。
- i：进度值。
- b：是否允许用户改变进度值，默认返回值为 true。

onTouchStart：当通过手指触发进度条时触发。

onTouchEnd：当通过手指离开时触发。

3.6.14 ScrollView 组件

ScrollView 是一种带滚动功能的组件，它采用滑动的方式在有限的区域内显示更多的内容。ScrollView 组件继承自 StackLayout，如图 3-82 所示。

```
public class ScrollView extends StackLayout {
    public ScrollView(Context context) {
        super((Context)null);
        throw new RuntimeException("Stub!");
    }

    public ScrollView(Context context, AttrSet attrSet) {
        super((Context)null);
        throw new RuntimeException("Stub!");
    }

    public ScrollView(Context context, AttrSet attrSet, String styleName) {
        super((Context)null);
        throw new RuntimeException("Stub!");
    }
```

图 3-82 ScrollView 组件继承自 StackLayout

可以直接把 ScrollView 当作顶级的布局组件来使用，如图 3-83 所示。

可以在 ScrollView 内放置一个方向布局组件，在方向组件内部放置多个 Image 组件，当所有 Image 组件的高度超过屏幕时，使用 ScrollView 就可以滚动查看了，如代码示例 3-50 所示。

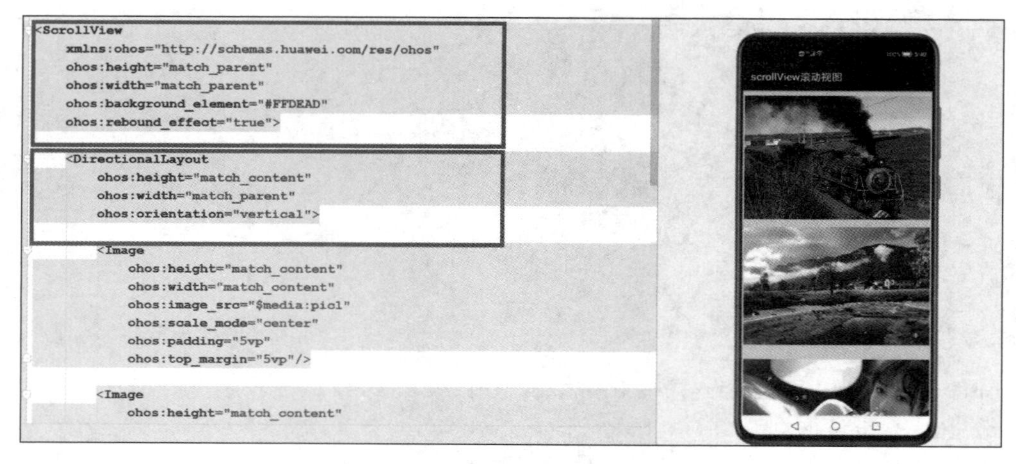

图 3-83 ScrollView 组件效果图

代码示例 3-50 ScrollView 布局用法

```
<?xml version = "1.0" encoding = "utf - 8"?>
< ScrollView
    xmlns:ohos = "http://schemas.huawei.com/res/ohos"
    ohos:height = "match_parent"
    ohos:width = "match_parent"
    ohos:background_element = "#FFDEAD"
    ohos:rebound_effect = "true">

< DirectionalLayout
   ohos:height = "match_content"
        ohos:width = "match_parent"
ohos:orientation = "vertical">

< Image
              ohos:height = "match_content"
              ohos:width = "match_content"
ohos:image_src = " $ media:pic1"
              ohos:padding = "5vp"
              ohos:top_margin = "5vp"/>

</DirectionalLayout >
</ScrollView >
```

根据像素数平滑滚动,代码如下:

```
btnScroll.setClickedListener(component -> {
    scrollView.fluentScrollByY(300);
});
```

平滑滚动到指定位置，代码如下：

```
scrollView.fluentScrollYTo(500);
```

设置布局方向：ScrollView 自身没有设置布局方向的属性，所以需要在其子布局中设置。以横向布局 horizontal 为例，代码如下：

```
<ScrollView>
<DirectionalLayout
        ohos:orientation="horizontal">
</DirectionalLayout>
</Scrollview>
```

在 xml 中设置回弹效果，代码如下：

```
<ScrollView
    ohos:rebound_effect="true">
</ScrollView>
```

在 Java 代码中设置回弹效果，代码如下：

```
scrollView.setReboundEffect(true);
```

在 xml 中设置缩放匹配效果，代码如下：

```
<ScrollView
    ohos:match_viewport="true">
</ScrollView>
```

在 Java 代码中设置缩放匹配效果，代码如下：

```
scrollView.setMatchViewportEnabled(true);
```

3.6.15 ListContainer 组件

ListContainer 是用来呈现连续、多行数据的组件，包含一系列相同类型的列表项。ListContainer 组件继承自 ComponentContainer 组件，是布局组件的一种，如图 3-84 所示。

下面通过 ListContainer 实现一个循环列表的页面，如图 3-85 所示。

在 layout 目录下，在 AbilitySlice 对应的布局文件 page_listcontainer.xml 中创建 ListContainer，如代码示例 3-51 所示。

```java
public class ListContainer extends ComponentContainer implements TextObserver {
    public static final int INVALID_INDEX = -1;

    public ListContainer(Context context) {
        super((Context)null);
        throw new RuntimeException("Stub!");
    }

    public ListContainer(Context context, AttrSet attrSet) {
        super((Context)null);
        throw new RuntimeException("Stub!");
    }

    public ListContainer(Context context, AttrSet attrSet, String styleName) {
        super((Context)null);
        throw new RuntimeException("Stub!");
    }
```

图 3-84 ListContainer 组件继承自 ComponentContainer 组件

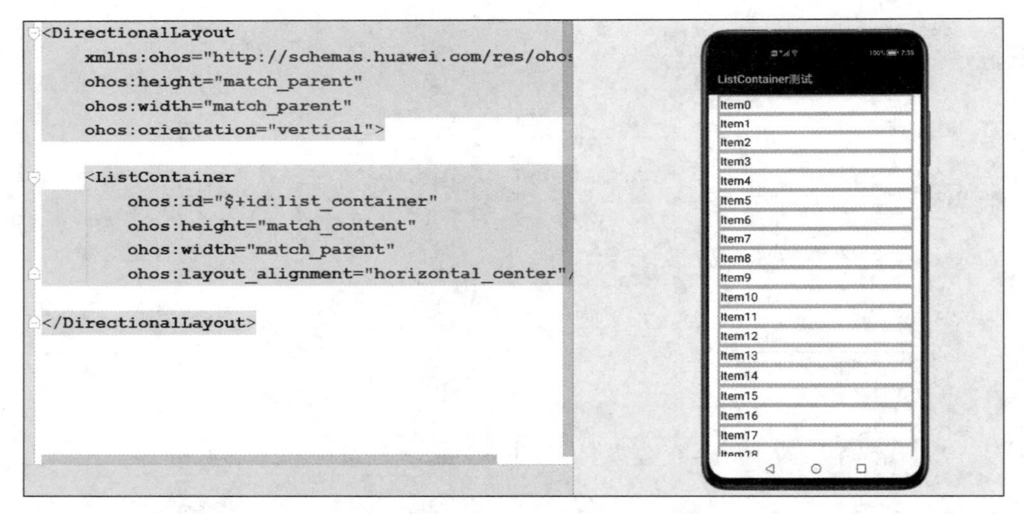

图 3-85 ListContainer 组件效果

代码示例 3-51 ListContainer 布局用法

```xml
<ListContainer
    ohos:id = " $ + id:list_container"
    ohos:height = "200vp"
    ohos:width = "300vp"
    ohos:layout_alignment = "horizontal_center"/>
```

在 layout 目录下新建 xml 文件(例如：item_sample.xml),作为 ListContainer 的子布局,如代码示例 3-52 所示。

代码示例 3-52　ListContainer 的子布局

```
<?xml version = "1.0" encoding = "utf - 8"?>
< DirectionalLayout
    xmlns:ohos = "http://schemas.huawei.com/res/ohos"
    ohos:height = "match_content"
    ohos:width = "match_parent"
    ohos:left_margin = "16vp"
    ohos:right_margin = "16vp"
    ohos:orientation = "vertical">
< Text
        ohos:id = " $ + id:item_index"
        ohos:height = "match_content"
        ohos:width = "match_content"
        ohos:padding = "4vp"
        ohos:text = "Item0"
        ohos:text_size = "20fp"
        ohos:layout_alignment = "center"/>
</DirectionalLayout >
```

创建 SampleItem. java 文件，作为 ListContainer 的数据包装类，如代码示例 3-53 所示。

代码示例 3-53　创建 ListContainer 的数据包装类

```
public class SampleItem {
    private String name;
    public SampleItem(String name) {
        this.name  =  name;
    }
    public String getName() {
        return name;
    }
    public void setName(String name) {
        this.name  =  name;
    }
}
```

ListContainer 每一行可以存放不同的数据，因此需要适配不同的数据结构，使其都能添加到 ListContainer 上。

创建 SampleItemProvider. java 文件，使其继承自 BaseItemProvider。必须重写的方法如表 3-9 所示。

表 3-9　继承 BaseItemProvider，必须重写的方法

方　　法	作　　用
int getCount()	返回填充的表项个数
Object getItem(int position)	根据 position 返回对应的数据

续表

方　　法	作　　用
long getItemId(int position)	返回某一项的 id
ComponentgetComponent(int position, Component covertComponent, ComponentContainer componentContainer)	根据 position 返回对应的界面组件

接下来为 ListContainer 组件提供数据源,如代码示例 3-54 所示。

代码示例 3-54　为 ListContainer 组件提供数据源

```
import ohos.aafwk.ability.AbilitySlice;
import ohos.agp.components.*;
import java.util.List;
public class SampleItemProvider extends BaseItemProvider{
    private List < SampleItem > list;
    private AbilitySlice slice;
    public SampleItemProvider(List < SampleItem > list, AbilitySlice slice) {
        this.list = list;
        this.slice = slice;
    }
    @Override
    public int getCount() {
        return list == null ? 0 : list.size();
    }
    @Override
    public Object getItem(int position) {
        if (list != null && position > 0 && position < list.size()){
            return list.get(position);
        }
        return null;
    }
    @Override
    public long getItemId(int position) {
        return position;
    }
    @Override
    public Component getComponent(int position, Component convertComponent, ComponentContainer
componentContainer) {
        final Component cpt;
        if (convertComponent == null) {
            cpt = LayoutScatter.getInstance(slice).parse(ResourceTable.Layout_item_
sample, null, false);
        } else {
            cpt = convertComponent;
        }
        SampleItem sampleItem = list.get(position);
        Text text = (Text) cpt.findComponentById(ResourceTable.Id_item_index);
```

```
        text.setText(sampleItem.getName());
        return cpt;
    }
}
```

在 Java 代码中向 ListContainer 添加数据，并适配其数据结构，如代码示例 3-55 所示。

代码示例 3-55 绑定 ListContainer 组件数据源

```
@Override
    public void onStart(Intent intent) {
        super.onStart(intent);
        super.setUIContent(ResourceTable.Layout_page_listcontainer);
        initListContainer();
    }
    private void initListContainer() {
ListContainer listContainer = (ListContainer) findComponentById(ResourceTable.Id_list_
container);
        List<SampleItem> list = getData();
        SampleItemProvider sampleItemProvider = new SampleItemProvider(list,this);
        listContainer.setItemProvider(sampleItemProvider);
    }
    private ArrayList<SampleItem> getData() {
        ArrayList<SampleItem> list = new ArrayList<>();
        for (int i = 0; i <= 8; i++) {
            list.add(new SampleItem("Item" + i));
        }
        return list;
    }
```

设置响应单击事件，代码如下：

```
listContainer.setItemClickedListener((container, component, position, id) ->{
    SampleItem item = (SampleItem) listContainer.getItemProvider().getItem(position);
    new ToastDialog(this)
            .setText("clicked:" + item.getName())
            //Toast 显示在界面中间
            .setAlignment(LayoutAlignment.CENTER)
            .show();
});
```

设置响应长按事件，代码如下：

```
listContainer.setItemLongClickedListener((container, component, position, id) ->{
        SampleItem item = (SampleItem)listContainer.getItemProvider().getItem(position);
        new ToastDialog(this)
             .setText("long clicked:" + item.getName())
```

```
                .setAlignment(LayoutAlignment.CENTER)
                .show();
            return false;
        }
    );
```

与 ListContainer 的样式设置相关的接口如表 3-10 所示。

表 3-10 与 ListContainer 的样式设置相关的接口

属　性	Java 方法	作　　用
orientation	setOrientation(int orientation)	设置布局方向
—	setContentStartOffSet(int startOffset) setContentEndOffSet(int endOffset) setContentOffSet(int startOffset, int endOffset)	设置列表容器的开始和结束偏移量
rebound_effect	setReboundEffect(boolean enabled)	设置是否启用回弹效果
—	setReboundEffectParams (int overscrollPercent, float overscrollRate, int remainVisiblePercent) setReboundEffectParams(ListContainer. ReboundEffectParams reboundEffectParams)	设置回弹效果参数
shader_color	setShaderColor(Color color)	设置着色器颜色

3.6.16 PageSlider 组件

PageSlider 组件继承自 StackLayout 布局组件，该组件提供页面向上下及左右滑动的功能，如图 3-86 所示。PageSlider 组件可以搭配 RadioContainer、PageSliderIndicator、TabList 等组件一起来使用。

```
public class PageSlider extends StackLayout {
    public static final int DEFAULT_CACHED_PAGES_LIMIT = 1;
    public static final int INVALID_INDEX = -1;
    public static final int SLIDING_STATE_DRAGGING = 1;
    public static final int SLIDING_STATE_IDLE = 0;
    public static final int SLIDING_STATE_SETTLING = 2;

    public PageSlider(Context context) {
        super((Context)null);
        throw new RuntimeException("Stub!");
    }

    public PageSlider(Context context, AttrSet attrSet) {
        super((Context)null);
        throw new RuntimeException("Stub!");
    }
```

图 3-86 PageSlider 组件继承自 StackLayout 布局组件

使用 PageSlider 搭配 RadioContainer 实现如图 3-87 所示效果，常用于导航页面，RadioContainer 用作 PageSlider 页码提示符。

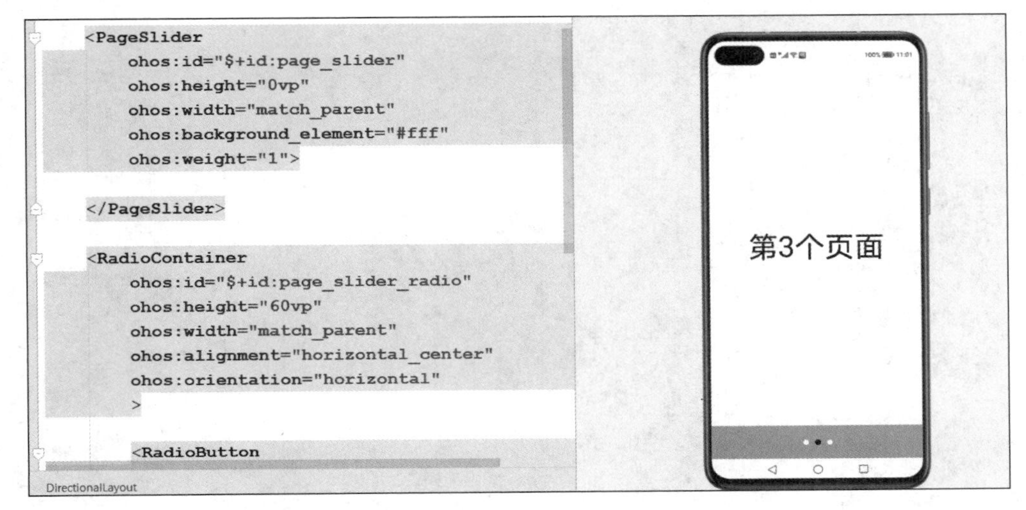

图 3-87　PageSlider 搭配 RadioContainer 效果图

使用 PageSlider 搭配 PageSliderIndicator 实现如图 3-88 所示效果，与图 3-87 所示的效果类似，默认这两个组件是配套使用的。

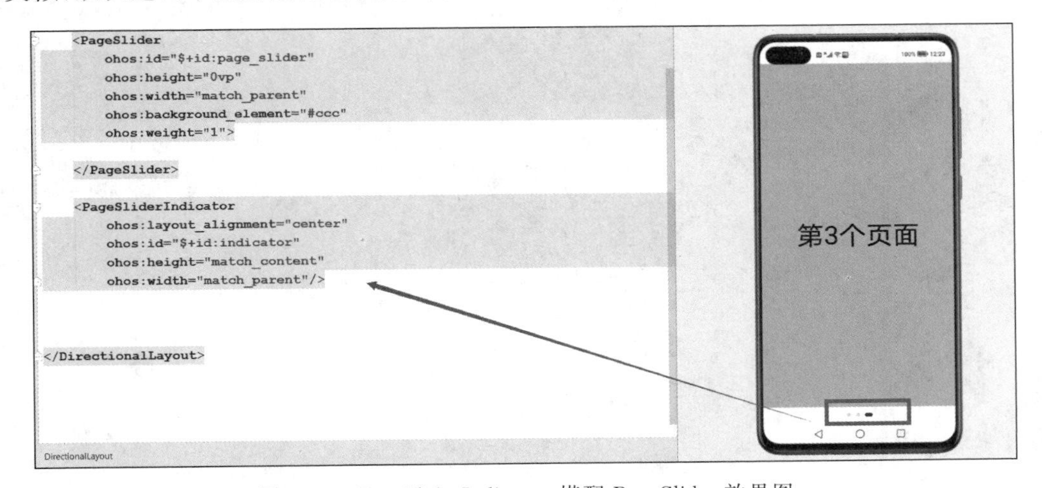

图 3-88　PageSliderIndicator 搭配 PageSlider 效果图

实现如图 3-87 所示效果，实现思路如下：

(1) 在垂直方向布局组件 DirectionalLayout 中添加 PageSlider 组件和 RadioContainer 组件，将 PageSlider 组件的高度设为 0vp，并将权重 weight 设为 1，PageSlider 组件所占空间等于屏幕高度减去 RadioContainer 高度后所剩余的空间。

(2) 如果 RadioButton 不设置文字，则显示的内容就是一个小点，利用这个小点作为翻页提示符。

实现 PageSlider 布局如代码示例 3-56 所示。

代码示例 3-56　PageSlider 布局实现

```
<?xml version = "1.0" encoding = "utf-8"?>
<DirectionalLayout
    xmlns:ohos = "http://schemas.huawei.com/res/ohos"
    ohos:height = "match_parent"
    ohos:width = "match_parent"
    ohos:background_element = "cyan"
    ohos:orientation = "vertical">


<PageSlider
        ohos:id = "$ + id:page_slider"
ohos:height = "0vp"
        ohos:width = "match_parent"
        ohos:background_element = "#fff"
 ohos:weight = "1">
</PageSlider>

<RadioContainer
        ohos:id = "$ + id:page_slider_radio"
 ohos:height = "60vp"
        ohos:width = "match_parent"
        ohos:alignment = "horizontal_center"
        ohos:orientation = "horizontal"
>

<RadioButton
            ohos:height = "match_parent"
            ohos:width = "match_content"
            ohos:text_size = "20fp"/>

<RadioButton
            ohos:height = "match_parent"
            ohos:width = "match_content"
            ohos:text_size = "20fp"/>

<RadioButton
            ohos:height = "match_parent"
            ohos:width = "match_content"
            ohos:text_size = "20fp"/>
</RadioContainer>
</DirectionalLayout>
```

有了上面的页面布局，我们看一看如何实现页面的左右滑动，实现思路如下：

（1）首先，需要单独创建3个页面布局文件，这里分别创建了slider_page0.xml、slider_page1.xml和slider_page2.xml共3个页面文件。

（2）定义一个容器列表，把上面创建好的布局页面添加到列表中。

（3）通过pageSlider.setProvider方法，向PageSlider添加页面数据源，具体的方法是通过重写createPageInContainer方法，通过该方法把页面按索引取出，并添加到pageSlider的组件容器中。

（4）通过监听addPageChangedListener页面变化的事件，重写onPageChosen方法，给RadioButton设置选中状态。

在base/element文件夹中创建一个json文件，用于设置Text组件中的文本信息，代码如下：

```
{
"string": [
    {
"name": "page1",
"value": "第 1 个页面"
    },
    {
"name": "page2",
"value": "第 2 个页面"
    },
    {
"name": "page3",
"value": "第 3 个页面"
    }
  ]
}
```

创建第1个页面，并将文件命名为slider_page0.xml，如代码示例3-57所示。

代码示例3-57　创建PageSlider第1个页面

```
<?xml version = "1.0" encoding = "utf - 8"?>
< DirectionalLayout
    xmlns:ohos = "http://schemas.huawei.com/res/ohos"
    ohos:height = "match_parent"
    ohos:width = "match_parent"
    ohos:alignment = "center"
    ohos:orientation = "vertical">

< Text
        ohos:text_size = "50fp"
```

```
            ohos:text = " $ string:page1"   //
            ohos:height = "match_content"
            ohos:width = "match_content"/>
</DirectionalLayout>
```

创建第 2 个页面,并将文件命名为 slider_page1. xml,如代码示例 3-58 所示。

代码示例 3-58　创建 PageSlider 第 2 个页面

```
<?xml version = "1.0" encoding = "utf - 8"?>
<DirectionalLayout
    xmlns:ohos = "http://schemas.huawei.com/res/ohos"
    ohos:height = "match_parent"
    ohos:width = "match_parent"
    ohos:alignment = "center"
    ohos:orientation = "vertical">

<Text
        ohos:text_size = "50fp"
        ohos:text = " $ string:page2"
        ohos:height = "match_content"
        ohos:width = "match_content"/>
</DirectionalLayout>
```

创建第 3 个页面,并将文件命名为 slider_page2. xml,如代码示例 3-59 所示。

代码示例 3-59　创建 PageSlider 第 3 个页面

```
<?xml version = "1.0" encoding = "utf - 8"?>
<DirectionalLayout
    xmlns:ohos = "http://schemas.huawei.com/res/ohos"
    ohos:height = "match_parent"
    ohos:width = "match_parent"
    ohos:alignment = "center"
    ohos:orientation = "vertical">

<Text
        ohos:text_size = "50fp"
        ohos:text = " $ string:page3"
        ohos:height = "match_content"
        ohos:width = "match_content"/>
</DirectionalLayout>
```

实现页面滑动,如代码示例 3-60 所示。

代码示例 3-60 实现 PageSlider 滑动

```
public class PageSliderAbilitySlice extends AbilitySlice {

    private PageSlider pageSlider;
    private RadioContainer radioContainer;

    @Override
    public void onStart(Intent intent) {
        super.onStart(intent);
        super.setUIContent(ResourceTable.Layout_ability_page_slider);

        pageSlider = (PageSlider)findComponentById(ResourceTable.Id_page_slider);
        radioContainer = (RadioContainer)findComponentById(ResourceTable.Id_page_slider_
radio);

        ((RadioButton)radioContainer.getComponentAt(0)).setChecked(true);

        LayoutScatter scatter = LayoutScatter.getInstance(getContext());
        DirectionalLayout page0 = (DirectionalLayout) scatter.parse(ResourceTable.Layout_
slider_page0,null,false);
        DirectionalLayout page1 = (DirectionalLayout) scatter.parse(ResourceTable.Layout_
slider_page1,null,false);
        DirectionalLayout page2 = (DirectionalLayout) scatter.parse(ResourceTable.Layout_
slider_page2,null,false);

        List<Component> pageArr = new ArrayList<>();
        pageArr.add(page0);
        pageArr.add(page1);
        pageArr.add(page2);


        //向 pageSlider 添加数据源
        pageSlider.setProvider(new PageSliderProvider() {
            @Override
            public int getCount() {
                return pageArr.size();
            }

            @Override
            public Object createPageInContainer(ComponentContainer componentContainer, int i) {
                componentContainer.addComponent(pageArr.get(i));
                return pageArr.get(i);
            }

            @Override
```

```
            public void destroyPageFromContainer(ComponentContainer componentContainer, int
i, Object o) {
                ((PageSlider)componentContainer).removeComponent(pageArr.get(i));
            }

            @Override
            public boolean isPageMatchToObject(Component component, Object o) {
                return component == o;
            }
        });



        //监听 page 滑动
        pageSlider.addPageChangedListener(new PageSlider.PageChangedListener() {
            @Override
            public void onPageSliding(int i, float v, int i1) {

            }

            @Override
            public void onPageSlideStateChanged(int i) {

            }

            @Override
            public void onPageChosen(int i) {
                ((RadioButton)radioContainer.getComponentAt(i)).setChecked(true);
            }
        });

        radioContainer.setMarkChangedListener(new RadioContainer.CheckedStateChangedListener() {
            @Override
            public void onCheckedChanged(RadioContainer radioContainer, int i) {
                new ToastDialog(PageSliderAbilitySlice.this)
                        .setText(i + "")
                        .show();

            }
        });
    }

    @Override
    public void onActive() {
        super.onActive();
```

```
    }

    @Override
    public void onForeground(Intent intent) {
        super.onForeground(intent);
    }
}
```

3.6.17　系统剪贴板服务

用户通过系统剪贴板服务，可实现应用之间的简单数据传递。例如：在应用 A 中复制的数据，可以在应用 B 中粘贴，反之亦可。

(1) HarmonyOS 提供了系统剪贴板服务的操作接口，支持用户程序从系统剪贴板中读取、写入和查询剪贴板数据，以及添加、移除系统剪贴板数据变化的回调。

(2) HarmonyOS 提供了剪贴板数据的对象定义，包含内容对象和属性对象。

1. 场景说明

同一设备的应用程序 A、B 之间可以借助系统剪贴板服务完成简单数据的传递，即应用程序 A 向剪贴板服务写入数据后，应用程序 B 可以从中读取数据，如图 3-89 所示。

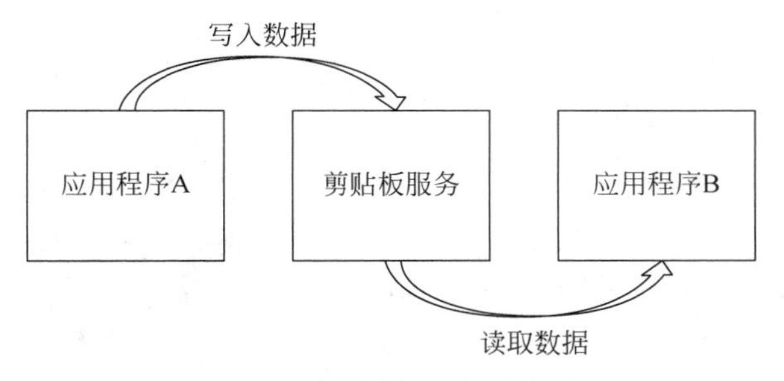

图 3-89　剪贴板服务示意图

在使用剪贴板服务时，需要注意以下几点：

(1) 只有在前台获取焦点的应用才有读取系统剪贴板的权限，但系统默认输入法应用除外。

(2) 写入剪贴板服务中的剪贴板数据不会随应用程序结束而被销毁。

(3) 对同一用户而言，写入剪贴板服务的数据会被下一次写入的剪贴板数据所覆盖。

(4) 在同一设备内，剪贴板单次传递内容不应超过 800KB。

2. 接口说明

SystemPasteboard 提供了系统剪贴板操作的相关接口，例如复制、粘贴、配置回调等。PasteData 是剪贴板服务操作的数据对象，一个 PasteData 由若干个内容节点(PasteData.Record)和一个属性集合对象(PasteData.DataProperty)组成。Record 是存放剪贴板数据信息的最小单位，每个 Record 都有其特定的 MIME 类型，如纯文本、HTML、URI、Intent

等。剪贴板数据的属性信息存放在 DataProperty 中，包括标签、时间戳等。

3. SystemPasteboard

SystemPasteboard 提供了系统剪贴板服务的操作接口，例如复制、粘贴、配置回调等，如表 3-11 所示。

表 3-11 SystemPasteboard 提供系统剪贴板操作的相关接口

接口名	描述
getSystemPasteboard(Context context)	获取系统剪切板服务的对象实例
getPasteData()	读取当前系统剪贴板中的数据
hasPasteData()	判断当前系统剪贴板中是否有内容
setPasteData(PasteData data)	将剪贴板数据写入系统剪贴板
clear()	清空系统剪贴板数据
addPasteDataChangedListener (IPasteDataChangedListener listener)	向用户程序添加系统剪贴板数据变化的回调，当系统剪贴板数据发生变化时，会触发用户程序的回调实现
removePasteDataChangedListener (IPasteDataChangedListener listener)	用户程序移除系统剪贴板数据变化的回调

4. 开发步骤

应用 A 获取系统剪贴板服务，代码如下：

```
SystemPasteboard pasteboard = SystemPasteboard.getSystemPasteboard(appContext);
```

应用 A 向系统剪贴板中写入一条纯文本数据，代码如下：

```
if (pasteboard != null) {
    pasteboard.setPasteData(PasteData.creatPlainTextData("Hello, world!"));
}
```

应用 B 从系统剪贴板读取数据，并将数据对象中的首个文本类型（纯文本/HTML）内容信息在控件中显示，但忽略其他类型内容，如代码示例 3-61 所示。

代码示例 3-61 应用 B 从系统剪贴板读取数据

```
PasteData pasteData = pasteboard.getPasteData();
if (pasteData == null) {
    return;
}
DataProperty dataProperty = pasteData.getProperty();
boolean hasHtml = dataProperty.hasMimeType(PasteData.MIMETYPE_TEXT_HTML);
boolean hasText = dataProperty.hasMimeType(PasteData.MIMETYPE_TEXT_PLAIN);
if (hasHtml || hasText) {
    Text text = (Text) findComponentById(ResourceTable.Id_text);
    for (int i = 0; i < pasteData.getRecordCount(); i++) {
```

```
            PasteData.Record record = pasteData.getRecordAt(i);
            String mimeType = record.getMimeType();
            if (mimeType.equals(PasteData.MIMETYPE_TEXT_HTML)) {
                text.setText(record.getHtmlText());
                break;
            } else if (mimeType.equals(PasteData.MIMETYPE_TEXT_PLAIN)) {
                text.setText(record.getPlainText().toString());
                break;
            } else {
                //skip records of other Mime type
            }
        }
    }
```

应用C注册添加系统剪贴板数据变化回调，当系统剪贴板数据发生变化时触发处理逻辑，如代码示例3-62所示。

代码示例3-62 应用C注册添加系统剪贴板数据变化回调

```
IPasteDataChangedListener listener = new IPasteDataChangedListener() {
    @Override
    public void onChanged() {
        PasteData pasteData = pasteboard.getPasteData();
        if (pasteData == null) {
            return;
        }
        //Operations to handle data change on the system pasteboard
    }
};
pasteboard.addPasteDataChangedListener(listener);
```

3.6.18 组件总结

本节介绍了15个鸿蒙ACE Java UI框架中的基础组件，通过这些基础组件，可以开发出更高级的业务组件及页面。

3.7 线程管理

不同应用在各自独立的进程中运行。当应用以任何形式启动时，系统为其创建进程，该进程将持续运行。当进程完成当前任务而处于等待状态且系统资源不足时，系统会自动回收此进程。

在启动应用时，系统会为该应用创建一个称为“主线程”的执行线程。该线程随着应用而创建或消失，是应用的核心线程。UI界面的显示和更新等操作，都在主线程上进行。主线程又称为 UI 线程,默认情况下,所有的操作都在主线程上执行。如果需要执行比较耗时的任务,如下载文件、查询数据库等,可创建其他线程来处理。

3.7.1 线程管理

如果应用的业务逻辑比较复杂,则可能需要创建多个线程来执行多个任务。这种情况下,代码复杂且难以维护,任务与线程的交互也会更加繁杂。要解决此问题,开发者可以使用 TaskDispatcher 来分发不同的任务。

1. 接口说明

TaskDispatcher 是一个任务分发器,它是 Ability 分发任务的基本接口,用于隐藏任务所在线程的实现细节。

为了保证应用有更好的响应性,我们需要设计任务的优先级。在 UI 线程上运行的任务默认以高优先级运行,如果某个任务无须等待结果,则可以采用低优先级,如表 3-12 所示。

表 3-12 UI 线程上运行的任务默认以高优先级运行

优先级	详细描述
HIGH	最高任务优先级,比默认优先级、低优先级的任务有更高概率得到执行
DEFAULT	默认任务优先级,比低优先级的任务有更高概率得到执行
LOW	低任务优先级,比高优先级、默认优先级的任务有更低概率得到执行

TaskDispatcher 具有多种实现,每种实现对应不同的任务分发器。在分发任务时可以指定任务的优先级,由同一个任务分发器分发出的任务具有相同的优先级。系统提供的任务分发器有 GlobalTaskDispatcher、ParallelTaskDispatcher、SerialTaskDispatcher、SpecTaskDispatcher。

- GlobalTaskDispatcher

全局并发任务分发器,由 Ability 执行 getGlobalTaskDispatcher()获取。适用于任务之间没有联系的情况。一个应用只有一个 GlobalTaskDispatcher,它在程序结束时才会被销毁。

```
TaskDispatcher globalTaskDispatcher = getGlobalTaskDispatcher(TaskPriority.DEFAULT);
```

- ParallelTaskDispatcher

并发任务分发器,由 Ability 执行 createParallelTaskDispatcher()创建并返回。与 GlobalTaskDispatcher 不同的是,ParallelTaskDispatcher 不具有全局唯一性,可以创建多个。开发者在创建或销毁 dispatcher 时,需要持有对应的对象引用。

```
String dispatcherName = "parallelTaskDispatcher";
TaskDispatcher parallelTaskDispatcher = createParallelTaskDispatcher ( dispatcherName,
TaskPriority.DEFAULT);
```

■ SerialTaskDispatcher

串行任务分发器，由 Ability 执行 createSerialTaskDispatcher()创建并返回。由该分发器分发的所有任务都是按顺序执行的，但执行这些任务的线程并不是固定的。如果要执行并行任务，则应使用 ParallelTaskDispatcher 或者 GlobalTaskDispatcher，而不是创建多个 SerialTaskDispatcher。如果任务之间没有依赖，则应使用 GlobalTaskDispatcher 实现。它的创建和销毁由开发者自己管理，开发者在使用期间需要持有该对象引用。

```
String dispatcherName = "serialTaskDispatcher";
TaskDispatcher serialTaskDispatcher = createSerialTaskDispatcher(dispatcherName, TaskPriority.
DEFAULT);
```

■ SpecTaskDispatcher

专有任务分发器，即绑定到专有线程上的任务分发器。目前已有的专有线程是主线程。UITaskDispatcher 和 MainTaskDispatcher 都属于 SpecTaskDispatcher。建议使用 UITaskDispatcher。

UITaskDispatcher：绑定到应用主线程的专有任务分发器，由 Ability 执行 getUITaskDispatcher()创建并返回。由该分发器分发的所有的任务都在主线程上按顺序执行，它在应用程序结束时被销毁。

```
TaskDispatcher uiTaskDispatcher = getUITaskDispatcher();
```

MainTaskDispatcher：由 Ability 执行 getMainTaskDispatcher()创建并返回。

```
TaskDispatcher mainTaskDispatcher = getMainTaskDispatcher()
```

2. 开发步骤

1）同步派发任务 syncDispatch

同步派发任务：派发任务并在当前线程等待任务执行完成。在返回前，当前线程会被阻塞。下面的代码示例展示了如何使用 GlobalTaskDispatcher 派发同步任务，如代码示例 3-63 所示。

代码示例 3-63 同步派发任务

```
TaskDispatcher globalTaskDispatcher = getGlobalTaskDispatcher(TaskPriority.DEFAULT);
globalTaskDispatcher.syncDispatch(new Runnable() {
    @Override
    public void run() {
        HiLog.info(label, "sync task1 run");
    }
});
HiLog.info(label, "after sync task1");
```

```
globalTaskDispatcher.syncDispatch(new Runnable() {
    @Override
    public void run() {
        HiLog.info(label, "sync task2 run");
    }
});
HiLog.info(label, "after sync task2");

globalTaskDispatcher.syncDispatch(new Runnable() {
    @Override
    public void run() {
        HiLog.info(label, "sync task3 run");
    }
});
HiLog.info(label, "after sync task3");

//执行结果如下:
//sync task1 run
//after sync task1
//sync task2 run
//after sync task2
//sync task3 run
//after sync task3
```

注意：如果 syncDispatch 使用不当，则会导致死锁。如下情形可能导致死锁发生：

(1) 在专有线程上，利用该专有任务分发器进行 syncDispatch。

(2) 在被某个串行任务分发器(dispatcher_a)派发的任务中，再次利用同一个串行任务分发器(dispatcher_a)对象派发任务。

(3) 在被某个串行任务分发器(dispatcher_a)派发的任务中，经过数次派发任务，最终又利用该(dispatcher_a)串行任务分发器派发任务。例如：dispatcher_a 派发的任务使用 dispatcher_b 进行任务派发，在 dispatcher_b 派发的任务中又利用 dispatcher_a 进行任务派发。

(4) 串行任务分发器(dispatcher_a)派发的任务中利用串行任务分发器(dispatcher_b)进行同步任务派发，同时 dispatcher_b 派发的任务中利用串行任务分发器(dispatcher_a)进行同步任务派发。在特定的线程执行顺序下将导致死锁。

2) 异步派发任务 asyncDispatch

异步派发任务：派发任务，并立即返回，返回值是一个可用于取消任务的接口。

如何使用 GlobalTaskDispatcher 派发异步任务，如代码示例 3-64 所示。

代码示例 3-64　异步派发任务

```
TaskDispatcher globalTaskDispatcher = getGlobalTaskDispatcher(TaskPriority.DEFAULT);
Revocable revocable = globalTaskDispatcher.asyncDispatch(new Runnable() {
    @Override
    public void run() {
        HiLog.info(label, "async task1 run");
    }
});
HiLog.info(label, "after async task1");

//执行结果可能如下
//after async task1
//async task1 run
```

3）异步延迟派发任务 delayDispatch

异步延迟派发任务：异步执行，函数立即返回，内部会在延时指定时间后将任务派发到相应队列中。延时时间参数仅代表在这段时间以后任务分发器会将任务加入队列中，任务的实际执行时间可能晚于这个时间。具体比这个数值晚多久，取决于队列及内部线程池的繁忙情况。

如何使用 GlobalTaskDispatcher 延迟派发任务，如代码示例 3-65 所示。

代码示例 3-65　异步延迟派发任务

```
final long callTime = System.currentTimeMillis();
final long delayTime = 50;
TaskDispatcher globalTaskDispatcher = getGlobalTaskDispatcher(TaskPriority.DEFAULT);
Revocable revocable = globalTaskDispatcher.delayDispatch(new Runnable() {
    @Override
    public void run() {
        HiLog.info(label, "delayDispatch task1 run");
        final long actualDelayMs = System.currentTimeMillis() - callTime;
        HiLog.info(label, "actualDelayTime >= delayTime: %{public}b", (actualDelayMs >=
delayTime));
    }
}, delayTime);
HiLog.info(label, "after delayDispatch task1");

//执行结果可能如下
//after delayDispatch task1
//delayDispatch task1 run
//actualDelayTime >= delayTime : true
```

4）任务组 Group

任务组表示一组任务，且该组任务之间有一定的联系，由 TaskDispatcher 执行

createDispatchGroup 创建并返回。将任务加入任务组，返回一个用于取消任务的接口。

任务组的使用方式是将一系列相关联的下载任务放入一个任务组，执行完下载任务后关闭应用，如代码示例 3-66 所示。

代码示例 3-66 任务组

```
String dispatcherName = "parallelTaskDispatcher";
TaskDispatcher dispatcher = createParallelTaskDispatcher(dispatcherName, TaskPriority.DEFAULT);
//创建任务组
Group group = dispatcher.createDispatchGroup();
//将任务 1 加入任务组,返回一个用于取消任务的接口
dispatcher.asyncGroupDispatch(group, new Runnable(){
    @Override
    public void run() {
        HiLog.info(label, "download task1 is running");
    }
});
//将与任务 1 相关联的任务 2 加入任务组
dispatcher.asyncGroupDispatch(group, new Runnable(){
    @Override
    public void run() {
        HiLog.info(label, "download task2 is running");
    }
});
//当任务组中的所有任务执行完成后执行指定任务
dispatcher.groupDispatchNotify(group, new Runnable(){
    @Override
    public void run() {
        HiLog. info ( label, " the close task is running after all tasks in the group are
completed");
    }
});

//可能的执行结果
//download task1 is running
//download task2 is running
//the close task is running after all tasks in the group are completed

//另外一种可能的执行结果
//download task2 is running
//download task1 is running
//the close task is running after all tasks in the group are completed
```

5）取消任务 Revocable

取消任务：Revocable 是取消一个异步任务的接口。异步任务包括通过 asyncDispatch、delayDispatch、asyncGroupDispatch 派发的任务。如果任务已经在执行中或执行完成，则会

返回取消失败。

如何取消一个异步延时任务,如代码示例 3-67 所示。

代码示例 3-67　取消任务

```
TaskDispatcher dispatcher = getUITaskDispatcher();
Revocable revocable = dispatcher.delayDispatch(new Runnable() {
    @Override
    public void run() {
        HiLog.info(label, "delay dispatch");
    }
}, 10);
boolean revoked = revocable.revoke();
HiLog.info(label, "%{public}b", revoked);

//一种可能的结果如下
//true
```

6）同步设置屏障任务 syncDispatchBarrier

同步设置屏障任务：在任务组上设立任务执行屏障,同步等待任务组中的所有任务执行完成,再执行指定任务。

在全局并发任务分发器(GlobalTaskDispatcher)上同步设置任务屏障,将不会起到屏障作用。如何同步设置屏障,如代码示例 3-68 所示。

代码示例 3-68　同步设置屏障任务

```
String dispatcherName = "parallelTaskDispatcher";
TaskDispatcher dispatcher = createParallelTaskDispatcher(dispatcherName, TaskPriority.DEFAULT);
//创建任务组
Group group = dispatcher.createDispatchGroup();
//将任务加入任务组,返回一个用于取消任务的接口
dispatcher.asyncGroupDispatch(group, new Runnable(){
    @Override
    public void run() {
        HiLog.info(label, "task1 is running");  //1
    }
});
dispatcher.asyncGroupDispatch(group, new Runnable(){
    @Override
    public void run() {
        HiLog.info(label, "task2 is running");  //2
    }
});

dispatcher.syncDispatchBarrier(new Runnable() {
    @Override
```

```
        public void run() {
            HiLog.info(label, "barrier");  //3
        }
    });
    HiLog.info(label, "after syncDispatchBarrier");  //4

    //1 和 2 的执行顺序不确定; 3 和 4 总是在 1 和 2 之后按顺序执行

    //可能的执行结果
    //task1 is running
    //task2 is running
    //barrier
    //after syncDispatchBarrier

    //另外一种执行结果
    //task2 is running
    //task1 is running
    //barrier
    //after syncDispatchBarrier
```

7）异步设置屏障任务 asyncDispatchBarrier

异步设置屏障任务：在任务组上设立任务执行屏障后直接返回，指定任务将在任务组中的所有任务执行完成后再执行。

在全局并发任务分发器(GlobalTaskDispatcher)上异步设置任务屏障，将不会起到屏障作用。可以使用并发任务分发器(ParallelTaskDispatcher)分离不同的任务组，达到微观并行、宏观串行的行为。

如何异步设置屏障，如代码示例 3-69 所示。

代码示例 3-69　异步设置屏障任务

```
TaskDispatcher dispatcher = createParallelTaskDispatcher("dispatcherName", TaskPriority.DEFAULT);
//创建任务组
Group group = dispatcher.createDispatchGroup();
//将任务加入任务组,返回一个用于取消任务的接口
dispatcher.asyncGroupDispatch(group, new Runnable(){
    @Override
    public void run() {
        HiLog.info(label, "task1 is running");  //1
    }
});
dispatcher.asyncGroupDispatch(group, new Runnable(){
    @Override
    public void run() {
        HiLog.info(label, "task2 is running");  //2
```

```
    }
});

dispatcher.asyncDispatchBarrier(new Runnable() {
    @Override
    public void run() {
        HiLog.info(label, "barrier");  //3
    }
});
HiLog.info(label, "after asyncDispatchBarrier");  //4

//1 和 2 的执行顺序不确定,但总在 3 和 4 之前执行; 4 可能在 3 之前执行

//可能的执行结果
//task1 is running
//task2 is running
//after syncDispatchBarrier
//barrier
```

8）执行多次任务 applyDispatch

执行多次任务：对指定任务执行多次。如何执行多次任务，如代码示例 3-70 所示。

代码示例 3-70 执行多次任务

```
final int total = 10;
final CountDownLatch latch = new CountDownLatch(total);
final ArrayList<Long> indexList = new ArrayList<>(total);
TaskDispatcher dispatcher = getGlobalTaskDispatcher(TaskPriority.DEFAULT);
//执行任务 total 次
dispatcher.applyDispatch((index) -> {
    indexList.add(index);
    latch.countDown();
}, total);

//设置任务超时
try {
    latch.await();
} catch (InterruptedException exception) {
    HiLog.error(label, "latch exception");
}
HiLog.info(label, "list size matches, %{public}b", (total == indexList.size()));

//执行结果
//list size matches, true
```

3.7.2 线程间通信

在开发过程中，开发者经常需要在当前线程中处理下载任务等较为耗时的操作，但是又不希望当前的线程受到阻塞。此时，就可以使用 EventHandler 机制。EventHandler 是 HarmonyOS 用于处理线程间通信的一种机制，可以通过 EventRunner 创建新线程，将耗时的操作放到新线程上执行。这样既不阻塞原来的线程，任务又可以得到合理处理。例如：主线程使用 EventHandler 创建子线程，子线程做耗时的下载图片操作，下载完成后，子线程通过 EventHandler 通知主线程，主线程再更新 UI。

EventRunner 是一种事件循环器，循环处理从该 EventRunner 创建的新线程的事件队列中获取 InnerEvent 事件或者 Runnable 任务。InnerEvent 是 EventHandler 投递的事件。

EventHandler 是一种用户在当前线程上将 InnerEvent 事件或者 Runnable 任务投递到异步线程上进行处理的机制。每个 EventHandler 和指定的 EventRunner 所创建的新线程绑定，并且该新线程内部有一个事件队列。EventHandler 可以将指定的 InnerEvent 事件或 Runnable 任务投递到这个事件队列。EventRunner 从事件队列里循环地取出事件，如果取出的事件是 InnerEvent 事件，则在 EventRunner 所在线程执行 processEvent 回调；如果取出的事件是 Runnable 任务，则在 EventRunner 所在线程执行 Runnable 的 run 回调。一般情况下，EventHandler 有两个主要作用：第 1 个作用，在不同线程间分发和处理 InnerEvent 事件或 Runnable 任务。第 2 个作用，延迟处理 InnerEvent 事件或 Runnable 任务。EventHandler 的运作机制如图 3-90 所示。

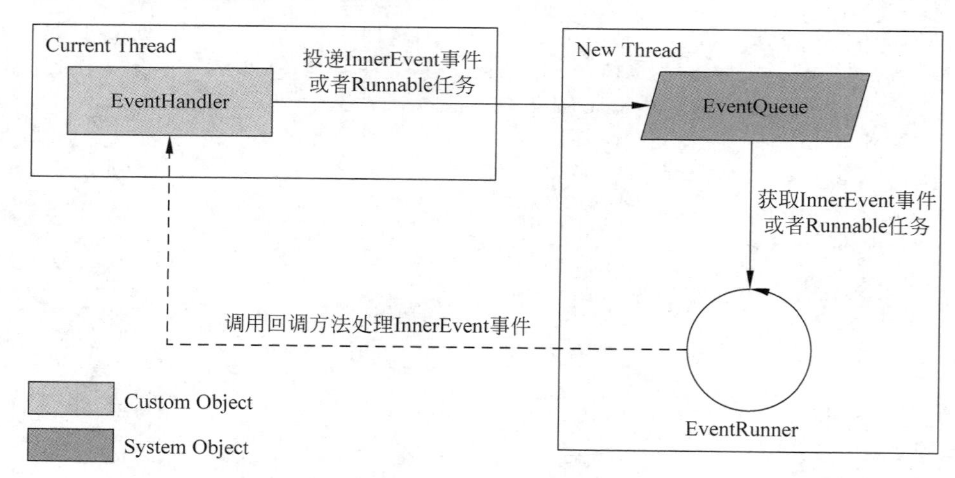

图 3-90 EventHandler 的运作机制

使用 EventHandler 实现线程间通信的主要流程如下：

(1) EventHandler 将具体的 InnerEvent 事件或者 Runnable 任务投递到 EventRunner 所创建的线程的事件队列。

(2) EventRunner 循环从事件队列中获取 InnerEvent 事件或者 Runnable 任务。

1. 处理事件或任务

如果 EventRunner 取出的事件为 InnerEvent 事件，则触发 EventHandler 的回调方法并触发 EventHandler 的处理方法，在新线程上处理该事件。

如果 EventRunner 取出的事件为 Runnable 任务，则 EventRunner 直接在新线程上处理 Runnable 任务。

2. 约束限制

在进行线程间通信的时候，EventHandler 只能和 EventRunner 所创建的线程进行绑定，EventRunner 创建时需要判断是否创建成功，只有确保获取的 EventRunner 实例非空时，才可以使用 EventHandler 绑定 EventRunner。

一个 EventHandler 只能同时与一个 EventRunner 绑定，一个 EventRunner 上可以创建多个 EventHandler。

1）EventHandler 开发场景

EventHandler 的主要功能是将 InnerEvent 事件或者 Runnable 任务投递到其他的线程进行处理，其使用的场景包括：

开发者需要将 InnerEvent 事件投递到新的线程，按照优先级和延时进行处理。投递时，EventHandler 的优先级可在 IMMEDIATE、HIGH、LOW、IDLE 中选择，并设置合适的 delayTime。

开发者需要将 Runnable 任务投递到新的线程，并按照优先级和延时进行处理。投递时，EventHandler 的优先级可在 IMMEDIATE、HIGH、LOW、IDLE 中选择，并设置合适的 delayTime。

开发者需要在新创建的线程里将事件投递到原线程进行处理。

2）EventRunner 工作模式

EventRunner 的工作模式可以分为托管模式和手动模式。两种模式是在调用 EventRunner 的 create()方法时，通过选择不同的参数实现的，默认为托管模式。

托管模式：不需要开发者调用 run()和 stop()方法去启动和停止 EventRunner。当 EventRunner 实例化时，系统调用 run()来启动 EventRunner；当 EventRunner 不被引用时，系统调用 stop()来停止 EventRunner。

手动模式：需要开发者自行调用 EventRunner 的 run()方法和 stop()方法来确保线程的启动和停止。

3. EventHandler 投递 InnerEvent 事件

EventHandler 投递 InnerEvent 事件，并按照优先级和延时进行处理，开发步骤如下。

(1) 创建 EventHandler 的子类，在子类中重写实现方法 processEvent()来处理事件，代码如下：

```
private class MyEventHandler extends EventHandler {
    private MyEventHandler(EventRunner runner) {
```

```
        super(runner);
    }
    //重写实现 processEvent 方法
    @Override
    public void processEvent(InnerEvent event) {
        super.processEvent(event);
        if (event == null) {
            return;
        }
        int eventId = event.eventId;
        switch (eventId) {
            case CASE1:
                //待执行的操作,由开发者定义
                break;
            case CASE2:
                //待执行的操作,由开发者定义
                break;
            default:
                break;
        }
    }
}
```

(2) 创建 EventRunner,以手动模式为例,代码如下:

```
EventRunner runner = EventRunner.create(false);  //如果 create()的参数是 true,则为托管模式
```

(3) 创建 EventHandler 子类的实例,代码如下:

```
MyEventHandler myHandler = new MyEventHandler(runner);
```

(4) 获取 InnerEvent 事件,代码如下:

```
//获取事件实例,其属性 eventId、param、object 由开发者确定,代码中只用于示例
int eventId1 = 0;
int eventId2 = 1;
long param = 0L;
Object object = null;
InnerEvent event1 = InnerEvent.get(eventId1, param, object);
InnerEvent event2 = InnerEvent.get(eventId2, param, object);
```

(5) 投递事件,投递的优先级以 IMMEDIATE 为例,延时选择 0ms 和 2ms,代码如下:

```
//优先级为 IMMEDIATE,投递之后立即处理,延时为 0ms,该语句等价于同步投递 sendSyncEvent
//(event1,EventHandler.Priority.IMMEDIATE);
```

```
myHandler.sendEvent(event1, 0, EventHandler.Priority.IMMEDIATE);
myHandler.sendEvent(event2, 2, EventHandler.Priority.IMMEDIATE); //延时 2ms 后立即处理
```

(6) 启动和停止 EventRunner，如果为托管模式，则不需要此步骤，代码如下：

```
runner.run();
//待执行操作
runner.stop();    //开发者根据业务需要在适当时机停止 EventRunner
```

4. EventHandler 投递 Runnable 任务

按照优先级和延时进行处理，开发步骤如下：

(1) 创建 EventHandler 的子类，创建 EventRunner，并创建 EventHandler 子类的实例，步骤与 EventHandler 投递 InnerEvent 场景的步骤 1～3 相同。

创建 Runnable 任务，代码如下：

```
Runnable task1 = new Runnable() {
    @Override
    public void run() {
        //待执行的操作，由开发者定义
    }
};
Runnable task2 = new Runnable() {
    @Override
    public void run() {
        //待执行的操作，由开发者定义
    }
};
```

(2) 投递 Runnable 任务，投递的优先级以 IMMEDIATE 为例，延时选择 0ms 和 2ms，代码如下：

```
//优先级为 immediate，延时 0ms，该语句等价于同步投递 myHandler.postSyncTask(task1,
//EventHandler.Priority.immediate);
myHandler.postTask(task1, 0, EventHandler.Priority.IMMEDIATE);

myHandler.postTask(task2, 2, EventHandler.Priority.IMMEDIATE);
//延时 2ms 后立即执行
```

(3) 启动和停止 EventRunner，如果是托管模式，则不需要此步骤，代码如下：

```
runner.run();
//待执行操作
runner.stop();    //停止 EventRunner
```

5. 在新创建的线程里将事件投递到原线程

EventHandler 从新创建的线程将事件投递到原线程并进行处理，开发步骤如下：

(1) 创建 EventHandler 的子类，在子类中重写实现方法 processEvent()来处理事件，代码如下：

```
private class MyEventHandler extends EventHandler {
    private MyEventHandler(EventRunner runner) {
        super(runner);
    }
    //重写实现 processEvent 方法
    @Override
    public void processEvent(InnerEvent event) {
        super.processEvent(event);
        if (event == null) {
            return;
        }
        int eventId = event.eventId;
        Object object = event.object;
        switch (eventId) {
            case CASE1:
                //待执行的操作，由开发者定义
                break;
            case CASE2:
                if (object instanceof EventRunner) {
                    //将原先线程的 EventRunner 实例投递给新创建的线程
                    EventRunner runner2 = (EventRunner) object;
                    //将原先线程的 EventRunner 实例与新创建的线程的 EventHandler 绑定
                    EventHandler myHandler2 = new EventHandler(runner2) {
                        @Override
                        public void processEvent(InnerEvent event) {
                            //需要在原先线程执行的操作
                        }
                    };
                    int eventId2 = 1;
                    long param2 = 0L;
                    Object object2 = null;
                    InnerEvent event2 = InnerEvent.get(eventId2, param2, object2);
                    myHandler2.sendEvent(event2); //将事件投递到原线程
                }
                break;
            default:
                break;
        }
    }
}
```

(2) 创建 EventRunner,以手动模式为例,代码如下:

```
EventRunner runner = EventRunner.create(false);
//如果 create()的参数是 true,则为托管模式
```

(3) 创建 EventHandler 子类的实例,代码如下:

```
MyEventHandler myHandler = new MyEventHandler(runner);
```

(4) 获取 InnerEvent 事件,代码如下:

```
//获取事件实例,其属性 eventId、param、object 由开发者确定,代码中只用于示例
int eventId = 0;
long param = 0L;
Object object = (Object) EventRunner.current();
InnerEvent event = InnerEvent.get(eventId, param, object);
```

(5) 投递事件,在新线程上直接处理,代码如下:

```
//将与当前线程绑定的 EventRunner 投递到由 runner 创建的新线程中
myHandler.sendEvent(event);
```

(6) 启动和停止 EventRunner,如果是托管模式,则不需要此步骤,代码如下:

```
runner.run();
//待执行操作
runner.stop();   //停止 EventRunner
```

3.8 网络媒体与设备

3.8.1 网络管理

HarmonyOS 网络管理模块主要提供以下功能。

- 数据连接管理:网卡绑定,打开 URL,数据链路参数查询。
- 数据网络管理:指定数据网络传输,获取数据网络状态变更,数据网络状态查询。
- 流量统计:获取蜂窝网络、所有网卡、指定应用或指定网卡的数据流量统计值。
- HTTP 缓存:有效管理 HTTP 缓存,减少数据流量。
- 创建本地套接字:实现本机不同进程间的通信,目前只支持流式套接字。

1. 约束与限制

使用网络管理模块的相关功能时,需要请求相应的权限,如表 3-13 所示。

表 3-13　需要请求相应的权限

权　限　名	权 限 描 述
ohos. permission. GET_NETWORK_INFO	获取网络连接信息
ohos. permission. SET_NETWORK_INFO	修改网络连接状态
ohos. permission. INTERNET	允许程序打开网络套接字,进行网络连接

1）场景介绍

应用使用当前的数据网络打开一个 URL 链接。

2）接口说明

应用使用当前网络打开一个 URL 链接,所使用的接口说明如表 3-14 所示。

表 3-14　网络管理功能的主要接口

类　　名	接　口　名	功 能 描 述
NetManager	getInstance(Context context)	获取网络管理的实例对象
	hasDefaultNet()	查询当前是否有默认可用的数据网络
	getDefaultNet()	获取当前默认的数据网络句柄
	addDefaultNetStatusCallback (NetStatusCallback callback)	获取当前默认的数据网络状态变化
	setAppNet(NetHandle netHandle)	应用绑定该数据网络
NetHandle	openConnection(URL URL, Proxy proxy) throws IOException	使用该网络打开一个 URL 链接

2. 开发步骤

(1) 调用 NetManager. getInstance(Context)获取网络管理的实例对象。

(2) 调用 NetManager. getDefaultNet()获取默认的数据网络。

(3) 调用 NetHandle. openConnection()打开一个 URL。

(4) 通过 URL 链接实例访问网站,如代码示例 3-71 所示。

代码示例 3-71　通过 URL 网址访问网站

```
NetManager netManager = NetManager.getInstance(null);

if (!netManager.hasDefaultNet()) {
    return;
}
NetHandle netHandle = netManager.getDefaultNet();

//可以获取网络状态的变化
NetStatusCallback callback = new NetStatusCallback() {
    //重写需要获取的网络状态变化的 override 函数
};
netManager.addDefaultNetStatusCallback(callback);
```

```
//通过 openConnection 获取 URLConnection
HttpURLConnection connection = null;
try {
    String URLString = "https://www.huawei.com/";
    URL URL = new URL(URLString);

    URLConnection URLConnection = netHandle.openConnection(URL,
                java.net.Proxy.NO_PROXY);
    if (URLConnection instanceof HttpURLConnection) {
        connection = (HttpURLConnection) URLConnection;
    }
    connection.setRequestMethod("GET");
    connection.connect();
    //之后可进行 URL 的其他操作
} catch(IOException e) {
} finally {
    connection.disconnect();
}
```

3.8.2　设备的位置信息

移动终端设备已经深入人们日常生活的方方面面，如查看所在城市的天气、新闻轶事、出行打车、旅行导航、运动记录等。这些习以为常的活动，都离不开定位用户终端设备的位置。

当用户处于这些丰富的使用场景中时，系统的位置能力可以提供实时准确的位置数据。对于开发者来讲，设计基于位置体验的服务，也可以使应用的使用体验更贴近每个用户。

当应用在实现基于设备位置的功能时，如驾车导航、记录运动轨迹等，可以调用该模块的 API，完成位置信息的获取。

1. 基本概念

位置能力用于确定用户设备在哪里，系统使用位置坐标标示设备的位置，并用多种定位技术提供服务，如 GNSS 定位、基站定位、WLAN/蓝牙定位(基站定位、WLAN/蓝牙定位后续统称“网络定位技术”)。通过这些定位技术，无论用户设备在室内或户外，都可以准确地确定设备位置。

1）坐标

系统以 1984 年建立的世界大地坐标系统为参考，使用经度、纬度数据描述地球上的一个位置。

2）GNSS 定位

基于全球导航卫星系统，包含 GPS、GLONASS、北斗、Galileo 等，通过导航卫星，由设备芯片提供的定位算法，用来确定设备的准确位置。定位过程具体使用哪些定位系统，取决于用户设备的硬件能力。

3）基站定位

根据设备当前驻网基站和相邻基站的位置，估算设备当前位置。此定位方式的定位结果精度相对较低，并且需要设备可以访问蜂窝网络。

4）WLAN、蓝牙定位

根据设备可搜索到的周围 WLAN、蓝牙设备位置，估算设备当前位置。此定位方式的定位结果精度依赖设备周围可见的固定 WLAN、蓝牙设备的分布，密度较高时，精度也相较于基站定位方式更高，同时也需要设备可以访问网络。

2. 运作机制

位置能力作为系统为应用提供的一种基础服务，需要应用在所使用的业务场景，向系统主动发起请求，并在业务场景结束时，主动结束此请求，在此过程中系统会将实时的定位结果上报给应用。

3. 约束与限制

使用设备的位置能力，需要用户进行确认并主动开启位置开关。如果位置开关没有开启，系统不会向任何应用提供位置服务。

设备位置信息属于用户敏感数据，所以即使用户已经开启位置开关，应用在获取设备位置前仍需向用户申请位置访问权限。在用户确认允许后，系统才会向应用提供位置服务。

4. 获取设备的位置信息

1）场景介绍

开发者可以调用 HarmonyOS 位置相关接口，获取设备实时位置，或者最近的历史位置。

对于位置敏感的应用业务，建议获取设备实时位置信息。如果不需要设备的实时位置信息，并且希望尽可能地节省耗电，则开发者可以考虑获取最近的历史位置。

2）接口说明

获取设备的位置信息，所使用的接口说明如表 3-15 所示。

表 3-15 获取设备的位置信息，所使用的接口说明

接口名	功能描述
Locator(Context context)	创建 Locator 实例对象
RequestParam(int scenario)	根据定位场景类型创建定位请求的 RequestParam 对象
onLocationReport(Location location)	获取定位结果
startLocating(RequestParam request, LocatorCallback callback)	向系统发起定位请求
requestOnce(RequestParam request, LocatorCallback callback)	向系统发起单次定位请求
stopLocating(LocatorCallback callback)	结束定位
getCachedLocation()	获取系统缓存的位置信息

3）开发步骤

应用在使用系统能力前，需要检查是否已经获取用户授权访问设备位置信息。如未获得授权，可以向用户申请需要的位置权限，申请方式参考动态申请权限开发步骤。

系统提供的定位权限如下：

```
ohos.permission.LOCATION
ohos.permission.LOCATION_IN_BACKGROUND
```

访问设备的位置信息，必须申请 ohos.permission.LOCATION 权限，并且获得用户授权。

如果应用在后台运行时也需要访问设备位置，除需要将应用声明为允许后台运行外，还必须申请 ohos.permission.LOCATION_IN_BACKGROUND 权限，这样应用在切入后台之后，系统可以继续上报位置信息。

开发者可以在应用 config.json 文件中声明所需要的权限，示例代码如下：

```
{
"reqPermissions": [{
"name": "ohos.permission.LOCATION",
"reason": "$string:reason_description",
"usedScene": {
"ability": ["com.myapplication.LocationAbility"],
"when": "inuse"
        }, {
        ...
        }
    ]
}
```

配置字段详细说明见权限开发指导。

实例化 Locator 对象，所有与基础定位能力相关的功能 API 都是通过 Locator 提供的。

```
Locator locator = new Locator(context);
```

其中入参需要提供当前应用程序的 AbilityInfo 信息，便于系统管理应用的定位请求。

实例化 RequestParam 对象，用于告知系统该向应用提供何种类型的位置服务，以及位置结果上报的频率。

方式一：

为了面向开发者提供贴近其使用场景的 API 使用方式，系统定义了几种常见的位置能力使用场景，并针对使用场景做了适当的优化处理，应用可以直接匹配使用，从而简化开发复杂度。系统当前支持的场景如表 3-16 所示。

表 3-16 系统当前支持的场景

场景名称	常量定义	说明
导航场景	SCENE_NAVIGATION	适用于在户外定位设备实时位置的场景，如车载、步行导航等。在此场景下，为了保证系统提供的位置结果精度最优，主要使用 GNSS 定位技术提供定位服务，结合场景特点，在导航启动之初，用户很可能在室内、车库等遮蔽环境，此时 GNSS 技术很难提供位置服务。为解决此问题，我们会在 GNSS 提供稳定位置结果之前，使用系统网络定位技术，向应用提供位置服务，以在导航初始阶段提升用户体验。 此场景默认以最短 1s 间隔上报定位结果，使用此场景的应用必须申请 ohos. permission. LOCATION 权限，同时获得用户授权
轨迹跟踪场景	SCENE_TRAJECTORY_TRACKING	适用于记录用户位置轨迹的场景，如运动类应用记录轨迹功能等。主要使用 GNSS 定位技术提供定位服务。 此场景默认以最短 1s 间隔上报定位结果，并且应用必须申请 ohos. permission. LOCATION 权限，同时获得用户授权
出行约车场景	SCENE_CAR_HAILING	适用于用户出行打车时定位当前位置的场景，如网约车类应用等。 此场景默认以最短 1s 间隔上报定位结果，并且应用必须申请 ohos. permission. LOCATION 权限，同时获得用户授权
生活服务场景	SCENE_DAILY_LIFE_SERVICE	生活服务场景，适用于不需要定位用户精确位置的使用场景，如新闻信息、网购、点餐类应用等，做推荐、推送时定位用户大致位置即可。 此场景默认以最短 1s 间隔上报定位结果，并且应用至少申请 ohos. permission. LOCATION 权限，同时获得用户授权
无功耗场景	SCENE_NO_POWER	无功耗场景，适用于不需要主动启动定位业务。系统在响应其他应用启动定位业务并上报位置结果时，会同时向请求此场景的应用程序上报定位结果，当前的应用程序不产生定位功耗。 此场景默认以最短 1s 间隔上报定位结果，并且应用需要申请 ohos. permission. LOCATION 权限，同时获得用户授权

以导航场景为例，实例化方式如下：

```
RequestParam requestParam = new
RequestParam(RequestParam.SCENE_NAVIGATION);
```

方式二：

如果定义的现有场景类型不能满足所需的开发场景，则可使用系统提供的基本的定位优先级策略类型，如表 3-17 所示。

表 3-17 系统提供的基本的定位优先级策略类型

策略类型	常量定义	说明
定位精度优先策略	PRIORITY_ACCURACY	定位精度优先策略主要以 GNSS 定位技术为主，在开阔场景下可以提供米级的定位精度，具体性能指标依赖用户设备的硬件定位能力，但在室内等强遮蔽定位场景下，无法提供准确的位置服务。 应用必须申请 ohos. permission. LOCATION 权限，同时获得用户授权
快速定位优先策略	PRIORITY_FAST_FIRST_FIX	快速定位优先策略会同时使用 GNSS 定位、基站定位和 WLAN、蓝牙定位技术，以便在室内和户外场景下，通过此策略可以获得位置结果，当各种定位技术都提供位置结果时，系统会选择其中精度较好的结果返回应用。因为各种定位技术同时使用，对设备的硬件资源消耗较大，功耗也较大。 应用必须申请 ohos. permission. LOCATION 权限，同时获得用户授权
低功耗定位优先策略	PRIORITY_LOW_POWER	低功耗定位优先策略主要使用基站定位和 WLAN、蓝牙定位技术，也可以同时提供室内和户外场景下的位置服务，因为其依赖周边基站、可见 WLAN、蓝牙设备的分布情况，定位结果的精度波动范围较大，如果对定位结果精度要求不高，或者使用场景多在有基站、可见 WLAN、蓝牙设备高密度分布的情况下，则推荐使用，可以有效节省设备功耗。 应用至少申请 ohos. permission. LOCATION 权限，同时获得用户授权

以定位精度优先策略为例，实例化方式如下：

```
RequestParam requestParam = new
RequestParam(RequestParam.PRIORITY_ACCURACY,0,0);
```

后两个入参用于限定系统向应用上报定位结果的频率，分别为位置上报的最短时间间隔和位置上报的最短距离间隔，开发者可以参考 API 具体说明进行开发。

实例化 LocatorCallback 对象，用于向系统提供位置上报的途径。

应用需要自行实现系统定义好的回调接口，并将其实例化。系统在定位成功并确定设备的实时位置结果时，会通过 onLocationReport 接口上报给应用。

应用程序可以在 onLocationReport 接口的实现中完成自己的业务逻辑，如代码示例 3-72 所示。

代码示例 3-72 onLocationReport 接口完成自己的业务逻辑

```
MyLocatorCallback locatorCallback = new MyLocatorCallback();

public class MyLocatorCallback implements LocatorCallback {
    @Override
    public void onLocationReport(Location location) {
    }

    @Override
    public void onStatusChanged(int type) {
    }

    @Override
    public void onErrorReport(int type) {
    }
}
```

启动定位，代码如下：

```
locator.startLocating(requestParam, locatorCallback);
```

如果应用不需要持续获取位置结果，则可以使用如下方式启动定位，系统会上报一次实时定位结果，然后自动结束应用的定位请求。应用不需要执行结束定位，代码如下：

```
locator.requestOnce(requestParam, locatorCallback);
```

（可选）结束定位，代码如下：

```
locator.stopLocating(locatorCallback);
```

如果应用使用场景不需要获取实时的设备位置，可以获取系统缓存的最近一次历史定位结果，代码如下：

```
locator.getCachedLocation();
```

此接口的使用需要应用向用户申请 ohos.permission.LOCATION 权限。

5. 地理编码转化成详细所在地信息

1）场景介绍

使用坐标描述一个位置，非常准确，但是并不直观，面向用户表达并不友好。

系统向开发者提供了地理编码转化能力（将坐标转化为地理编码信息），以及逆地理编码转化能力（将地理描述转化为具体坐标）。其中地理编码包含多个属性来描述位置，包括国家、行政区、街道、门牌号、地址描述等，这样的信息更便于用户理解。

2）接口说明

进行坐标和地理编码信息的相互转化，所使用的接口说明如表3-18所示。

表3-18 进行坐标和地理编码信息的相互转化，所使用的接口说明

接口名	功能描述
GeoConvert()	创建GeoConvert实例对象
GeoConvert(Locale locale)	根据自定义参数创建GeoConvert实例对象
getAddressFromLocation(double latitude, double longitude, int maxItems)	根据指定的经纬度坐标获取地理位置信息
getAddressFromLocationName(String description, int maxItems)	根据地理位置信息获取相匹配的包含坐标数据的地址列表
getAddressFromLocationName(String description, double minLatitude, double minLongitude, double maxLatitude, double maxLongitude, int maxItems)	根据指定的位置信息和地理区域获取相匹配的包含坐标数据的地址列表

3）开发步骤

GeoConvert需要访问后端服务，需要确保设备联网，以进行信息获取。

实例化GeoConvert对象，所有与（逆）地理编码转化能力相关的功能API都是通过GeoConvert提供的，代码如下：

```
GeoConvert geoConvert = new GeoConvert();
```

如果需要根据自定义参数实例化GeoConvert对象，如语言、地区等，则可以使用GeoConvert(Locale locale)。

获取转化结果。

坐标转化地理位置信息，代码如下：

```
geoConvert.getAddressFromLocation(纬度值, 经度值, 1);
```

参考接口API说明，应用可以获得与此坐标相匹配的GeoAddress列表，应用可以根据实际使用需求，读取相应的参数数据。

位置描述转化为坐标，代码如下：

```
geoConvert.getAddressFromLocationName("北京大兴国际机场", 1);
```

参考接口 API 说明，应用可以获得与位置描述相匹配的 GeoAddress 列表，其中包含对应的坐标数据，参考 API 使用。

如果需要查询的位置描述可能出现多地重名的请求，则可以通过设置一个经纬度范围，以便高效获取期望的准确结果，代码如下：

```
geoConvert.getAddressFromLocationName("北京大兴国际机场", 纬度下限, 经度下限, 纬度上限,
经度上限, 1);
```

3.8.3 视频

HarmonyOS 视频模块支持视频业务的开发和生态开放，开发者可以通过已开放的接口很容易地实现视频媒体的播放、操作和新功能开发。视频媒体的常见操作有视频编解码、视频合成、视频提取、视频播放及视频录制等。

视频播放包括播放控制、播放设置和播放查询，如播放的开始/停止、播放速度设置和是否循环播放等。视频播放类 Player 的主要接口如表 3-19 所示。

表 3-19 视频播放类 Player 的主要接口

接口名	功能描述
Player(Context context)	创建 Player 实例
setSource(Source source)	设置媒体源
prepare()	准备播放
play()	开始播放
pause()	暂停播放
stop()	停止播放
rewindTo(long microseconds)	拖曳播放
setVolume(float volume)	调节播放音量
setVideoSurface(Surface surface)	设置视频播放的窗口
enableSingleLooping(boolean looping)	设置为单曲循环
isSingleLooping()	检查是否单曲循环播放
isNowPlaying()	检查是否播放
getCurrentTime()	获取当前播放位置
getDuration()	获取媒体文件总时长
getVideoWidth()	获取视频宽度
getVideoHeight()	获取视频高度
setPlaybackSpeed(float speed)	设置播放速度
getPlaybackSpeed()	获取播放速度
setAudioStreamType(int type)	设置声频类型
getAudioStreamType()	获取声频类型

续表

接 口 名	功 能 描 述
setNextPlayer(Player next)	设置当前播放结束后的下一个播放器
reset()	重置播放器
release()	释放播放资源
setPlayerCallback(IPlayerCallback callback)	注册回调,接收播放器的事件通知或异常通知

创建 Player 实例,可调用 Player(Context context),创建本地播放器,用于在本设备播放。

```
Player mPlayer = new Player(VideoAbilitySlice.this);
```

构造数据源对象,并调用 Player 实例的 setSource(Source source)方法,设置媒体源,播放本地视频,如代码示例 3-73 所示。

代码示例 3-73 播放本地视频

```
private void playLocalFile(Surface surface) {
    try {
        RawFileDescriptor filDescriptor = getResourceManager().getRawFileEntry("resources/
rawfile/test.mp4").openRawFileDescriptor();
        Source source = new Source(filDescriptor.getFileDescriptor(), filDescriptor.
getStartPosition(), filDescriptor.getFileSize());
        mPlayer.setSource(source);
        mPlayer.setVideoSurface(surface);
        mPlayer.setPlayerCallback(new VideoPlayerCallback());
        mPlayer.prepare();

        mPlayer.play();
    } catch (Exception e) {
        System.out.println("playURL Exception:" + e.getMessage());
    }
}
```

播放网络视频,如代码示例 3-74 所示。

代码示例 3-74 播放网络视频

```
private void playURL(Surface surface) {
    System.out.println("playURL called:");

    try {
        Source source = new Source("https://media.w3.org/2010/05/sintel/trailer.mp4");
        mPlayer.setSource(source);
```

```java
            mPlayer.setVideoSurface(surface);
            mPlayer.setPlayerCallback(new VideoPlayerCallback());
            mPlayer.prepare();
            mSurfaceProvider.setTop(0);
            mPlayer.play();
        } catch (Exception e) {
            System.out.println("playURL Exception:" + e.getMessage());
        }
    }
```

调用 prepare()方法，准备播放。

构造 IPlayerCallback，IPlayerCallback 需要实现 onPlayBackComplete 和 onError(int errorType, int errorCode)两种方法，实现播放完成和播放异常时做相应的操作，如代码示例 3-75 所示。

代码示例 3-75　播放视频回调

```java
    class VideoPlayerCallback implements Player.IPlayerCallback {
        @Override
        public void onPrepared() {
            System.out.println("onPrepared");
        }

        @Override
        public void onMessage(int i, int i1) {
            System.out.println("onMessage");
        }

        @Override
        public void onError(int i, int i1) {
            System.out.println("onError: i=" + i + ", i1=" + i1);
        }

        @Override
        public void onResolutionChanged(int i, int i1) {
            System.out.println("onResolutionChanged");
        }

        @Override
        public void onPlayBackComplete() {
            System.out.println("onPlayBackComplete");
            if (mPlayer != null) {
                mPlayer.stop();
                mPlayer = null;
            }
```

```
    }

    @Override
    public void onRewindToComplete() {
        System.out.println("onRewindToComplete");
    }

    @Override
    public void onBufferingChange(int i) {
        System.out.println("onBufferingChange");
    }

    @Override
    public void onNewTimedMetaData(Player.MediaTimedMetaData mediaTimedMetaData) {
        System.out.println("onNewTimedMetaData");
    }

    @Override
    public void onMediaTimeIncontinuity(Player.MediaTimeInfo mediaTimeInfo) {
        System.out.println("onMediaTimeIncontinuity");
    }
}
```

调用 play()方法，开始播放。

调用 pause()方法和 play()方法，可以实现暂停和恢复播放。

调用 rewindTo(long microseconds)方法实现播放中的拖曳功能。

调用 getDuration()方法和 getCurrentTime()方法，可以实现获取总播放时长及当前播放位置功能。

调用 stop()方法，停止播放。

播放结束后，调用 release()释放资源。

利用 SurfaceProvider 控件实现了一个视频播放的 Demo，包括由开始、暂停和循环播放功能构建的播放页面，如代码示例 3-76 所示。

代码示例 3-76 利用 SurfaceProvider 控件实现了一个视频播放的 Demo

```
<StackLayout
    xmlns:ohos="http://schemas.huawei.com/res/ohos"
    ohos:width="match_parent"
    ohos:height="match_parent">
<ohos.agp.components.surfaceprovider.SurfaceProvider
        ohos:id="$+id:surfaceProvider"
        ohos:width="match_parent"
        ohos:height="match_parent"/>
<Image
```

```
        ohos:id = "$ + id:img"
        ohos:height = "match_content"
        ohos:width = "match_content"
        ohos:image_src = "$media:icon"
        ohos:layout_alignment = "center"
        ohos:visibility = "hide"/>
</StackLayout>
```

在 Slice 代码 onStart 方法中初始化，如代码示例 3-77 所示。

代码示例 3-77　实现了一个视频播放

```
private static Player player;
    private SurfaceProvider sfProvider;
    private Image image;
    static final HiLogLabel logLabel = new HiLogLabel(HiLog.LOG_APP, 0x00201,
"VideoAbilitySlice");

    @Override
    public void onStart(Intent intent) {
        super.onStart(intent);
        super.setUIContent(ResourceTable.Layout_ability_video);
        initPlayer();
    }
```

实现播放功能的代码如代码示例 3-78 所示。

代码示例 3-78　实现播放功能

```
private void initPlayer() {
sfProvider = (SurfaceProvider) findComponentById(ResourceTable.Id_surfaceProvider);
        image = (Image) findComponentById(ResourceTable.Id_img);
        sfProvider.getSurfaceOps().get().addCallback(new VideoSurfaceCallback());
        //sfProvider.pinToZTop(boolean) -- 如果设置为 true,则视频控件会在最上层展示,但是
//当设置为 false 时,虽然不在最上层展示,却会出现黑屏
        //需加上一行代码 WindowManager.getInstance().getTopWindow().get().setTransparent
(true);
        sfProvider.pinToZTop(false);

WindowManager.getInstance().getTopWindow().get().setTransparent(true);
        player = new Player(getContext());
//sfProvider 添加监听事件
        sfProvider.setClickedListener(new Component.ClickedListener() {
            @Override
            public void onClick(Component component) {
                if(player.isNowPlaying()){
                    //如果正在播放,则暂停
                    player.pause();
```

```
                    //播放按钮可见
                    image.setVisibility(Component.VISIBLE);
                }else {
                    //如果暂停,则单击后继续播放
                    player.play();
                    //播放按钮隐藏
                    image.setVisibility(Component.HIDE);
                }
            }
        });
    }
    private void playLocalFile(Surface surface) {
        try {
            RawFileDescriptor filDescriptor = getResourceManager().getRawFileEntry("
resources/rawfile/test.mp4").openRawFileDescriptor();
            Source source = new Source(filDescriptor.getFileDescriptor(),filDescriptor.
getStartPosition(),filDescriptor.getFileSize());
            player.setSource(source);
            player.setVideoSurface(surface);
            player.setPlayerCallback(new VideoPlayerCallback());
            player.prepare();
            sfProvider.setTop(0);
            player.play();
        } catch (Exception e) {
            HiLog.info(logLabel,"playURL Exception:" + e.getMessage());
        }
    }
    private class VideoSurfaceCallback implements SurfaceOps.Callback {
        @Override
        public void surfaceCreated(SurfaceOps surfaceOps) {
            HiLog.info(logLabel,"surfaceCreated() called.");
            if (sfProvider.getSurfaceOps().isPresent()) {
                Surface surface = sfProvider.getSurfaceOps().get().getSurface();
                playLocalFile(surface);
            }
        }
        @Override
        public void surfaceChanged(SurfaceOps surfaceOps, int i, int i1, int i2) {
            HiLog.info(logLabel,"surfaceChanged() called.");
        }
        @Override
        public void surfaceDestroyed(SurfaceOps surfaceOps) {
            HiLog.info(logLabel,"surfaceDestroyed() called.");
        }
    }
```

```
private class VideoPlayerCallback implements Player.IPlayerCallback {
    @Override
    public void onPrepared() {
        HiLog.info(logLabel,"onPrepared");
    }
    @Override
    public void onMessage(int i, int i1) {
        HiLog.info(logLabel,"onMessage");
    }
    @Override
    public void onError(int i, int i1) {
        HiLog.info(logLabel,"onError: i=" + i + ", i1=" + i1);
    }
    @Override
    public void onResolutionChanged(int i, int i1) {
        HiLog.info(logLabel,"onResolutionChanged");
    }
    @Override
    public void onPlayBackComplete() {
        //播放完成回调,重新播放
        if (player != null) {
            player.prepare();
            player.play();
        }
    }
    @Override
    public void onRewindToComplete() {
        HiLog.info(logLabel,"onRewindToComplete");
    }
    @Override
    public void onBufferingChange(int i) {
        HiLog.info(logLabel,"onBufferingChange");
    }
    @Override
    public void onNewTimedMetaData(Player.MediaTimedMetaData mediaTimedMetaData) {
        HiLog.info(logLabel,"onNewTimedMetaData");
    }
    @Override
    public void onMediaTimeIncontinuity(Player.MediaTimeInfo mediaTimeInfo) {
        HiLog.info(logLabel,"onMediaTimeIncontinuity");
    }
}
```

在 Slice 的其他生命周期对 player 进行资源管理,代码如下:

```
@Overridepublic void onActive() {
    super.onActive();
    player.play();
}@Overrideprotected void onBackground() {
    super.onBackground();
    player.pause();
}@Overrideprotected void onStop() {
    super.onStop();
    player.stop();
    player.release();
}
```

3.8.4 图像

HarmonyOS 图像模块支持图像业务的开发,常见功能如图像解码、图像编码、基本的位图操作、图像编辑等。当然,也支持通过接口组合实现更复杂的图像处理逻辑。

1. 基本概念

1) 图像解码

图像解码就是将不同的存档格式图片(如 JPEG、PNG 等)解码为无压缩的位图格式,以方便在应用或者系统中进行相应处理。

2) PixelMap

PixelMap 是图像解码后无压缩的位图格式,用于图像显示或者进一步处理。

3) 渐进式解码

渐进式解码是在无法一次性提供完整图像文件数据的场景下,随着图像文件数据的逐步增加,通过多次增量解码逐步完成图像解码的模式。

4) 预乘

预乘时,RGB 各通道的值被替换为原始值乘以 Alpha 通道不透明的比例(0~1)后的值,方便后期直接合成叠加。不预乘指 RGB 各通道的数值是图像的原始值,与 Alpha 通道的值无关。

5) 图像编码

图像编码就是将无压缩的位图格式,编码成不同格式的存档格式图片(JPEG、PNG 等),以方便在应用或者系统中进行相应处理。

2. 约束与限制

为及时释放本地资源,建议在图像解码的 ImageSource 对象、位图图像 PixelMap 对象或图像编码的 ImagePacker 对象使用完成后,主动调用 ImageSource、PixelMap 和 ImagePacker 的 release()方法。

3. 普通解码开发步骤

创建图像数据源 ImageSource 对象，可以通过 SourceOptions 指定数据源的格式信息，此格式信息仅为给解码器的提示，正确提供信息能帮助提高解码效率，如果不设置或设置不正确，则会自动检测正确的图像格式。不使用该选项时，可以将 create 接口传入的 SourceOptions 设置为 null。

设置解码参数，解码获取 PixelMap 图像对象，解码过程中同时支持图像处理操作。设置 desiredRegion 支持按矩形区域裁剪，如果设置为全 0，则不进行裁剪。设置 desiredSize 支持按尺寸缩放，如果设置为全 0，则不进行缩放。设置 rotateDegrees 支持旋转角度，以图像中心点为圆心顺时针旋转。如果只需解码原始图像，不使用该选项时，则可将向 createPixelMap 传入的 DecodingOptions 设置为 null。

解码完成并获取 PixelMap 对象后，可以进行后续处理，例如渲染显示等。

我们通过一个解析远程图片的例子，了解图像的解码过程，如代码示例 3-79 所示。

代码示例 3-79 解析远程图片的例子

```
public void loadNetImageURL() {
    String URLImage = "https://www.baidu.com/img/flexible/logo/plus_logo_web_white_2.png";
    HttpURLConnection connection = null;

    try {
        URL URL = new URL(URLImage);
        URLConnection URLConnection = URL.openConnection();
        if (URLConnection instanceof HttpURLConnection) {
            connection = (HttpURLConnection) URLConnection;
        }
        if (connection != null) {
            connection.connect();
            //之后可进行 URL 的其他操作
            //得到服务器返回的流对象
            InputStream inputStream = URLConnection.getInputStream();
            ImageSource imageSource = ImageSource.create(inputStream, new ImageSource.
SourceOptions());
            ImageSource.DecodingOptions decodingOptions = new ImageSource.DecodingOptions();
            decodingOptions.desiredPixelFormat = PixelFormat.ARGB_8888;
            //普通解码叠加旋转、缩放、裁剪
            PixelMap pixelMap = imageSource.createPixelmap(decodingOptions);
            //普通解码
            getUITaskDispatcher().syncDispatch(() -> {
                new ToastDialog(MainAbilitySlice.this)
                        .setText("333")
                        .show();
                img.setPixelMap(pixelMap);
                img.release();
```

```
            });
        }

    } catch (Exception e) {
        System.out.println("----------------------------------");
        System.out.println(e.fillInStackTrace());
        System.out.println("----------------------------------");
    }
}
```

3.8.5 相机

HarmonyOS 相机模块支持相机业务的开发,开发者可以通过已开放的接口实现相机硬件的访问、操作和新功能开发,最常见的操作如:预览、拍照、连拍和录像等。

1. 基本概念

1) 相机静态能力

用于描述相机固有能力的一系列参数,例如朝向、支持的分辨率等信息。

2) 物理相机

物理相机就是独立的实体摄像头设备。物理相机 ID 是用于标识每个物理摄像头的唯一字符串。

3) 逻辑相机

逻辑相机是由多个物理相机组合出来的抽象设备,逻辑相机通过同时控制多个物理相机设备来完成相机某些功能,如大光圈、变焦等功能。逻辑摄像机 ID 是一个唯一的字符串,用于标识多个物理摄像机的抽象能力。

4) 帧捕获

相机启动后对帧的捕获动作统称为帧捕获。主要包含单帧捕获、多帧捕获、循环帧捕获。

5) 单帧捕获

指的是相机启动后,在帧数据流中捕获一帧数据,常用于普通拍照。

6) 多帧捕获

指的是相机启动后,在帧数据流中连续捕获多帧数据,常用于连拍。

7) 循环帧捕获

指的是相机启动后,在帧数据流中一直捕获帧数据,常用于预览和录像。

2. 约束与限制

在同一时刻只能有一个相机应用在运行。

相机模块内部有状态控制,开发者必须按照指导文档的流程进行接口的顺序调用,否则可能会出现调用失败等问题。

为了使所开发的相机应用拥有更好的兼容性,在创建相机对象或者进行参数设置前务必进行能力查询。

3. 相机开发流程

相机模块的主要工作是给相机应用开发者提供基本的相机 API,用于使用相机系统的功能,进行相机硬件的访问、操作和新功能开发。相机的开发流程如图 3-91 所示。

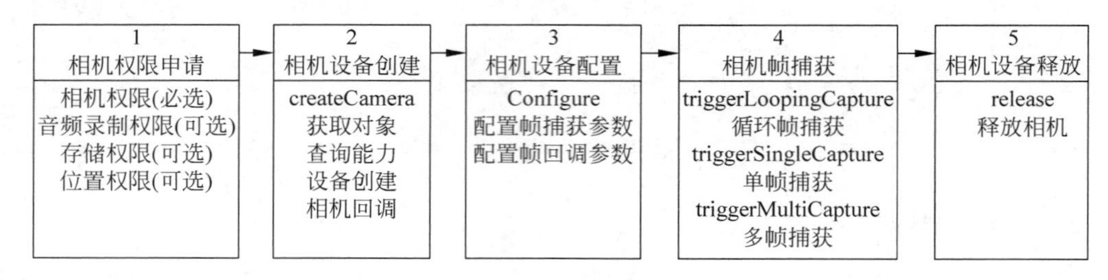

图 3-91 相机开发流程

1) 接口说明

相机模块为相机应用开发者提供了 3 个开发包,如表 3-20 所示。包括方法、枚举及常量/变量,方便开发者更容易地实现相机功能。详情可查阅对应的开发场景。

表 3-20 相机模块为相机应用开发者提供了 3 个开发包

包 名	功 能
ohos. media. camera. CameraKit	相机功能入口类。获取当前支持的相机列表及其静态能力信息,创建相机对象
ohos. media. camera. device	相机设备操作类。提供相机能力查询、相机配置、相机帧捕获、相机状态回调等功能
ohos. media. camera. params	相机参数类。提供相机属性、参数和操作结果的定义

2) 相机权限申请

在使用相机之前,需要申请相机的相关权限,保证应用拥有相机硬件及其他功能权限,应用权限的介绍可参考权限章节,相机涉及权限如表 3-21 所示。

表 3-21 在使用相机之前,需要申请相机的相关权限

权限名称	权限属性值	是否必选
相机权限	ohos. permission. CAMERA	必选
录音权限	ohos. permission. MICROPHONE	可选(需要录像时申请)
存储权限	ohos. permission. WRITE_USER_STORAGE	可选(需要将图像及视频保存到设备的外部存储时申请)
位置权限	ohos. permission. LOCATION	可选(需要保存图像及视频位置信息时申请)

3) 相机设备创建

CameraKit 类是相机的入口 API 类,用于获取相机设备特性、打开相机,其接口如

表 3-22 所示。

表 3-22 CameraKit 类是相机的入口 API 类

接口名	描述
createCamera(String cameraId, CameraStateCallback callback, EventHandler handler)	创建相机对象
getCameraAbility(String cameraId)	获取指定逻辑相机或物理相机的静态能力
getCameraIds()	获取当前逻辑相机列表
getCameraInfo(String cameraId)	获取指定逻辑相机的信息
getInstance(Context context)	获取 CameraKit 实例
registerCameraDeviceCallback (CameraDeviceCallback callback, EventHandler handler)	注册相机使用状态回调
unregisterCameraDeviceCallback (CameraDeviceCallback callback)	注销相机使用状态回调

基于 HarmonyOS 实现一个相机应用,无论将来想应用到哪个或者哪些设备上,都必须先创建一个独立的相机设备,然后才能继续相机的其他操作。创建相机设备的建议步骤如下:

通过 CameraKit. getInstance(Context context)方法获取唯一的 CameraKit 对象,如代码示例 3-80 所示。

代码示例 3-80 创建 CameraKit 对象

```
private void openCamera(){
    //获取 CameraKit 对象
    CameraKit cameraKit = CameraKit.getInstance(context);
    if (cameraKit == null) {
        //处理 cameraKit 获取失败的情况
    }
}
```

如果此步骤操作失败,则说明相机可能被占用或无法使用。如果被占用,则必须等到相机释放后才能重新获取 CameraKit 对象。

通过 getCameraIds()方法,获取当前使用的设备所支持的逻辑相机列表。逻辑相机列表中存储了当前设备拥有的所有逻辑相机 ID,如果列表不为空,则列表中的每个 ID 都支持独立创建相机对象。否则,说明正在使用的设备无可用的相机,不能继续后续的操作,如代码示例 3-81 所示。

代码示例 3-81 获取当前使用的设备所支持的逻辑相机列表

```
try {
    //获取当前设备的逻辑相机列表
    String[] cameraIds = cameraKit.getCameraIds();
```

```
        if (cameraIds.length <= 0) {
            HiLog.error(LABEL, "cameraIds size is 0");
        }
    } catch (IllegalStateException e) {
        //处理异常
    }
```

还可以继续查询指定相机 ID 的静态信息，如表 3-23 和表 3-24 所示。

表 3-23 CameraInfo 的主要接口

接口名	描述
getDeviceLinkType(String physicalId)	获取物理相机连接方式
getFacingType()	获取相机朝向信息
getLogicalId()	获取逻辑相机 ID
getPhysicalIdList()	获取对应的物理相机 ID 列表

表 3-24 CameraAbility 的主要接口

接口名	描述
getSupportedSizes(int format)	根据格式查询输出图像的分辨率列表
getSupportedSizes(Class<T> clazz)	根据 Class 类型查询分辨率列表
getParameterRange(ParameterKey.Key<T> parameter)	获取指定参数能够设置的值范围
getPropertyValue(PropertyKey.Key<T> property)	获取指定属性对应的值
getSupportedAeMode()	获取当前相机支持的自动曝光模式
getSupportedAfMode()	获取当前相机支持的自动对焦模式
getSupportedFaceDetection()	获取相机支持的人脸检测类型范围
getSupportedFlashMode()	当前相机支持的闪光灯取值范围
getSupportedParameters()	当前相机支持的参数设置
getSupportedProperties()	获取当前相机的属性列表
getSupportedResults()	获取当前相机支持的参数，设置可返回的结果列表
getSupportedZoom()	获取相机支持的变焦范围

调用 getDeviceLinkType(String physicalId)方法获取物理相机连接方式；

调用 getCameraInfo(String cameraId)方法查询相机硬件朝向等信息；

调用 getCameraAbility(String cameraId)方法查询相机能力信息(例如支持的分辨率列表等)。

通过 createCamera(String cameraId, CameraStateCallback callback, EventHandler handler)方法，创建相机对象，此步骤执行成功意味着相机系统的硬件已经完成了上电。

```
//创建相机设备
cameraKit.createCamera(cameraIds[0], cameraStateCallback, eventHandler);
```

第 1 个参数 cameraIds 可以是上一步获取的逻辑相机列表中的任何一个相机 ID。

第 2 个和第 3 个参数负责相机创建和相机运行时的数据和状态检测，务必保证在整个相机运行周期内有效，如代码示例 3-82 所示。

代码示例 3-82 相机设备的创建及完成回调

```
private final class CameraStateCallbackImpl extends CameraStateCallback {
    @Override
    public void onCreated(Camera camera) {
        //创建相机设备
    }

    @Override
    public void onConfigured(Camera camera) {
        //配置相机设备
    }

    @Override
    public void onPartialConfigured(Camera camera) {
        //当使用了 addDeferredSurfaceSize 配置了相机，会接到此回调
    }

    @Override
    public void onReleased(Camera camera) {
        //释放相机设备
    }
}

//相机创建和相机运行时的回调
CameraStateCallbackImpl cameraStateCallback = new
CameraStateCallbackImpl();
if(cameraStateCallback == null) {
    HiLog.error(LABEL, "cameraStateCallback is null");
}
import ohos.eventhandler.EventHandler;
import ohos.eventhandler.EventRunner;

//执行回调的 EventHandler
EventHandler eventHandler = new
EventHandler(EventRunner.create("CameraCb"));
if(eventHandler == null) {
    HiLog.error(LABEL, "eventHandler is null");
}
```

至此，相机设备的创建已经完成。相机设备创建成功会在 CameraStateCallback 中触发 onCreated(Camera camera)回调。在进入相机设备配置前，需要确保相机设备已经创建成功。否则会触发相机设备创建失败的回调，并返回错误码，进行错误处理后，重新执行相机设备的创建。

4）相机设备配置

创建相机设备成功后，在 CameraStateCallback 中会触发 onCreated(Camera camera)回调，并且返回 Camera 对象，用于执行相机设备的操作。

当一个新的相机设备成功创建后，首先需要对相机进行配置，调用 configure(CameraConfig)方法实现配置。相机配置主要是设置预览、拍照、录像所用到的 Surface(详见 ohos. agp. graphics. Surface)，如果不配置 Surface，则相应的功能不能使用。

为了获取相机帧捕获结果的数据和进行状态检测，还需要在相机配置时调用 setFrameStateCallback(FrameStateCallback, EventHandler)方法设置帧回调，如代码示例 3-83 所示。

代码示例 3-83　对相机进行配置

```
private final class CameraStateCallbackImpl extends CameraStateCallback {
    @Override
    public void onCreated(Camera camera) {
        cameraConfigBuilder = camera.getCameraConfigBuilder();
        if (cameraConfigBuilder == null) {
            HiLog.error(LABEL, "onCreated cameraConfigBuilder is null");
            return;
        }
        //配置预览的 Surface
        cameraConfigBuilder.addSurface(previewSurface);
        //配置拍照的 Surface

        cameraConfigBuilder.addSurface(imageReceiver.getRecevingSurface());
        //配置帧结果的回调
        cameraConfigBuilder.setFrameStateCallback(frameStateCallbackImpl, handler);
        try {
            //相机设备配置
            camera.configure(cameraConfigBuilder.build());
        } catch (IllegalArgumentException e) {
            HiLog.error(LABEL, "Argument Exception");
        } catch (IllegalStateException e) {
            HiLog.error(LABEL, "State Exception");
        }
    }
}
```

相机配置成功后，在 CameraStateCallback 中会触发 onConfigured(Camera camera)回调，然后才可以执行与相机帧捕获相关的操作，如表 3-25 所示。

表 3-25　CameraConfig. Builder 的主要接口

接　口　名	描　　述
addSurface(Surface surface)	在相机配置中增加 Surface
build()	相机配置的构建类
removeSurface(Surface surface)	移除先前添加的 Surface
setFrameStateCallback(FrameStateCallback callback, EventHandler handler)	设置用于相机帧结果返回的 FrameStateCallback 和 Handler
addDeferredSurfaceSize(Size surfaceSize, Class<T> clazz)	添加延迟 Surface 的尺寸、类型
addDeferredSurface(Surface surface)	设置延迟的 Surface,此 Surface 的尺寸和类型必须和使用 addDeferredSurfaceSize 配置的一致

5）相机帧捕获

Camera 操作类,包括相机预览、录像、拍照等功能接口,如表 3-26 所示。

表 3-26　Camera 的主要接口

接　口　名	描　　述
triggerSingleCapture(FrameConfig frameConfig)	启动相机帧的单帧捕获
triggerMultiCapture(List<FrameConfig> frameConfigs)	启动相机帧的多帧捕获
configure(CameraConfig config)	配置相机
flushCaptures()	停止并清除相机帧的捕获,包括循环帧/单帧/多帧捕获
getCameraConfigBuilder()	获取相机配置构造器对象
getCameraId()	获取当前相机的 ID
getFrameConfigBuilder(int type)	获取指定类型的相机帧,配置构造器对象
release()	释放相机对象及资源
triggerLoopingCapture(FrameConfig frameConfig)	启动或者更新相机帧的循环捕获
stopLoopingCapture()	停止当前相机帧的循环捕获

6）启动预览(循环帧捕获)

用户一般先看见预览画面才执行拍照或者其他功能,所以对于一个普通的相机应用,预览是必不可少的。启动预览的建议步骤如下。

通过 getFrameConfigBuilder(FRAME_CONFIG_PREVIEW)方法获取预览配置模板,常用帧配置项见表 3-27,更多的帧配置项及详细使用方法可参考 API 说明的 FrameConfig. Builder 部分,如表 3-27 所示。

表 3-27 常用帧配置项

接口名	描述	是否必选
addSurface(Surface surface)	配置预览 Surface 和帧的绑定	是
setAfMode(int afMode, Rect rect)	配置对焦模式	否
setAeMode(int aeMode, Rect rect)	配置曝光模式	否
setZoom(float value)	配置变焦值	否
setFlashMode(int flashMode)	配置闪光灯模式	否
setFaceDetection(int type, boolean isEnable)	配置人脸检测或者笑脸检测	否
setParameter(Key<T> key, T value)	配置其他属性(如自拍镜像等)	否
setMark(Object mark)	配置一个标签,后续可以从 FrameConfig 中通过 Object getMark()得到标签,判断两个是否相等,如果相等就说明是同一个配置	否
setCoordinateSurface(Surface surface)	配置坐标系基准 Surface,后续计算 Ae/Af 等区域都会基于此 Surface 为基本的中心坐标系,不设置则默认使用添加的第一个 Surface	否

通过 triggerLoopingCapture(FrameConfig)方法实现循环帧捕获(如预览/录像),如代码示例 3-84 所示。

代码示例 3-84 实现循环帧捕获

```
private final class CameraStateCallbackImpl extends CameraStateCallback {
    @Override
    public void onConfigured(Camera camera) {
        //获取预览配置模板
        frameConfigBuilder = camera.getFrameConfigBuilder(FRAME_CONFIG_PREVIEW);
        //配置预览 Surface
        frameConfigBuilder.addSurface(previewSurface);
        previewFrameConfig = frameConfigBuilder.build();
        try {
            //启动循环帧捕获
            int triggerId = camera.triggerLoopingCapture(previewFrameConfig);
        } catch (IllegalArgumentException e) {
            HiLog.error(LABEL, "Argument Exception");
        } catch (IllegalStateException e) {
            HiLog.error(LABEL, "State Exception");
        }
    }
}
```

经过以上操作,相机应用已经可以正常进行实时预览了。在预览状态下,开发者还可以执行其他操作,例如当预览帧配置更改时,可以通过 triggerLoopingCapture(FrameConfig)

方法实现预览帧配置的更新，代码如下：

```
//预览帧变焦值变更
frameConfigBuilder.setZoom(1.2f);
//调用 triggerLoopingCapture 方法实现预览帧配置更新
triggerLoopingCapture(frameConfigBuilder.build());
//通过 stopLoopingCapture()方法停止循环帧捕获(停止预览)
//停止预览帧捕获
camera.stopLoopingCapture(frameConfigBuilder.build())
```

7）实现拍照（单帧捕获）

拍照功能属于相机应用的最重要功能之一，而且照片质量对用户至关重要。相机模块基于相机复杂的逻辑，从应用接口层到器件驱动层都已经做好了最适合用户的默认配置，这些默认配置尽可能地保证用户拍出的每张照片的质量。实现拍照的建议步骤如下。

通过 getFrameConfigBuilder(FRAME_CONFIG_PICTURE)方法获取拍照配置模板，并且设置拍照帧配置，如表 3-28 所示。

表 3-28 常用帧配置项

接 口 名	描 述	是否必选
FrameConfig.Builder addSurface(Surface)	实现拍照 Surface 和帧的绑定	必选
FrameConfig.Builder setImageRotation(int)	设置图片旋转角度	可选
FrameConfig.Builder setLocation(Location)	设置图片地理位置信息	可选
FrameConfig.Builder setParameter(Key<T>,T)	配置其他属性（如自拍镜像等）	可选

拍照前准备图像帧数据的接收，如代码示例 3-85 所示。

代码示例 3-85 拍照前准备图像帧数据的接收

```
//图像帧数据接收处理对象
private ImageReceiver imageReceiver;
//执行回调的 EventHandler
private EventHandler eventHandler = new
EventHandler(EventRunner.create("CameraCb"));
//拍照支持分辨率
private Size pictureSize;

//单帧捕获生成图像回调 Listener
private final ImageReceiver.IImageArrivalListener imageArrivalListener = new ImageReceiver.
IImageArrivalListener() {
    @Override
    public void onImageArrival(ImageReceiver imageReceiver) {
        StringBuffer fileName = new StringBuffer("picture_");
        fileName.append(UUID.randomUUID()).append(".jpg");   //定义生成图片文件名
        File myFile = new File(dirFile, fileName.toString()); //创建图片文件
```

```
            imageSaver = new ImageSaver(imageReceiver.readNextImage(), myFile); //创建一个读写
//线程任务用于保存图片
            eventHandler.postTask(imageSaver);            //执行读写线程任务生成图片
        }
    };

    //保存图片、图片数据读写,以及图像生成见 run 方法
    class ImageSaver implements Runnable {
        private final Image myImage;
        private final File myFile;

        ImageSaver(Image image, File file) {
            myImage = image;
            myFile = file;
        }

        @Override
        public void run() {
            Image.Component component = myImage.getComponent(ImageFormat.ComponentType.JPEG);
            Byte[] Bytes = new Byte[component.remaining()];
            component.read(Bytes);
            FileOutputStream output = null;
            try {
                output = new FileOutputStream(myFile);
                output.write(Bytes);                          //写图像数据
            } catch (IOException e) {
                HiLog.error(LABEL, "save picture occur exception!");
            } finally {
                myImage.release();
                if (output != null) {
                    try {
                        output.close();                       //关闭流
                    } catch (IOException e) {
                        HiLog.error(LABEL, "image release occur exception!");
                    }
                }
            }
        }
    }
    private void takePictureInit() {
        List<Size> pictureSizes = cameraAbility.getSupportedSizes(ImageFormat.JPEG);
//获取拍照支持分辨率列表
        pictureSize = getpictureSize(pictureSizes) //根据拍照要求选择合适的分辨率
        imageReceiver = ImageReceiver.create(Math.max(pictureSize.width, pictureSize.height),
            Math.min(pictureSize.width, pictureSize.height), ImageFormat.JPEG, 5); //创建
//ImageReceiver 对象,注意 create 函数中宽度要大于高度; 5 为最大支持的图像素,需要根据实际设置
```

```
    imageReceiver.setImageArrivalListener(imageArrivalListener);
}
```

通过 triggerSingleCapture(FrameConfig)方法实现单帧捕获(如拍照),如代码示例 3-86 所示。

代码示例 3-86 实现单帧捕获(如拍照)

```
private void capture() {
    //获取拍照配置模板
    framePictureConfigBuilder = cameraDevice.getFrameConfigBuilder(FRAME_CONFIG_PICTURE);
    //配置拍照 Surface

    framePictureConfigBuilder.addSurface(imageReceiver.getRecevingSurface());
    //配置拍照其他参数
    framePictureConfigBuilder.setImageRotation(90);
    try {
        //启动单帧捕获(拍照)
        camera.triggerSingleCapture(framePictureConfigBuilder.build());
    } catch (IllegalArgumentException e) {
        HiLog.error(LABEL, "Argument Exception");
    } catch (IllegalStateException e) {
        HiLog.error(LABEL, "State Exception");
    }
}
```

为了捕获到质量更高和效果更好的图片,还可以在帧结果中实时监测自动对焦和自动曝光的状态,一般而言,在自动对焦完成且自动曝光收敛后的瞬间是发起单帧捕获的最佳时机。

实现连拍(多帧捕获):连拍功能方便用户一次拍照获取多张照片,用于捕捉精彩瞬间。同普通拍照的实现流程一致,但连拍需要使用 triggerMultiCapture(List < FrameConfig > frameConfigs)方法。

启动录像(循环帧捕获):启动录像和启动预览类似,但需要另外配置录像 Surface 才能使用。录像前需要进行音视频模块的配置,如代码示例 3-87 所示。

代码示例 3-87 进行音视频模块的配置

```
private Source source;                                        //音视频源
private AudioProperty.Builder audioPropertyBuilder;           //声频属性
private VideoProperty.Builder videoPropertyBuilder;           //视频属性
private StorageProperty.Builder storagePropertyBuilder;       //音视频存储属性
private Recorder mediaRecorder;                               //录像操作对象
private String recordName;                                    //音视频文件名

private void initMediaRecorder() {
    HiLog.info(LABEL, "initMediaRecorder begin");
```

```
    videoPropertyBuilder.setRecorderBitRate(10000000);          //设置录制比特率
    int rotation = DisplayManager.getInstance().getDefaultDisplay(this).get().getRotation();
    videoPropertyBuilder.setRecorderDegrees(getOrientation(rotation));   //设置录像方向
    videoPropertyBuilder.setRecorderFps(30);                    //设置录制采样率
    videoPropertyBuilder.setRecorderHeight(Math.min(recordSize.height, recordSize.width));
    //设置录像支持的分辨率,需保证 width > height
    videoPropertyBuilder.setRecorderWidth(Math.max(recordSize.height, recordSize.width));

    videoPropertyBuilder.setRecorderVideoEncoder(Recorder.VideoEncoder.H264);
    //设置视频编码方式
    videoPropertyBuilder.setRecorderRate(30);                   //设置录制帧率
    source.setRecorderAudioSource(Recorder.AudioSource.MIC);    //设置录制声频源
    source.setRecorderVideoSource(Recorder.VideoSource.SURFACE); //设置视频窗口
    mediaRecorder.setSource(source);                            //设置音视频源
    mediaRecorder.setOutputFormat(Recorder.OutputFormat.MPEG_4); //设置音视频输出格式
    StringBuffer fileName = new StringBuffer("record_");        //生成随机文件名
    fileName.append(UUID.randomUUID()).append(".mp4");
    recordName = fileName.toString();
    File file = new File(dirFile, fileName.toString());         //创建录像文件对象
    storagePropertyBuilder.setRecorderFile(file);               //设置存储音视频文件名
    mediaRecorder.setStorageProperty(storagePropertyBuilder.build());
    audioPropertyBuilder.setRecorderAudioEncoder(Recorder.AudioEncoder.AAC);
    //设置声频编码格式
    mediaRecorder.setAudioProperty(audioPropertyBuilder.build());
    //设置声频属性
    mediaRecorder.setVideoProperty(videoPropertyBuilder.build());
    //设置视频属性
    mediaRecorder.prepare();                                    //准备录制
    HiLog.info(LABEL, "initMediaRecorder end");
}
```

配置录像帧,启动录像,如代码示例3-88所示。

代码示例 3-88 配置录像帧,启动录像

```
private final class CameraStateCallbackImpl extends CameraStateCallback {
    @Override
    public void onConfigured(Camera camera) {
        //获取预览配置模板
        frameConfigBuilder = camera.getFrameConfigBuilder(FRAME_CONFIG_PREVIEW);
        //配置预览 Surface
        frameConfigBuilder.addSurface(previewSurface);
        //配置录像的 Surface
        mRecorderSurface = mediaRecorder.getVideoSurface();
        cameraConfigBuilder.addSurface(mRecorderSurface);
        previewFrameConfig = frameConfigBuilder.build();
        try {
```

```
                //启动循环帧捕获
                int triggerId = camera.triggerLoopingCapture(previewFrameConfig);
            } catch (IllegalArgumentException e) {
                HiLog.error(LABEL, "Argument Exception");
            } catch (IllegalStateException e) {
                HiLog.error(LABEL, "State Exception");
            }
        }
    }
```

通过 camera.stopLoopingCapture()方法停止循环帧捕获(录像)。

8) 相机设备释放

使用完相机后,必须通过 release()来关闭相机和释放资源,否则可能导致其他相机应用无法启动。一旦相机被释放,它所提供的操作就不能再被调用,否则会导致不可预期的结果,或者会引发状态异常。相机设备释放的示例如代码示例 3-89 所示。

代码示例 3-89 设备释放

```
private void releaseCamera() {
    if (camera != null) {
        //关闭相机和释放资源
        camera.release();
        camera = null;
    }
    //拍照配置模板置空
    framePictureConfigBuilder = null;
    //预览配置模板置空
    previewFrameConfig = null;
}
```

3.8.6 声频

HarmonyOS 声频模块支持声频业务的开发,提供与声频相关的功能,主要包括声频播放、声频采集、音量管理和短音播放等。

在使用完 AudioRenderer 声频播放类和 AudioCapturer 声频采集类后,需要调用 release()方法进行资源释放。

声频采集所使用的最终采样率与采样格式取决于输入设备,不同设备支持的格式及采样率范围不同,可以通过 AudioManager 类的 getDevices 接口查询。

在进行声频采集之前,需要申请话筒权限 ohos.permission.MICROPHONE。

1) 场景介绍

声频播放的主要工作是将声频数据转码为可听见的声频模拟信号并通过输出设备进行播放,同时对播放任务进行管理。声频播放类 AudioRenderer 的主要接口如表 3-29 所示。

表 3-29 声频播放类 AudioRenderer 的主要接口

接口名	描述
AudioRenderer(AudioRendererInfo audioRendererInfo, PlayMode pm)	构造函数，设置与播放相关的声频参数和播放模式，使用默认播放设备
AudioRenderer(AudioRendererInfo audioRendererInfo, PlayMode pm, AudioDeviceDescriptor outputDevice)	构造函数，设置与播放相关的声频参数、播放模式和播放设备
start()	播放声频流
write(Byte[] data, int offset, int size)	将声频数据以 Byte 流写入声频接收器以进行播放
write(short[] data, int offset, int size)	将声频数据以 short 流写入声频接收器以进行播放
write(float[] data, int offset, int size)	将声频数据以 float 流写入声频接收器以进行播放
write(java.nio.ByteBuffer data, int size)	将声频数据以 ByteBuffer 流写入声频接收器以进行播放
pause()	暂停播放声频流
stop()	停止播放声频流
release()	释放播放资源
getCurrentDevice()	获取当前工作的声频播放设备
setPlaybackSpeed(float speed)	设置播放速度
setPlaybackSpeed(AudioRenderer.SpeedPara speedPara)	设置播放速度与音调
setVolume(ChannelVolume channelVolume)	设置指定声道上的输出音量
setVolume(float vol)	设置所有声道上的输出音量
getMinBufferSize(int sampleRate, AudioStreamInfo.EncodingFormat format, AudioStreamInfo.ChannelMask channelMask)	获取 Stream 播放模式所需的 buffer 大小
getState()	获取声频播放的状态
getRendererSessionId()	获取声频播放的 sessionID
getSampleRate()	获取采样率
getPosition()	获取声频播放的帧数位置
setPosition(int position)	设置起始播放帧位置
getRendererInfo()	获取声频渲染信息
duckVolume()	降低音量并将声频与另一个拥有声频焦点的应用程序混合
unduckVolume()	恢复音量
getPlaybackSpeed()	获取播放速度、音调参数
setSpeed(SpeedPara speedPara)	设置播放速度、音调参数
getAudioTime()	获取播放时间戳信息
flush()	刷新当前的播放流数据队列
getMaxVolume()	获取播放流可设置的最大音量
getMinVolume()	获取播放流可设置的最小音量
getStreamType()	获取播放流的声频流类型

2）开发步骤

构造声频流参数的数据结构 AudioStreamInfo，推荐使用 AudioStreamInfo. Builder 类来构造，模板如下，模板中所设置的值均为 AudioStreamInfo. Builder 类的默认值，根据声频流的具体规格设置具体参数。

使用创建的声频流构建声频播放的参数结构 AudioRendererInfo，推荐使用 AudioRendererInfo. Builder 类来构造，模板如下，模板中所设置的值均为 AudioRendererInfo. Builder 类的默认值，根据声频播放的具体规格设置具体参数，如代码示例 3-90 所示。

代码示例 3-90 本地声频播放案例封装

```
import ohos.app.AbilityContext;
import ohos.app.dispatcher.TaskDispatcher;
import ohos.app.dispatcher.task.Revocable;
import ohos.app.dispatcher.task.TaskPriority;
import ohos.media.audio.*;
import java.io.*;

public class PlaySoundUtil{

    private AudioStreamInfo audioStreamInfo = null;
    private AudioRendererInfo audioRendererInfo = null;
    private AudioRenderer.PlayMode playMode = AudioRenderer.PlayMode.MODE_STREAM;
    private AudioRenderer audioRenderer = null;
    private AudioManager audioManager = null;
    private AudioInterrupt audioInterrupt = null;
    private InputStream soundInputStream = null;
    private String fileName = null;


    public PlaySoundUtil() throws IOException {
        System.out.println("音乐播放器初始化");
        audioStreamInfo = new AudioStreamInfo.Builder().sampleRate(44100) //44.1kHz
                .audioStreamFlag(AudioStreamInfo.AudioStreamFlag.AUDIO_STREAM_FLAG_MAY_
DUCK) //混音
                .encodingFormat(AudioStreamInfo.EncodingFormat.ENCODING_PCM_16 位) //16 -
bit PCM
                .channelMask(AudioStreamInfo.ChannelMask.CHANNEL_OUT_STEREO) //双声道输出
                .streamUsage(AudioStreamInfo.StreamUsage.STREAM_USAGE_MEDIA) //媒体类声频
                .build();

        audioRendererInfo = new AudioRendererInfo.Builder().audioStreamInfo(audioStreamInfo)
                  .audioStreamOutputFlag(AudioRendererInfo.AudioStreamOutputFlag.AUDIO_
STREAM_OUTPUT_FLAG_DIRECT_PCM) //pcm 格式的输出流
```

```
                .bufferSizeInBytes(1024)
                .isOffload(false) //false 表示分段传输 buffer 并播放,true 表示将整个声频
                                  //流一次性传输到 HAL 层进行播放
                .build();
        audioRenderer = new AudioRenderer(audioRendererInfo,playMode);
    }

    public void loadSound(String fileName){
        this.fileName = fileName;
        String filePath = String.format("assets/entry/resources/rawfile/music/%s", fileName);
        soundInputStream = this.getClass().getClassLoader().getResourceAsStream(filePath);

        int bufSize = audioRenderer.getBufferFrameSize();

        System.out.println("bufSize" + bufSize);

        Byte[] buffer = new Byte[1024];
        int len;
        try {
            audioRenderer.start();
            while((len = soundInputStream.read(buffer,0,buffer.length)) != -1){
                audioRenderer.write(buffer,0,buffer.length) ;
            }
            soundInputStream.close();
        } catch (Exception e) {
            e.printStackTrace();
        }
    }

    public AudioRenderer getAudioRenderer(){
        return this.audioRenderer;
    }


    private TaskDispatcher globalTaskDispatcher = null;

    public void asyncLoadSound(String fileName, AbilityContext cxt){
        globalTaskDispatcher = cxt.getGlobalTaskDispatcher(TaskPriority.DEFAULT);
        System.out.println("音乐播放状态: " + audioRenderer.getState());
        if(audioRenderer.getState() == AudioRenderer.State.STATE_PLAYING){
            audioRenderer.pause();
            return;
        }

        Revocable revocable = globalTaskDispatcher.asyncDispatch(new Runnable() {
```

```
                @Override
                public void run() {
                    loadSound(fileName);
                }
            });
        }
    }
```

第4章 面向 Ability 开发

鸿蒙操作系统中软件的基础单位是 Ability(能力,可以是界面能力,也可以是服务能力)。Ability 是应用所具备能力的抽象,也是应用程序的重要组成部分。一个应用可以具备多种能力,即可以包含多个 Ability,鸿蒙 HarmonyOS 支持应用以 Ability 为单位进行部署。

4.1 Ability 分类

Ability 可以分为 FA(Feature Ability)和 PA(Particle Ability)两种类型,每种类型为开发者提供了不同的模板,以便实现不同的业务功能。

(1) FA 支持 Page Ability。

Page 模板是 FA 唯一支持的模板,用于提供与用户交互的能力。一个 Page 实例可以包含一组相关页面,每个页面用一个 AbilitySlice 实例表示。

(2) PA 支持 Service Ability 和 Data Ability。

Service 模板:用于提供后台运行任务的能力。

Data 模板:用于对外部提供统一的数据访问抽象。

在配置文件(config.json)中注册 Ability 时,可以通过配置 Ability 元素中的 type 属性来指定 Ability 模板类型。

其中,type 的取值可以为 page、service 或 data,分别代表 Page 模板、Service 模板、Data 模板。为了便于表述,后文中将基于 Page 模板、Service 模板、Data 模板实现的 Ability 分别简称为 Page、Service、Data。示例代码如下:

```
{
"module": {
"abilities": [
            {
"type": "page"
   }
        ]
```

```
    }
}
```

接下来，分别介绍不同的 Ability 的用法。

4.2 Page Ability

Page 模板（以下简称 Page）是 FA 唯一支持的模板，用于提供与用户交互的能力。一个 Page 可以由一个或多个 AbilitySlice 构成，AbilitySlice 是指应用的单个页面及其控制逻辑的总和。

当一个 Page 由多个 AbilitySlice 共同构成时，这些 AbilitySlice 页面提供的业务能力应具有高度相关性。例如，新闻浏览功能可以通过一个 Page 实现，其中包含了两个 AbilitySlice：一个 AbilitySlice 用于展示新闻列表，另一个 AbilitySlice 用于展示新闻详情。Page 和 AbilitySlice 的关系如图 4-1 所示。

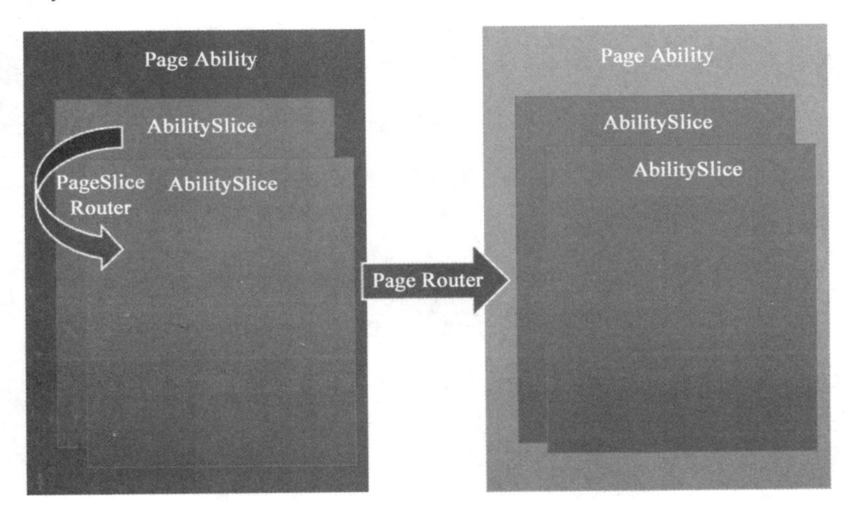

图 4-1　Page Ability 与 Slice Ability 的关系

相比于桌面场景，移动场景下应用之间的交互更为频繁。通常，单个应用专注于某个方面的能力开发，当它需要其他能力辅助时，会调用其他应用提供的能力。例如，外卖应用提供了联系商家的业务功能入口，当用户在使用该功能时，会跳转到通话应用的拨号页面。与此类似，HarmonyOS 支持不同 Page 之间的跳转，并可以指定跳转到目标 Page 中某个具体的 AbilitySlice。

4.2.1 Page Ability 的创建

通过 DevEco Studio 创建一个 Page Ability，如图 4-2 所示。

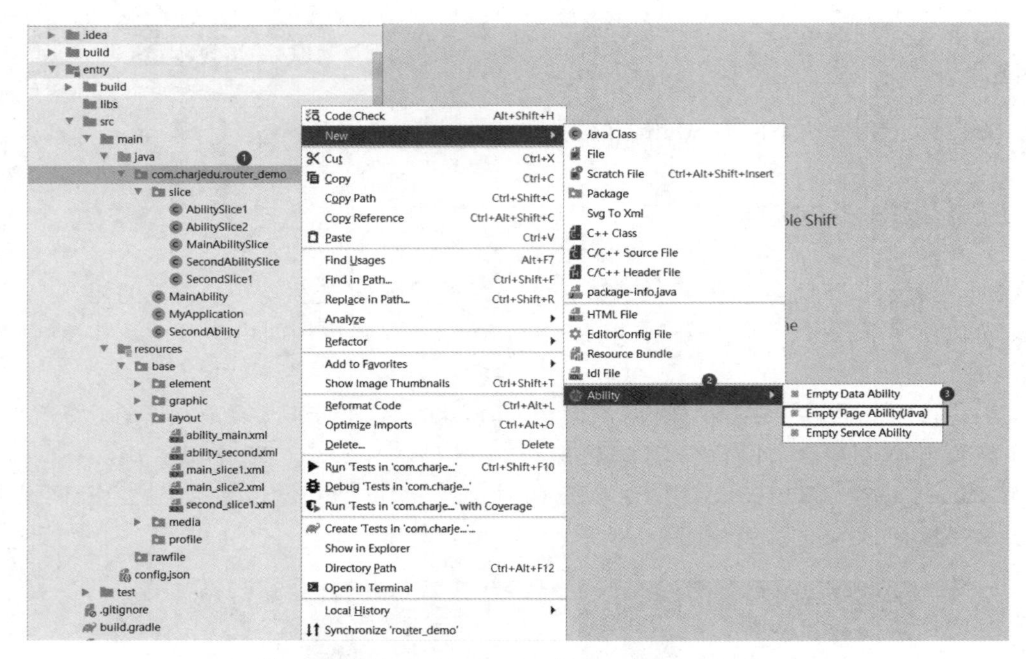

图 4-2 通过 DevEco Studio 创建一个 Page Ability

DevEco Studio 在创建一个 Page Ability 时，会同时修改 config.json 文件，在 abilities 数组中添加这个 Page Ability 的配置，如图 4-3 所示。

```
"abilities": [
  {"name": "com.charjedu.router_demo.MainAbility"...},
  {
    "skills": [
      {
        "entities": [
        ],
        "actions": [
          "secondslice1"
        ]
      }
    ],
    "orientation": "unspecified",
    "name": "com.charjedu.router_demo.SecondAbility",
    "icon": "$media:icon",
    "description": "Java_Phone_Empty Feature Ability",
    "label": "第2个Page",
    "type": "page",
    "launchType": "standard"
  }
]
```

图 4-3 修改 config.json 文件

Skills 项默认为没有，当需要添加子路由跳转的时候，再添加就可以了。

配置项中的配置说明如下。

(1) "orientation": "unspecified": 这里可以设置横屏、竖屏，unspecified 是默认的模式。

(2) "name": "com.charjedu.router_demo.SecondAbility": Page Ability 的名称。

(3) "icon": "$media:icon": /默认图标。

(4) "description": "$string:secondability_description": 描述。

(5) "label": "第 2 个 Page": 默认标题栏上的文字。

(6) "type": "page": 类型为 page、service、data 共 3 种，这里是 page。

(7) "launchType": "standard": 启动模式 standard、singleton(单例模式)。

DevEco Studio 为每个 Page Ability，创建一个 Page Ability 代码文件和一个 Slice 代码文件。一个 Page Ability 会有一个默认的 Page Slice Ability，如图 4-4 所示。

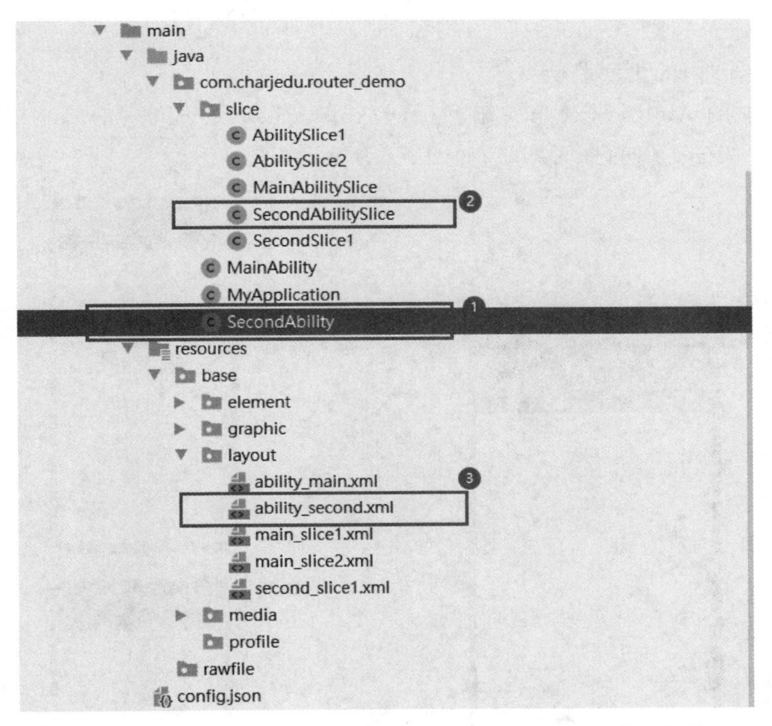

图 4-4　创建一个 Page Ability 代码文件和一个 Slice 代码文件

一个新的 Page Ability 类继承自 Ability 类，同时需要覆盖 Ability 中的 onStart 方法，我们需要在 onStart 方法中设置默认启动的主路由，默认为 Slice。如果有多个 Slice 子页面，则可以通过 addActionRoute 来配置，如图 4-5 所示。

```
public class SecondAbility extends Ability {
    @Override
    public void onStart(Intent intent) {
        super.onStart(intent);
        super.setMainRoute(SecondAbilitySlice.class.getName());

        super.addActionRoute( action: "secondslice1", SecondSlice1.class.getName());
    }
}
```

图 4-5　一个新的 Page Ability 类继承自 Ability 类

4.2.2　Page Ability 页面导航

Page Ability 页面导航分为两类：同一 Page 内的多个 Ability Slice 的跳转和不同 Page 间的跳转。同一 Page 内的 Slice 跳转使用 present 方法，不同 Page 间的跳转使用 startAbility 方法。

1. 同一 Page 内的页面跳转

当发起导航的 AbilitySlice 和导航目标的 AbilitySlice 处于同一个 Page 时，这种类型的导航被称为同一 Page 内的页面导航，如图 4-6 所示。

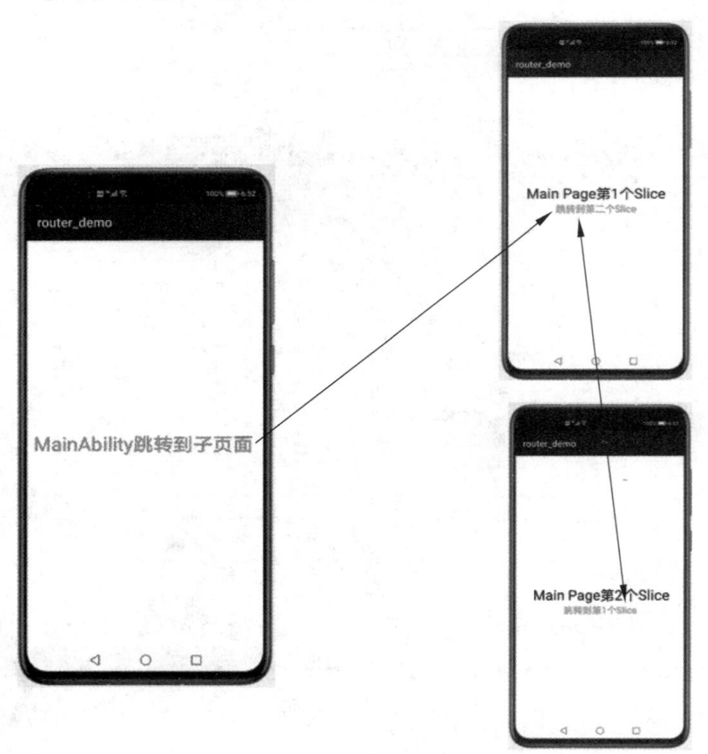

图 4-6　同一 Page 内的页面跳转

首先，在 layout 目录下添加两个布局文件，将这两个文件分别命名为 main_slice1 和 main_slice2，如图 4-7 所示。

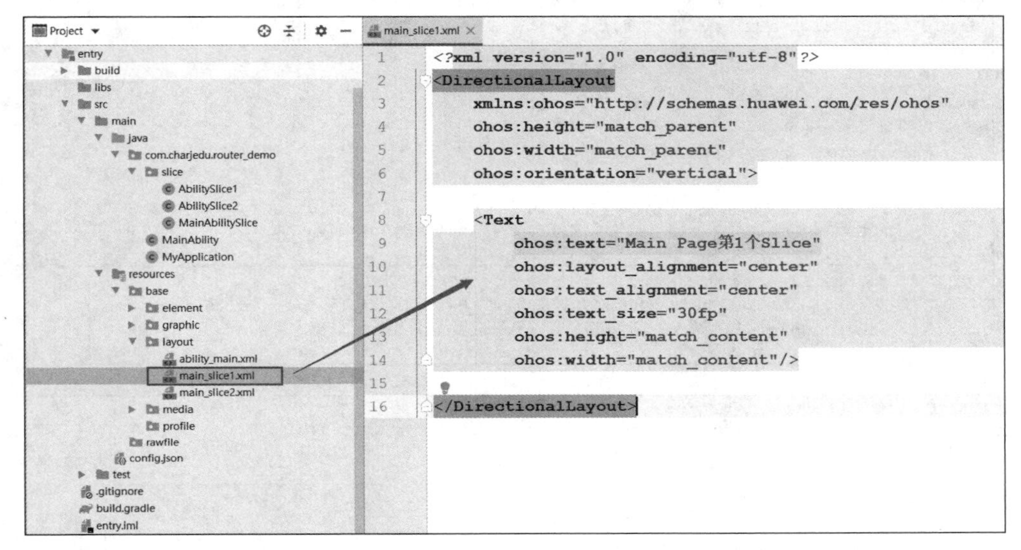

图 4-7　在 layout 目录下添加两个布局文件

在 slice 目录下添加两个 AbilitySlice，并分别命名为 AbilitySlice1 和 AbilitySlice2，如图 4-8 所示。

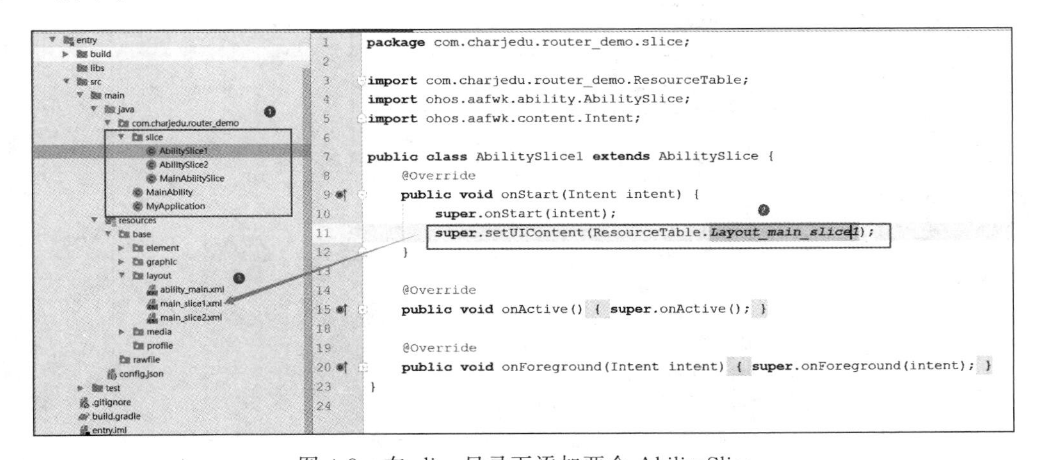

图 4-8　在 slice 目录下添加两个 AbilitySlice

在 Page Ability 代码中注册子路由，通过 super. addActionRoute 方法把两个 AblitySlice 添加到路由中，action 中的值是开发者任意设置的，这个值需要在 config. json 中进行注册才可以使用，如图 4-9 所示。

在 config. json 中注册 Ability Slice 的 action 名称，action 名称是注册子路由和子路由跳转的标记，如图 4-10 所示。

```java
import com.charjedu.router_demo.slice.AbilitySlice1;
import com.charjedu.router_demo.slice.AbilitySlice2;
import com.charjedu.router_demo.slice.MainAbilitySlice;
import ohos.aafwk.ability.Ability;
import ohos.aafwk.content.Intent;

public class MainAbility extends Ability {
    @Override
    public void onStart(Intent intent) {
        super.onStart(intent);
        //主页面路由
        super.setMainRoute(MainAbilitySlice.class.getName());

        //向Page添加多个子Ability Slice
        super.addActionRoute( action: "abilityslice1", AbilitySlice1.class.getName());
        super.addActionRoute( action: "abilityslice2", AbilitySlice2.class.getName());
    }
}
```

图 4-9　在 Page Ability 代码中注册子路由

```json
"abilities": [
  {
    "skills": [
      {
        "entities": [
          "entity.system.home"
        ],
        "actions": [
          "action.system.home",
          "abilityslice1", ❶
          "abilityslice2"  ❷
        ]
      }
    ],
    "orientation": "unspecified",
    "name": "com.charjedu.router_demo.MainAbility",
    "icon": "$media:icon",
    "description": "Java_Phone_Empty Feature Ability",
    "label": "router_demo",
    "type": "page",
    "launchType": "standard"
  }
]
```

图 4-10　在 config.json 中注册 Ability Slice 的 action 名称

向 main_slice1.xml，main_slice2.xml 布局文件添加两个 Button 按钮，在 AbilitySlice1 和 AbilitySlice2 代码中添加单击事件监听，单击跳转按钮进行导航。

main_slice1.xml 布局需进行修改，如代码示例 4-1 所示。

代码示例 4-1 main_slice1. xml 布局

```
<?xml version = "1.0" encoding = "utf - 8"?>
< DirectionalLayout
    xmlns:ohos = "http://schemas.huawei.com/res/ohos"
    ohos:height = "match_parent"
    ohos:width = "match_parent"
    ohos:alignment = "center"
    ohos:orientation = "vertical">

< Text
        ohos:text = "Main Page 第 1 个 Slice"
        ohos:layout_alignment = "center"
        ohos:text_alignment = "center"
        ohos:text_size = "30fp"
        ohos:height = "match_content"
        ohos:width = "match_content"/>

< Button
        ohos:id = " $ + id:btn1"
        ohos:text_size = "20fp"
        ohos:text_color = "cyan"
        ohos:text = "跳转到第 2 个 Slice"
        ohos:height = "match_content"
        ohos:width = "match_content"/>

</DirectionalLayout >
```

向 AbilitySlice1 添加跳转事件，通过 preset()方法进行跳转，如代码示例 4-2 所示。

代码示例 4-2 向 AbilitySlice1 添加跳转事件

```
public class AbilitySlice1 extends AbilitySlice {
    @Override
    public void onStart(Intent intent) {
        super.onStart(intent);
        super.setUIContent(ResourceTable.Layout_main_slice1);

        Button btn  =  (Button)findComponentById(ResourceTable.Id_btn1);
        btn.setClickedListener(new Component.ClickedListener() {
            @Override
            public void onClick(Component component) {
                Intent intent1  =  new Intent();
                present(new AbilitySlice2(),intent1);

            }
```

```
        });
    }

    @Override
    public void onActive() {
        super.onActive();
    }

    @Override
    public void onForeground(Intent intent) {
        super.onForeground(intent);
    }
}
```

main_slice2.xml 布局需进行修改，如代码示例 4-3 所示。

代码示例 4-3 main_slice2.xml 布局

```
<?xml version = "1.0" encoding = "utf - 8"?>
< DirectionalLayout
    xmlns:ohos = "http://schemas.huawei.com/res/ohos"
    ohos:height = "match_parent"
    ohos:width = "match_parent"
    ohos:alignment = "center"
    ohos:orientation = "vertical">

< Text
        ohos:text = "Main Page 第 2 个 Slice"
        ohos:layout_alignment = "center"
        ohos:text_alignment = "center"
        ohos:text_size = "30fp"
        ohos:height = "match_content"
        ohos:width = "match_content"/>

< Button
        ohos:id = " $ + id:btn2"
        ohos:text_size = "20fp"
        ohos:text_color = "cyan"
        ohos:text = "跳转到第 1 个 Slice"
        ohos:height = "match_content"
        ohos:width = "match_content"/>
</DirectionalLayout >
```

向 AbilitySlice2 添加跳转事件，如代码示例 4-4 所示。

代码示例 4-4 向 AbilitySlice2 添加跳转事件

```
public class AbilitySlice2 extends AbilitySlice {
    @Override
    public void onStart(Intent intent) {
        super.onStart(intent);
        super.setUIContent(ResourceTable.Layout_main_slice2);

        Button btn = (Button)findComponentById(ResourceTable.Id_btn2);
        btn.setClickedListener(new Component.ClickedListener() {
            @Override
            public void onClick(Component component) {
                Intent intent1 = new Intent();
                present(new AbilitySlice1(),intent1);
            }
        });
    }

    @Override
    public void onActive() {
        super.onActive();
    }

    @Override
    public void onForeground(Intent intent) {
        super.onForeground(intent);
    }
}
```

ability_main.xml 布局需进行修改，如代码示例 4-5 所示。

代码示例 4-5 ability_main.xml 布局

```
<?xml version = "1.0" encoding = "utf - 8"?>
< DirectionalLayout
    xmlns:ohos = "http://schemas.huawei.com/res/ohos"
    ohos:height = "match_parent"
    ohos:width = "match_parent"
    ohos:alignment = "center"
    ohos:orientation = "vertical">

< Text
        ohos:id = " $ + id:txt"
        ohos:height = "match_content"
        ohos:width = "match_content"
        ohos:text_color = "cyan"
        ohos:background_element = " $ graphic:background_ability_main"
```

```
        ohos:layout_alignment = "center"
        ohos:text = "跳转到子页面"
        ohos:text_size = "30fp"
        />

</DirectionalLayout>
```

同样向 MainAbilitySlice 添加跳转事件，如代码示例 4-6 所示。

代码示例 4-6　从 MainAbility 调整到子 Slice 页面

```
public class MainAbilitySlice extends AbilitySlice {
    @Override
    public void onStart(Intent intent) {
        super.onStart(intent);
        super.setUIContent(ResourceTable.Layout_ability_main);
        Text btn = (Text)findComponentById(ResourceTable.Id_txt);
        btn.setClickedListener(new Component.ClickedListener() {
            @Override
            public void onClick(Component component) {
                Intent intent1 = new Intent();
  present(new AbilitySlice1(),intent1);

            }
        });
    }

    @Override
    public void onActive() {
        super.onActive();
    }

    @Override
    public void onForeground(Intent intent) {
        super.onForeground(intent);
    }
}
```

2. 同一 Page 内的页面有返回值的跳转

如果开发者希望在用户从导航目标 AbilitySlice 返回时，能够获得其返回结果，则应当使用 presentForResult()实现导航。用户从导航目标 AbilitySlice 返回时，系统将回调 onResult()来接收和处理返回结果，开发者需要重写该方法。返回结果由导航目标 AbilitySlice 在其生命周期内通过 setResult()进行设置。

presentForResult：第 1 个参数是跳转的页面；第 2 个参数为 Intent：用来传递值；第 3 个参数用来作为识别的数字，可以是任意数，在这个数字后面在 onResult 方法中，通过这个数字来判断是哪个页面发起的请求，代码如下：

```
btn2 = (Text)findComponentById(ResourceTable.Id_txt2);
btn2.setClickedListener(new Component.ClickedListener() {
    @Override
    public void onClick(Component component) {
        Intent intent1 = new Intent();
        presentForResult(new AbilitySlice1(),intent1,100);

    }
});
```

在跳转到的目标页面中，单击返回按钮，即可返回发起跳转的页面。通过 setResult 方法把 intent 的参数值返回，然后调用 terminate 方法关闭页面，即可返回发起跳转的页面，代码如下：

```
//单击带参数返回,并关闭页面
Button btn_back = (Button)findComponentById(ResourceTable.Id_btn_back);
btn_back.setClickedListener(new Component.ClickedListener() {
    @Override
    public void onClick(Component component) {
        Intent intent1 = new Intent();
        intent1.setParam("back","51itcto.com");
        setResult(intent1);
        //关闭本页面
        terminate();
    }
});
```

在发起跳转的页面接收返回值，需要重写 onResult 方法，通过 onResult 的 requestCode 判断，获取返回参数值，代码如下：

```
//接收其他子页面的返回值的方法
@Override
protected void onResult(int requestCode, Intent resultIntent) {
    super.onResult(requestCode, resultIntent);
//这里的 100 是由上面的跳转方法所设置的数字
    if(requestCode == 100) {
        btn2.append(resultIntent.getStringParam("back"));
    }
}
```

系统为每个 Page 维护了一个 AbilitySlice 实例的栈，每个进入前台的 AbilitySlice 实例均会入栈。如果开发者在调用 present()或 presentForResult()时指定的 AbilitySlice 实例已经在栈中存在，则栈中位于此实例之上的 AbilitySlice 均会出栈并终止其生命周期。在前面的示例代码中，导航时指定的 AbilitySlice 实例均是新建的，即便重复执行此代码（此时作为导航目标的这些实例是同一个类），也不会导致任何 AbilitySlice 出栈。

3. 不同 Page 间的页面跳转

不同 Page 中的 AbilitySlice 相互不可见，因此无法通过 present()或 presentForResult()方法直接导航到其他 Page 的 AbilitySlice。AbilitySlice 作为 Page 的内部单元，以 Action 的形式对外暴露，因此可以通过配置 Intent 的 Action 导航到目标 AbilitySlice。Page 间的导航可以使用 startAbility()或 startAbilityForResult()方法，获得返回结果的回调为 onAbilityResult()。在 Ability 中调用 setResult()可以设置返回结果。

从 A 页面的主 Slice 页面跳转到 B 页面的主 Slice 页面，示例代码如下：

```
Text btn3 = (Text)findComponentById(ResourceTable.Id_txt_nav_page);
btn3.setClickedListener(new Component.ClickedListener() {
    @Override
    public void onClick(Component component) {

        Intent intent = new Intent();
//动作参数
        Operation operation = new Intent.OperationBuilder()
                .withDeviceId("")  //设备号,设备号为空则表示是本地
                .withBundleName("com.charjedu.router_demo")
                .withAbilityName(".SecondAbility")
                .build();
//通过 intent 携带这些 action 参数
        intent.setOperation(operation);
//启动指定的 bundlename 的 abiltiy
        startAbility(intent);

    }
});
```

withBundleName 对应 config.json 中的 ability 的配置，如图 4-11 所示。

从 A 页面的主 Slice 跳转到 B 页面的子 Slice 页面的第 1 步，注册 B 页面的子 Slice，并将 action 定义为 secondslice1，如图 4-12 所示。

第 2 步，在 config.json 中找到 abilities 数组中对应的 ability 配置项，添加 actions，如图 4-13 所示。

第 3 步，跳转页面时，这里可以不使用 Operation 对象设置 bundleName，而直接使用 setAction 指定需要跳转的 action 名就可以了，这个 action 名已经被全局注册了，所以可以直接指定，如图 4-14 所示。

```
{
  "skills": [
    {
      "entities": [
      ],
      "actions": [
        "secondslice1"
      ]
    }
  ],
  "orientation": "unspecified",
  "name": "com.charjedu.router_demo.SecondAbility",
  "icon": "$media:icon",
  "description": "Java_Phone_Empty Feature Ability",
  "label": "第2个Page",
  "type": "page",
  "launchType": "standard"
}
```

图 4-11 withBundleName 对应 config.json 中的 ability 的配置

```
public class SecondAbility extends Ability {
    @Override
    public void onStart(Intent intent) {
        super.onStart(intent);
        super.setMainRoute(SecondAbilitySlice.class.getName());

        super.addActionRoute( action: "secondslice1", SecondSlice1.class.getName());

    }
}
```

图 4-12 注册 B 页面的子 Slice

```
{
  "skills": [
    {
      "entities": [
      ],
      "actions": [
        "secondslice1"
      ]
    }
  ],
  "orientation": "unspecified",
  "name": "com.charjedu.router_demo.SecondAbility",
  "icon": "$media:icon",
  "description": "$string:secondability_description",
  "label": "第2个Page",
  "type": "page",
  "launchType": "standard"
}
```


图 4-13 在 ability 配置中添加 actions

```
Text btn3 = (Text)findComponentById(ResourceTable.Id_txt_nav_page);
btn3.setClickedListener(new Component.ClickedListener() {
    @Override
    public void onClick(Component component) {
        Intent intent = new Intent();
        //跳转到第2个页面的slice页面，可直接通过action名跳转
        intent.setAction("secondslice1");
        startAbility(intent);
    }
});
```

图 4-14　使用 setAction 指定需要跳转的 action

4.2.3　Page Ability 的生命周期

系统管理或用户操作等行为均会引起 Page 实例在其生命周期的不同状态之间进行转换。Ability 类提供的回调机制能够让 Page 及时感知外界变化，从而正确地应对状态的变化，例如释放资源，这有助于提升应用的性能和稳健性。

Page 生命周期的不同状态转换及其对应的回调，如图 4-15 所示。

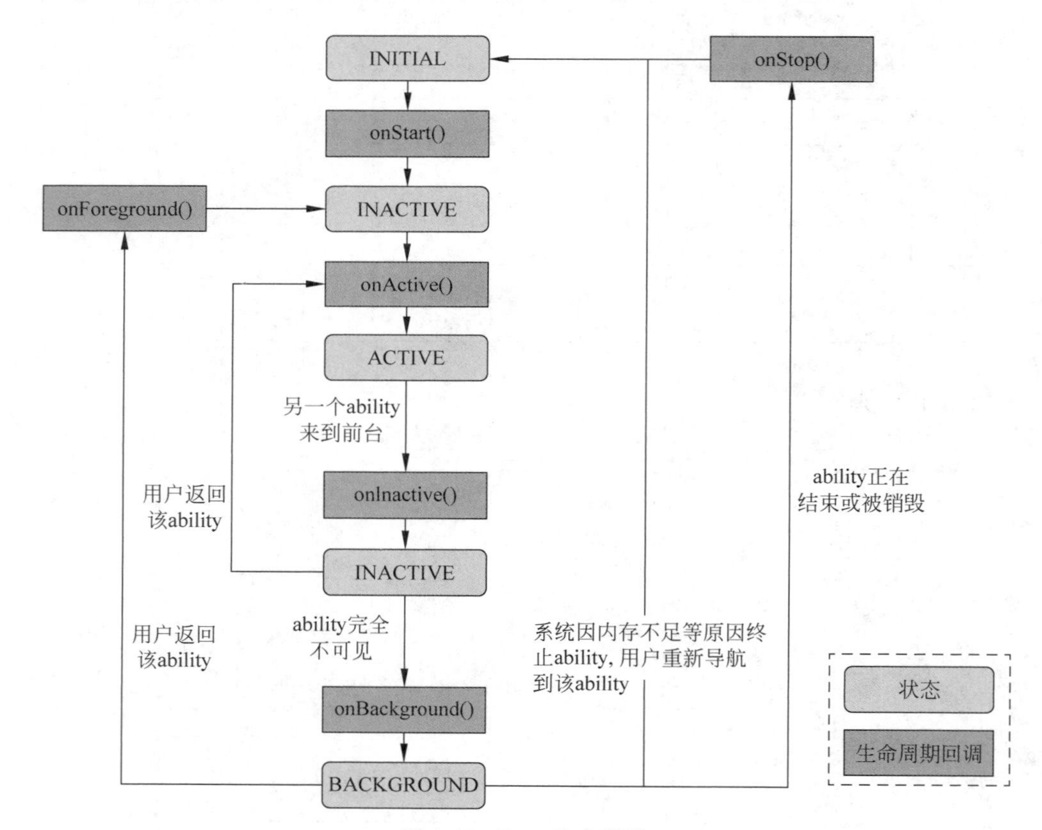

图 4-15　Page 生命周期

1. onStart()

当系统首次创建Page实例时，触发该回调。对于一个Page实例，该回调在其生命周期过程中仅触发一次，Page在该逻辑后将进入INACTIVE状态。开发者必须重写该方法，并在此配置默认展示的AbilitySlice。

```
@Override
    public void onStart(Intent intent) {
        super.onStart(intent);
        super.setMainRoute(FooSlice.class.getName());
    }
```

2. onActive()

Page会在进入INACTIVE状态后来到前台，然后由系统调用此回调。Page在此之后进入ACTIVE状态，该状态是应用与用户交互的状态。Page将保持在此状态，除非某类事件发生导致Page失去焦点，例如用户单击返回键或导航到其他Page。当此类事件发生时，会触发Page回到INACTIVE状态，系统将调用onInactive()回调。此后，Page可能重新回到ACTIVE状态，系统将再次调用onActive()回调，因此，开发者通常需要成对实现onActive()和onInactive()，并在onActive()中获取在onInactive()中被释放的资源。

3. onInactive()

当Page失去焦点时，系统将调用此回调，此后Page进入INACTIVE状态。开发者可以在此回调中实现Page失去焦点时应表现的恰当行为。

4. onBackground()

如果Page不再对用户可见，系统将调用此回调通知开发者用户进行相应的资源释放，此后Page进入BACKGROUND状态。开发者应该在此回调中释放Page不可见时无用的资源，或在此回调中执行较为耗时的状态保存操作。

5. onForeground()

处于BACKGROUND状态的Page仍然驻留在内存中，当重新回到前台时，例如用户重新导航到此Page，系统将先调用onForeground()回调通知开发者，而后Page的生命周期状态回到INACTIVE状态。开发者应当在此回调中重新申请在onBackground()中释放的资源，最后Page的生命周期状态进一步回到ACTIVE状态，系统将通过onActive()回调通知开发者用户。

6. onStop()

系统将要销毁Page时，将会触发此回调函数，通知用户进行系统资源的释放。销毁Page的可能原因包括以下几个方面：

(1) 用户通过系统管理能力关闭指定Page，例如使用任务管理器关闭Page。

(2) 用户行为触发Page的terminateAbility()方法调用，例如使用应用的退出功能。

(3) 配置变更导致系统暂时销毁Page并重建。

(4) 系统出于资源管理目的，自动触发对处于 BACKGROUND 状态的 Page 的销毁。

4.2.4 Ability Slice 的生命周期

AbilitySlice 作为 Page 的组成单元，其生命周期依托于其所属 Page 的生命周期。AbilitySlice 和 Page 具有相同的生命周期状态和同名的回调，当 Page 生命周期发生变化时，它的 AbilitySlice 也会发生相同的生命周期变化。此外，AbilitySlice 还具有独立于 Page 的生命周期变化，这发生在同一 Page 中的 AbilitySlice 之间进行导航时，此时 Page 的生命周期状态不会改变。

AbilitySlice 生命周期回调与 Page 的相应回调类似，因此不再赘述。由于 AbilitySlice 承载具体的页面，开发者必须重写 AbilitySlice 的 onStart() 回调，并在此方法中通过 setUIContent()方法设置页面，代码如下：

```
@Override
    protected void onStart(Intent intent) {
        super.onStart(intent);
        setUIContent(ResourceTable.Layout_main_layout);
    }
```

AbilitySlice 实例的创建和管理通常由应用负责，系统仅在特定情况下会创建 AbilitySlice 实例。例如，通过导航启动某个 AbilitySlice 时，是由系统负责实例化，但是在同一个 Page 中不同的 AbilitySlice 间进行导航时则由应用负责实例化。

当 AbilitySlice 处于前台且具有焦点时，其生命周期状态随着所属 Page 的生命周期状态的变化而变化。当一个 Page 拥有多个 AbilitySlice 时，例如：MyAbility 下有 FooAbilitySlice 和 BarAbilitySlice，当前 FooAbilitySlice 处于前台并获得焦点，并即将导航到 BarAbilitySlice，在此期间的生命周期状态变化顺序如下：

(1) FooAbilitySlice 从 ACTIVE 状态变为 INACTIVE 状态。

(2) BarAbilitySlice 则从 INITIAL 状态首先变为 INACTIVE 状态，然后变为 ACTIVE 状态(假定此前 BarAbilitySlice 未曾启动)。

(3) FooAbilitySlice 从 INACTIVE 状态变为 BACKGROUND 状态。

对应两个 slice 的生命周期方法回调顺序为

```
FooAbilitySlice.onInactive() --> BarAbilitySlice.onStart() --> BarAbilitySlice.onActive()
--> FooAbilitySlice.onBackground()
```

在整个流程中，MyAbility 始终处于 ACTIVE 状态，但是，当 Page 被系统销毁时，其所有已实例化的 AbilitySlice 将联动被销毁，而不仅是处于前台的 AbilitySlice。

4.3 Service Ability

Service Ability 是一种运行在后台的 Ability。主要用于运行后台任务，如执行音乐播放、文件下载等，但不提供用户交互界面。Service 可由其他应用或 Ability 启动，即使用户切换到其他应用，Service 仍将在后台继续运行。

4.3.1 Service Ability 概述

Service 是单实例的。在一个设备上，相同的 Service 只会存在一个实例。如果多个 Ability 共用这个实例，只有当与 Service 绑定的所有 Ability 都退出后，Service 才能够退出。由于 Service 是在主线程里执行的，因此，如果在 Service 里面的操作时间过长，则开发者必须在 Service 里创建新的线程来处理（详见线程间通信部分内容），以防止造成主线程阻塞，从而导致应用程序无响应。

Service Ability 需要在应用配置文件中进行注册，需要将注册类型 type 设置为 service。

Service 也是一种 Ability，创建 Ability 的子类，实现 Service 相关的生命周期方法。Ability 为 Service 提供了以下生命周期方法。Service Ability 在 config.json 中定义的类型为 Service，Page Ability 的类型为 page，代码如下：

```
{
"name": "com.cangjie.myapplication.LocalServiceAbility",
"icon": "$media:icon",
"description": "$string:localserviceability_description",
"type": "service"
}
```

4.3.2 Service Ability 生命周期

说明： 与 Page 类似，Service 也拥有生命周期，如图 4-16 所示。根据调用方法的不同，其生命周期有以下两种路径。

（1）启动 Service：该 Service 在其他 Ability 调用 startAbility()时创建，然后保持运行。其他 Ability 通过调用 stopAbility()来停止 Service，Service 停止后，系统会将其销毁。

（2）连接 Service：该 Service 在其他 Ability 调用 connectAbility()时创建，客户端可通过调用 disconnectAbility()断开连接。多个客户端可以绑定到相同的 Service，而且当所有绑定全部取消后，系统才会销毁该 Service。

4.3.3 创建 Service Ability

创建 Ability 的子类，实现 Service 相关的生命周期方法。Service 也是一种 Ability，

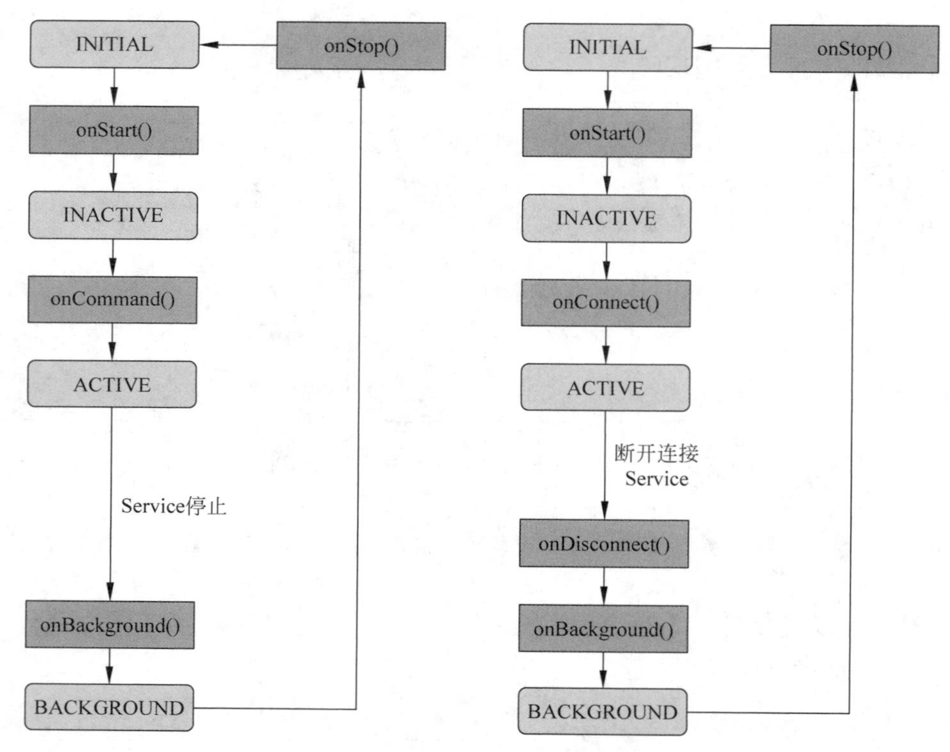

图 4-16 Service Ability 生命周期

Ability 为 Service 提供以下生命周期方法，用户可以重写这些方法来添加自己的处理。

通过 DevEco Studio 自动生成 Service Ability，具体的步骤如下。

第 1 步：在项目的目录上右击新建一个空的 Service Ability，如图 4-17 所示。

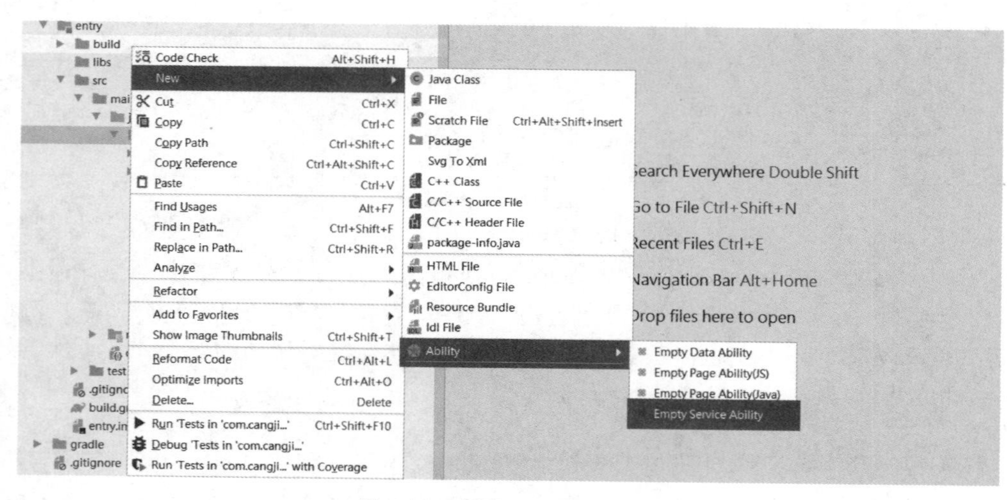

图 4-17 创建 Service Ability

第 2 步：配置 Service Ability。Service Ability 可以分为本地和远程，在代码上两者并无区别，只是在 FA 调用上有区别。Service Ability 在创建的时候是没有模板选项的，如图 4-18 所示。

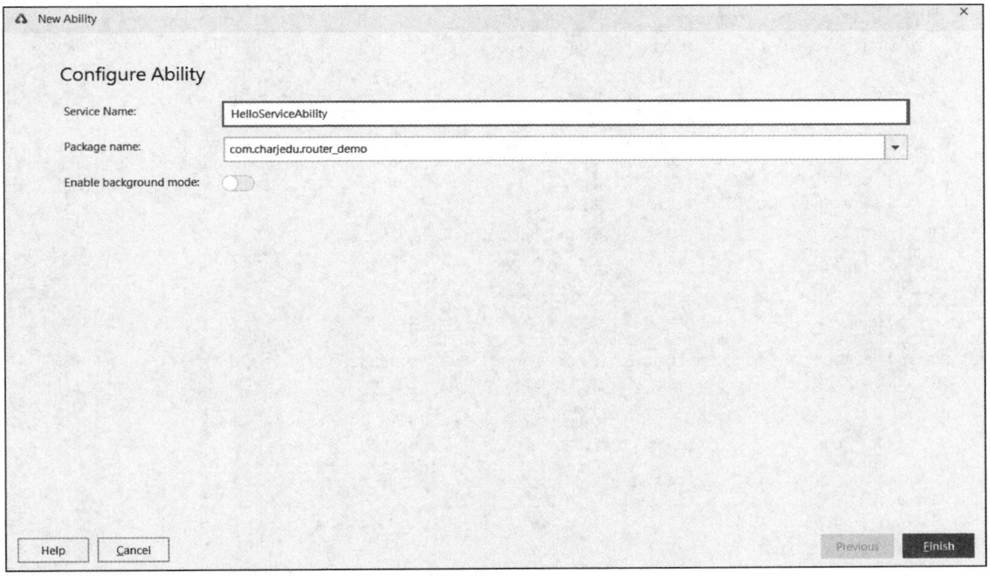

图 4-18　配置 Service Ability

创建完成后，DevEco Studio 会自动在 config.json 添加一个 ability 的配置，其 type 为 service，如图 4-19 所示。

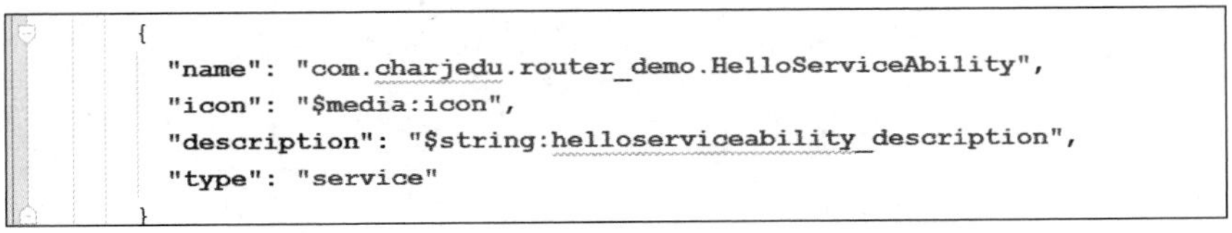

```
{
  "name": "com.charjedu.router_demo.HelloServiceAbility",
  "icon": "$media:icon",
  "description": "$string:helloserviceability_description",
  "type": "service"
}
```

图 4-19　在 config.json 添加一个 ability 的配置

第 3 步：重写 Ability 方法，来添加自己的处理逻辑，如代码示例 4-7 所示。

代码示例 4-7　添加自己的处理逻辑

```java
public class HelloServiceAbility extends Ability {
    private static final HiLogLabel LABEL_LOG = new HiLogLabel(3, 0xD001100, "Demo");

    @Override
    public void onStart(Intent intent) {
        HiLog.error(LABEL_LOG, "HelloServiceAbility::onStart");
        super.onStart(intent);
```

```
    }

    @Override
    public void onBackground() {
        super.onBackground();
        HiLog.info(LABEL_LOG, "HelloServiceAbility::onBackground");
    }

    @Override
    public void onStop() {
        super.onStop();
        HiLog.info(LABEL_LOG, "HelloServiceAbility::onStop");
    }

    @Override
    public void onCommand(Intent intent, boolean restart, int startId) {
        //restart : ability 的启动方式.true:销毁后重启,false:正常重启
        //startId: 计数器 onCommand 被调用一次,累加 1
        new ToastDialog(this)
                .setText("启动了" + startId + "次")
                .show();
    }

    @Override
    public IRemoteObject onConnect(Intent intent) {
        return null;
    }

    @Override
    public void onDisconnect(Intent intent) {
    }
```

- onStart 方法：启动时调用一次；
- onCommand 方法：每次调用都会执行；
- onConnect 方法：在本地调用中不需要处理此方法，但是远程调用时需要重写此方法。

4.3.4 启动 Service Ability

说明：Service Ability 的启动需要区分本地和远程，启用和关闭远程 Service Ability 需要设置支持分布式调度系统多设备启动的标识：Intent.FLAG_ABILITYSLICE_MULTI_DEVICE。

启动本地设备 Service 的代码如代码示例 4-8 所示。

代码示例 4-8 启动本地设备 Service

```
Intent intent = new Intent();
Operation operation = new Intent.OperationBuilder()
        .withDeviceId("")           //为空则表示本地
        .withBundleName("com.huawei.himusic")
        .withAbilityName("com.huawei.himusic.ServiceAbility")
        .build();
intent.setOperation(operation);
startAbility(intent);
```

(1) DeviceId：表示设备 ID。如果是本地设备，则可以直接留空；如果是远程设备，则可以通过 ohos.distributedschedule.interwork.DeviceManager 提供的 getDeviceList 获取设备列表。

(2) BundleName：表示包名称。

(3) AbilityName：表示待启动的 Ability 名称。

启动远程设备 Service 的代码如代码示例 4-9 所示。

代码示例 4-9 启动远程设备 Service

```
Operation operation = new Intent.OperationBuilder()
        .withDeviceId("deviceId")
        .withBundleName("com.huawei.himusic")
        .withAbilityName("com.huawei.hiworld.ServiceAbility")
        //设置支持分布式调度系统多设备启动的标识
        .withFlags(Intent.FLAG_ABILITYSLICE_MULTI_DEVICE)
        .build();
Intent intent = new Intent();
intent.setOperation(operation);
startAbility(intent);
```

说明：DeviceId 为空时表示本地调用，可以通过以下代码获取 DeviceId 列表。

执行上述代码后，Ability 将通过 startAbility()方法来启动 Service。

(1) 如果 Service 尚未运行，则系统会先调用 onStart()来初始化 Service，再回调 Service 的 onCommand()方法来启动 Service。

(2) 如果 Service 正在运行，则系统会直接回调 Service 的 onCommand()方法来启动 Service。

4.3.5 关闭 Service Ability

Service 一旦创建就会一直保持在后台运行，除非必须回收内存资源，否则系统不会停止或销毁 Service。开发者可以在 Service 中通过 terminateAbility()停止本 Service 或在其他 Ability 调用 stopAbility()来停止 Service。

关闭本地的 Service Ability,如代码示例 4-10 所示。

代码示例 4-10 关闭本地的 Service Ability

```
Operation operation = new Intent.OperationBuilder()
        .withDeviceId("")
        .withBundleName("com.huawei.hiworld.himusic")
        .withAbilityName("com.huawei.hiworld.himusic.ServiceAbility")
        //设置支持分布式调度系统多设备启动的标识
        .withFlags(Intent.FLAG_ABILITYSLICE_MULTI_DEVICE)
        .build();
Intent intent = new Intent();
intent.setOperation(operation);
stopAbility(intent);
```

关闭远程的 Service Ability,如代码示例 4-11 所示。

代码示例 4-11 关闭远程的 Service Ability

```
Intent intent = new Intent();
String deviceid = ""; //此次代码为封装获取 DeviceId
if (deviceid != null){
    ElementName ele = new ElementName(deviceid, "com.cangjie.myapplication", "com.cangjie.
myapplication.RemoteServiceAbility");
    intent.setElement(ele);
    intent.addFlags(Intent.FLAG_ABILITYSLICE_MULTI_DEVICE);
 stopAbility(intent);
}
```

这里需要获取远程设备的编号,调用 DeviceManager 的 getDeviceList 接口,通过 FLAG_GET_ONLINE_DEVICE 标记获得在线设备列表。获取远程在线设备的信息将在 4.4.4 节进行详细讲解。

注意:停止 Service 同样支持停止本地设备 Service 和停止远程设备 Service,使用方法与启动 Service 一样。一旦调用停止 Service 的方法,系统便会尽快销毁 Service。

4.3.6 连接远程 Service Ability

通过连接 Service Ability,可以在 FA 与 Service Ability 间建立通道,进行数据传输。这个与上面讲到的启动和关闭 Service Ability 是不一样的,连接后可以传输数据。

如果 Service 需要与 Page Ability 或其他应用的 Service Ability 进行交互,则应先创建用于连接的 Connection。Service 支持其他 Ability 通过 connectAbility()方法与其进行连接。

要连接远程机器安装的 Service Ability，需要先能够获取远程在线设备的设备信息，这样才可以发起连接。鸿蒙操作系统提供了分布式软总线，帮助我们自动发现，以便组网附近的在线设备。

1. 连接与启动/关闭的不同

仅通过启动/关闭这两种方式对 PA 进行调度无法应对需长期交互的场景，因此，分布式任务调度平台向开发者提供了跨设备 PA 连接及断开连接的能力。

启动/关闭这两种方式适用于执行时间不长的场景，如打开或者关闭电视，不相互通信。对于需要长期交互的场景，如图 4-20 所示的加减音量，就需要以连接的方式调用 Server Ability 进行数据交互了。

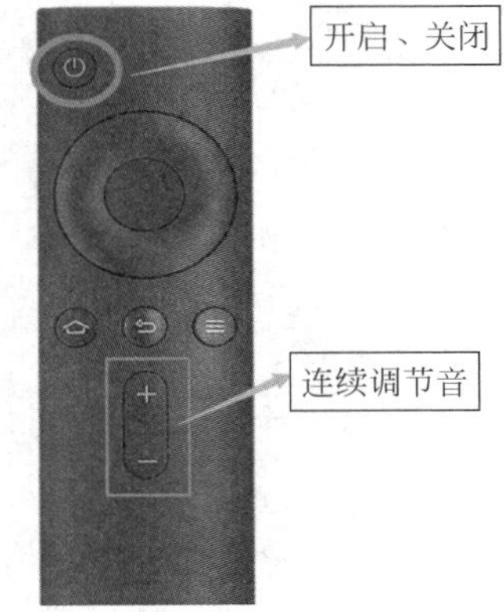

图 4-20　遥控器效果图

2. 申请访问权限

首先需要在 config.json 中申请权限，将 reqPermissions 添加到 module 配置中，如图 4-21 所示。

```
"module": {
  "package": "com.cangjie.jsabilitydemo",
  "name": ".MyApplication",
  "deviceType": [...],
  "distro": {"deliveryWithInstall": true...},
  "abilities": [...],
  "js": [...],
  "reqPermissions": [...]
}
```

图 4-21　在 config.json 中申请权限

reqPermissions 申请的权限如下：

```
"reqPermissions": [
  {
"name": "ohos.permission.DISTRIBUTED_DATASYNC"
  },
  {
"name": "ohos.permission.servicebus.ACCESS_SERVICE"
  },
  {
"name": "ohos.permission.servicebus.BIND_SERVICE"
  },
  {
"name": "ohos.permission.DISTRIBUTED_DEVICE_STATE_CHANGE"
  },
  {
```

```
"name": "ohos.permission.GET_DISTRIBUTED_DEVICE_INFO"
    },
    {
"name": "ohos.permission.GET_BUNDLE_INFO"
    }
]
```

- "ohos. permission. DISTRIBUTED_DATASYNC"：访问分布式数据同步权限；
- "ohos. permission. servicebus. BIND_SERVICE"：访问服务总线的绑定服务权限；
- "ohos. permission. servicebus. ACCESS_SERVICE"：访问服务总线的服务接入权限；

上面这3个和服务访问相关的权限是必须的权限，下面的这3个权限与设备相关，也是必须的权限。

- "ohos. permission. DISTRIBUTED_DEVICE_STATE_CHANGE"：与访问分布式设备状态相关的权限；
- "ohos. permission. GET_DISTRIBUTED_DEVICE_INFO"：分布式设备信息获取权限；
- "ohos. permission. GET_BUNDLE_INFO"：获取 Bundle 信息的权限。

上面我们通过在 config. json 中申请了权限，同时需要在 MainAbility 代码中获取权限，如代码示例 4-12 所示。

代码示例 4-12　在 MainAbility 代码中获取权限

```
@Override
public void onStart(Intent intent) {
    requestPermissionsFromUser(new String[]{
"ohos.permission.DISTRIBUTED_DATASYNC",
"ohos.permission.servicebus.BIND_SERVICE",
"ohos.permission.servicebus.ACCESS_SERVICE"
    },0);
    super.onStart(intent);
    super.setMainRoute(MainAbilitySlice.class.getName());
}
```

3. 获取在线设备的信息

获得设备列表，开发者可在得到的在线设备列表中选择目标设备执行操作。调用 DeviceManager 的 getDeviceList 接口，通过 FLAG_GET_ONLINE_DEVICE 标记获得在线设备列表，如代码示例 4-13 所示。

代码示例 4-13　获得设备列表

```
public class DeviceUtil {
    //ISelectResult 是一个自定义接口，用来处理指定设备 deviceId 后执行的行为
```

```
    public interface ISelectResult {
        void onSelectResult(String deviceId);
    }
    //获得设备列表,开发者可在得到的在线设备列表中选择目标设备执行相关操作
    public static void getOnlineDeviceInfo(ISelectResult listener) {
        //调用 DeviceManager 的 getDeviceList 接口,通过 FLAG_GET_ONLINE_DEVICE 标记获得在线
        //设备列表
         List < DeviceInfo > onlineDevices = DeviceManager.getDeviceList(DeviceInfo.FLAG_
GET_ONLINE_DEVICE);
        //判断组网设备是否为空
        if (onlineDevices.isEmpty()) {
            listener.onSelectResult(null);
            return;
        }
        int numDevices = onlineDevices.size();
        ArrayList < String > deviceIds = new ArrayList <>(numDevices);
        ArrayList < String > deviceNames = new ArrayList <>(numDevices);
        onlineDevices.forEach((device) -> {
            deviceIds.add(device.getDeviceId());
            deviceNames.add(device.getDeviceName());
        });
        //以选择首个设备作为目标设备为例
        //开发者也可按照具体场景,通过别的方式进行设备选择
        String selectDeviceId = deviceIds.get(0);
        listener.onSelectResult(selectDeviceId);
    }

}
```

上面的方法所使用的调用方式,代码如下:

```
DeviceUtil.getOnlineDeviceInfo(new DeviceUtil.ISelectResult() {
    @Override
    public void onSelectResult(String deviceId) {
        Intent intent = new Intent();
        Operation operation = new Intent.OperationBuilder()
                .withDeviceId(deviceId)                //为空则表示本地
                .withBundleName("com.charjedu.ability_demo")
                .withAbilityName(".RemoteServiceAbility")
                .build();
        intent.setOperation(operation);
        connectAbility(intent,conn);
    }
});
```

FA 与远程 PA 的调用流程如图 4-22 所示。

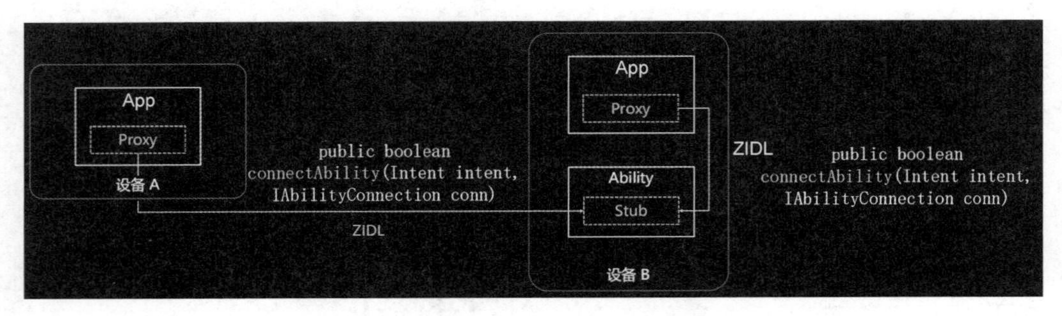

图 4-22　FA 与 PA 的调用关系

下面介绍如何创建远程端 ability，通过 FA 操作远程的 PA。

4. 创建本地代理

在发起端创建远程端 Service Ability 的本地代理，本地代理类需要实现 IRemoteBroker 接口，实现接口的 asObject 方法。给这个类添加一个 IRemoteObject 属性，这个属性就是远程代理对象，通过本地代理类的构造方法进行赋值，同时作为 asObject 方法的返回值。这个类的构造赋值，是通过 FA 调用 connectAbility 后的连接成功的回调中赋值，如代码示例 4-14 所示。

代码示例 4-14　创建本地代理

```
import ohos.rpc.*;

//发起端创建的远程端的 ability 本地代理
public class MyRemoteProxy implements IRemoteBroker {

    private static final int COMMAND_PLUS = IRemoteObject.MAX_TRANSACTION_ID;
    private static final int ERR_OK = 0;                //表示通信成功的 code
    private final IRemoteObject remote;

    public MyRemoteProxy(IRemoteObject remote) {
        this.remote = remote;
    }

    //返回远程端的对象
    @Override
public IRemoteObject asObject() {
        return remote;
    }

 //自定义的 Service Ability 的方法
 public int plus(int a,int b){

        //需要传递的数据
```

```
        //MessageParcel 类似数组的格式
        MessageParcel data = MessageParcel.obtain();
        data.writeInt(a);
        data.writeInt(b);

        //远程端返回的响应数据
        MessageParcel reply = MessageParcel.obtain();

        //信息传输的模式 option
        MessageOption option = new MessageOption(MessageOption.TF_SYNC);
        //最核心方法,实现调用
        int result = 0;
        try {
            remote.sendRequest(COMMAND_PLUS,data,reply,option);
            //通信结束后,响应的结果存放在 reply 中
            //判断是否成功
            int ec = reply.readInt();
            if(ec != ERR_OK){
                throw new RemoteException();
            }
            result = reply.readInt();

        } catch (RemoteException e) {
            e.printStackTrace();
        }
        return result;
    }
}
```

5. 创建远程端代理

创建远程端代理，用于接收发起端发送的数据，然后进行处理，并返回响应结果。这个类同样需要实现 IRemoteBroker 接口，同时需要是 RemoteObject 子类，在 asObject 方法中把当前类返回。通过重写 onRemoteRequest 方法接收发送端的信息，处理并返回，这种方法和发送端的 onSendRequest 方法是配套使用的，该方法如代码示例 4-15 所示。

代码示例 4-15 创建远程端代理

```
import ohos.rpc.*;

//远程端代理
public class MyRemote extends RemoteObject implements IRemoteBroker {

    private static final int COMMAND_PLUS = IRemoteObject.MAX_TRANSACTION_ID;
    private static final int ERR_OK = 0;    //表示通信成功的 cod
    private static final int ERROR = -1; //表示出错的 code
```

```
    public MyRemote() {
        super("my_remote");
    }

    @Override
    public IRemoteObject asObject() {
        return this;                        //返回的是自己的对象
    }

    //核心方法,接收发起端发送的数据,然后进行处理,并返回响应结果
    @Override
    public boolean onRemoteRequest(int code, MessageParcel data, MessageParcel reply, MessageOption
option) throws RemoteException {
        if (code != COMMAND_PLUS){
            reply.writeInt(ERR_OK);
            return false;
        }
        int v1 = data.readInt();
        int v2 = data.readInt();
        int rs = v1 + v2;
        reply.writeInt(ERR_OK);   //状态值
        reply.writeInt(rs);
        return true;
    }
}
```

6. 创建远程的 Service Ability

创建远程的 Service Ability,把上面的远程代理 MyRemote 类作为私有属性通过重写 onConnect 方法返回,当发起端调用 connectAbility 方法时,会调用 Service Ability 的 onConnect 方法,onConnect 成功返回 MyRemote 对象,connectAbility 绑定的回调方法将触发,通过在成功回调的方法中获取这个远程代理对象,如代码示例 4-16 所示。

代码示例 4-16　创建远程的 Service Ability

```
public class RemoteServiceAbility extends Ability {
    private static final HiLogLabel LABEL_LOG = new HiLogLabel(3, 0xD001100, "Demo");
    //调用 MyRemote
    private MyRemote remote = new MyRemote();
    @Override
    public void onStart(Intent intent) {
        HiLog.error(LABEL_LOG, "RemoteServiceAbility::onStart");
        super.onStart(intent);
        new ToastDialog(this)
                .setText("onStart")
```

```
                .show();
    }

    @Override
    public void onBackground() {
        super.onBackground();
        HiLog.info(LABEL_LOG, "RemoteServiceAbility::onBackground");
    }

    @Override
    public void onStop() {
        super.onStop();
        HiLog.info(LABEL_LOG, "RemoteServiceAbility::onStop");
        new ToastDialog(this)
                .setText("onStop")
                .show();
    }

    @Override
    public void onCommand(Intent intent, boolean restart, int startId) {
        new ToastDialog(this)
                .setText("remote onCommand!")
                .show();
    }

    @Override
    public IRemoteObject onConnect(Intent intent) {
        //返回 remote 对象,并返回到发起的 FA
        return remote;
    }

    @Override
    public void onDisconnect(Intent intent) {
    }
}
```

7. 在 FA 中连接远端 Service Ability

connectAbility 方法的第 1 个参数是 Intent 对象,第 2 个参数是连接回调对象,连接是否成功通过 IAbilityConnection 的回调方法返回远程代理对象,如代码示例 4-17 所示。

IAbilityConnection 提供了两种方法供开发者实现:

(1) onAbilityConnectDone()方法用来处理连接的回调。

(2) onAbilityDisconnectDone()方法用来处理断开连接的回调。

代码示例 4-17　FA 中连接远端 Service Ability

```
Intent intent = new Intent();
String deviceid = DeviceUtils.getDeviceId(); //获取远程设备的编号
if (deviceid != null) {
      ElementName ele = new ElementName(deviceid, "com.cangjie.myapplication", "com.
cangjie.myapplication.RemoteServiceAbility");
      intent.setElement(ele);
      intent.addFlags(Intent.FLAG_ABILITYSLICE_MULTI_DEVICE);
 connectAbility(intent, conn);
}
```

调用 connectAbility 方法，通过 Page 中的事件触发，代码如下：

```
public class MainAbilitySlice extends AbilitySlice {


   private MyRemoteProxy proxy = null;
   private IAbilityConnection conn = new IAbilityConnection() {
        @Override
        public void onAbilityConnectDone(ElementName elementName, IRemoteObject iRemoteObject,
int i) {
              new ToastDialog(MainAbilitySlice.this)
                      .setText("onAbilityConnectDone")
                      .show();
              proxy = new MyRemoteProxy(iRemoteObject);
        }

        @Override
        public void onAbilityDisconnectDone(ElementName elementName, int i) {
            new ToastDialog(MainAbilitySlice.this)
                    .setText("onAbilityDisconnectDone:")
                    .show();
        }
    };

    @Override
    public void onStart(Intent intent) {
        super.onStart(intent);
        super.setUIContent(ResourceTable.Layout_ability_main);

        //连接远程 Service Ability
        Button btn3 = (Button)findComponentById(ResourceTable.Id_btn_connect);
        btn3.setClickedListener(component -> {
String deviceId = DeviceUtils.getDeviceId();
            new ToastDialog(MainAbilitySlice.this)
```

```
                    .setText("deviceId:" + deviceId)
                    .show();

            Intent intent5 = new Intent();
            String deviceid = DeviceUtils.getDeviceId();
            if (deviceid != null) {
                ElementName ele = new ElementName(deviceid, "com.charjedu.ability_demo",
"com.charjedu.ability_demo.sa.RemoteServiceAbility");
                intent5.setElement(ele);
                intent5.addFlags(Intent.FLAG_ABILITYSLICE_MULTI_DEVICE);
                connectAbility(intent5, conn);
            }

        });

    }

}
```

8. 在FA中调用远程Service Ability

connectAbility方法调用成功后，通过onAbilityConnectDone回调方法获取远程代理对象，这样就可以通过这个代理对象进行服务调用了，如代码示例4-18所示。

代码示例4-18　连接远程Service Ability

```
public class MainAbilitySlice extends AbilitySlice {


    private MyRemoteProxy proxy = null;
    private IAbilityConnection conn = new IAbilityConnection() {
        @Override
        public void onAbilityConnectDone(ElementName elementName, IRemoteObject iRemoteObject,
int i) {
            new ToastDialog(MainAbilitySlice.this)
                    .setText("onAbilityConnectDone")
                    .show();
            proxy = new MyRemoteProxy(iRemoteObject);
        }

        @Override
        public void onAbilityDisconnectDone(ElementName elementName, int i) {
            new ToastDialog(MainAbilitySlice.this)
                    .setText("onAbilityDisconnectDone:")
                    .show();
        }
    };
```

```
    @Override
    public void onStart(Intent intent) {
        super.onStart(intent);
        super.setUIContent(ResourceTable.Layout_ability_main);

        //连接远程 Service Ability
        Button btn3 = (Button)findComponentById(ResourceTable.Id_btn_connect);
        btn3.setClickedListener(component -> {
            String deviceId = DeviceUtils.getDeviceId();
            new ToastDialog(MainAbilitySlice.this)
                    .setText("deviceId:" + deviceId)
                    .show();

            Intent intent5 = new Intent();
            String deviceid = DeviceUtils.getDeviceId();
            if (deviceid != null) {
                ElementName ele = new ElementName(deviceid, "com.charjedu.ability_demo",
"com.charjedu.ability_demo.sa.RemoteServiceAbility");
                intent5.setElement(ele);
                intent5.addFlags(Intent.FLAG_ABILITYSLICE_MULTI_DEVICE);
                connectAbility(intent5, conn);
            }

        });

        //调用远程 SA 方法
        Button btn4 = (Button)findComponentById(ResourceTable.Id_btn_call);
        btn4.setClickedListener(component -> {
 if (proxy != null) {
int result = proxy.plus(100,200);
                new ToastDialog(MainAbilitySlice.this)
                        .setText("res:" + result)
                        .show();
   }
        });

        //关闭远程
        Button btn5 = (Button)findComponentById(ResourceTable.Id_btn_close_remote);
        btn5.setClickedListener(component -> {
            disconnectAbility(conn);
        });


    }
```

```
    @Override
    public void onActive() {
        super.onActive();
    }

    @Override
    public void onForeground(Intent intent) {
        super.onForeground(intent);
    }
}
```

onAbilityConnectDone方法被回调，表明成功连接了远程Service Ability，回调成功后把连接好的远程代理对象赋值给本地代理，此时就可以通过这个连接好的Proxy进行远程调用了，如代码示例4-19所示。

代码示例4-19 调用远程方法

```
if(myproxy != null){
   int num = myproxy.plus(100,200);
     new ToastDialog(MusicAbilitySlice.this)
             .setText("remote service result:" + num)
             .show();
 }
```

断开连接：触发远程Service Ability中的onDisconnect方法。

```
disconnectAbility(conn);
```

4.3.7 前台Service Ability

一般情况下，Service都是在后台运行的，并且后台Service的优先级都是比较低的，当资源不足时，系统有可能回收正在运行的后台Service。

在一些场景下，如播放音乐，用户希望应用能够一直保持运行，此时就需要使用前台Service。前台Service会始终保持正在运行的图标在系统状态栏显示。

说明：使用前台Service并不复杂，开发者只需要在Service创建的方法里调用keepBackgroundRunning()并将Service与通知绑定。调用keepBackgroundRunning()方法前需要在配置文件中声明ohos.permission.KEEP_BACKGROUND_RUNNING权限，同时还需要在配置文件中添加对应的backgroundModes参数。在onStop()方法中调用cancelBackgroundRunning()方法可停止前台Service。

通过 DevEco Studio 创建一个新的前台 ServiceAbility，如图 4-23 所示。

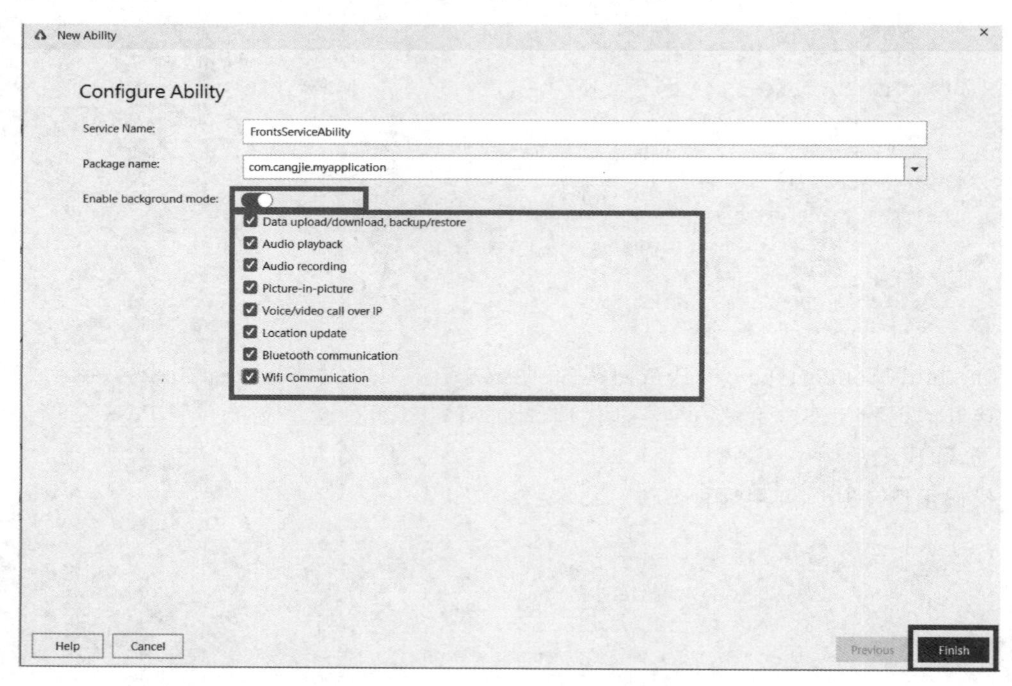

图 4-23　创建前台 Service Ability

通过 DevEco Studio 创建的 Service Ability 会自动在 config.json 中添加如下配置：

```
{
"backgroundModes": [
"dataTransfer",
"audioPlayback",
"audioRecording",
"pictureInPicture",
"voip",
"location",
"bluetoothInteraction",
"WiFiInteraction"
        ],
"name": "com.cangjie.myapplication.FrontServiceAbility",
"icon": "$media:icon",
"description": "$string:frontserviceability_description",
"type": "service"
 }
```

在创建的 Service Ability 的 onStart 方法中添加启动前台服务，如代码示例 4-20 所示。

代码示例 4-20 启动前台服务

```
@Override
public void onStart(Intent intent) {
        HiLog.error(LABEL_LOG, "FrontServiceAbility::onStart");
        super.onStart(intent);

        NotificationRequest request = new NotificationRequest(1005);
        NotificationRequest.NotificationNormalContent content = new NotificationRequest.
NotificationNormalContent();
        content.setTitle("title").setText("前台服务通知");
        NotificationRequest.NotificationContent notificationContent = new NotificationRequest.
NotificationContent(content);
        request.setContent(notificationContent);
        keepBackgroundRunning(1005,request);
    }

    @Override
    public void onStop() {
        super.onStop();
        HiLog.info(LABEL_LOG, "FrontServiceAbility::onStop");
        cancelBackgroundRunning();
    }
```

4.4 Data Ability

使用Data模板的Ability(以下简称Data)有助于应用管理其自身和其他应用存储数据的访问,并提供与其他应用共享数据的方法。Data既可用于同设备不同应用的数据共享,也支持跨设备不同应用的数据共享。

4.4.1 DataAbility 概述

数据的存放形式多样,可以是数据库形式,也可以是磁盘上的文件形式。Data对外提供对数据的增、删、改、查操作,以及打开文件等接口,这些接口的具体实现由开发者提供。

1. 数据 URI

通过URI(Uniform Resource Identifier)来标识一个具体的数据,例如数据库中的某个表或磁盘上的某个文件。HarmonyOS的URI仍基于URI通用标准,格式如图4-24所示。

Scheme://[authority]/[path][?query][#fragment]

Scheme	authority	path	?query	#fragment
协议方案名	设备ID	资源路径	查询参数	访问的子资源

图 4-24 URI组成部分

scheme：协议方案名，固定为 dataability，代表 Data Ability 所使用的协议类型。

authority：设备 ID。如果为跨设备场景，则为目标设备的 ID；如果为本地设备场景，则不需要填写。

path：资源的路径信息，代表特定资源的位置信息。

query：查询参数。

fragment：可以用于指示要访问的子资源。

URI 示例：

跨设备场景：dataability://device_id/com.huawei.dataability.persondata/person/10

本地设备场景：dataability:///com.huawei.dataability.persondata/person/10

注意：访问本地 DataAbility，如果没有设备编号，则 dataability:///需要使用 ///才可以，否则在转换时，会报格式转换错误。

---

2. DataAbilityHelper 工具类

开发者可以通过 DataAbilityHelper 类访问当前应用或其他应用提供的共享数据。DataAbilityHelper 作为客户端，与提供方的 Data 进行通信。Data 接收到请求后，执行相应的处理，并返回结果。DataAbilityHelper 提供了一系列与 Data Ability 对应的方法。

下面介绍 DataAbilityHelper 具体的使用步骤。

1）声明使用权限

如果待访问的 Data 声明了访问需要权限，则访问此 Data 时需要在配置文件中声明需要此权限。声明参考权限申请字段，代码如下：

```
"reqPermissions": [
    //访问文件还需要添加访问存储读写权限
    {
"name": "ohos.permission.READ_USER_STORAGE"
    },
    {
"name": "ohos.permission.WRITE_USER_STORAGE"
    }
]
```

2）创建 DataAbilityHelper

DataAbilityHelper 为开发者提供了 creator()方法来创建 DataAbilityHelper 实例。该方法为静态方法，有多个重载。最常见的方法是通过传入一个 context 对象来创建 DataAbilityHelper 对象。

```
DataAbilityHelper helper = DataAbilityHelper.creator(this);
```

3）数据接口

DataAbilityHelper 为开发者提供了一系列接口用于访问不同类型的数据，例如文件、数据库等。

DataAbilityHelper 为开发者提供了 FileDescriptor openFile(Uri uri，String mode)方法来操作文件。此方法需要传入两个参数，其中 uri 用来确定目标资源路径，mode 用来指定打开文件的方式，可选方式包含 r(读)、w(写)、rw(读写)、wt(覆盖写)、wa(追加写)、rwt(覆盖写且可读)。

该方法返回一个目标文件的 FD(文件描述符)，把文件描述符封装成流，开发者就可以对文件流进行自定义处理。

访问文件示例：

```
//读取文件描述符
FileDescriptor fd = helper.openFile(uri, "r");
FileInputStream fis = new FileInputStream(fd);
//使用文件描述符封装成的文件流,进行文件操作
```

DataAbilityHelper 为开发者提供了增、删、改、查及批量处理等方法来操作数据库，如表 4-1 所示。

表 4-1　DataAbilityHelper 为开发者提供的操作数据库的方法

方　　法	描　　述
ResultSet query(Uri uri，String[] columns，DataAbilityPredicates predicates)	查询数据库
int insert(Uri uri，ValuesBucket value)	向数据库中插入单条数据
int batchInsert(Uri uri，ValuesBucket[] values)	向数据库中插入多条数据
int delete(Uri uri，DataAbilityPredicates predicates)	删除一条或多条数据
int update(Uri uri，ValuesBucket value，DataAbilityPredicates predicates)	更新数据库
DataAbilityResult[] executeBatch(ArrayList <DataAbilityOperation> operations)	批量操作数据库

4.4.2　DataAbility 创建本地数据库

通过 DataAbility 创建本地数据库，可以通过以下步骤实现。

1. 创建 DataAbility

通过 DataAbility 模板生成 DataAbility 代码文件及配置，这是最简单实用的方法。生成的 DataAbility 代码定义了 6 种方法供用户对数据库表数据进行增、删、改、查操作。这 6 种方法在 Ability 中已默认实现，开发者可按需重写。

创建完成 DataAbility 代码文件后，会自动在 config.json 中的 abilities 中添加该 ability

的配置，如图 4-25 和图 4-26 所示。

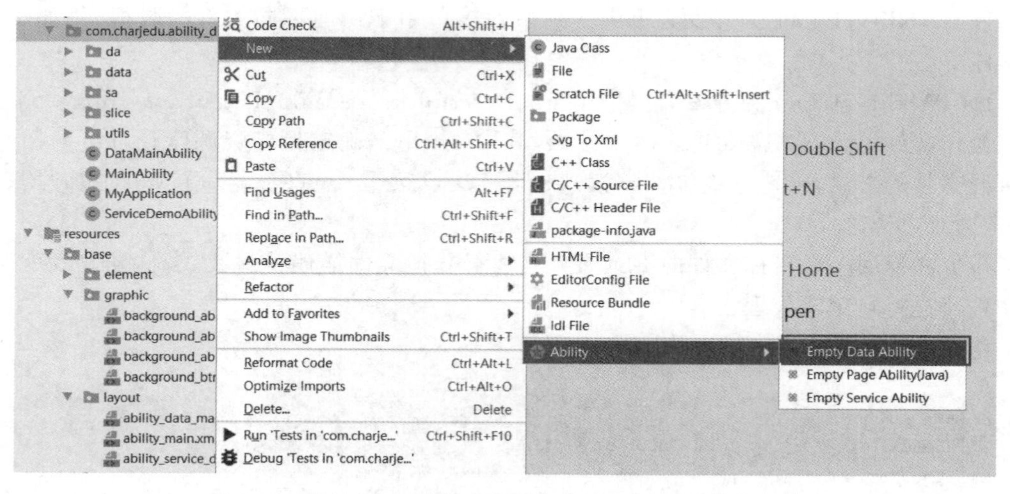

图 4-25　通过 DataAbility 模板生成 DataAbility 代码文件

```
{
  "permissions": [
    "com.charjedu.ability_demo.DataAbilityShellProvider.PROVIDER"
  ],
  "name": "com.charjedu.ability_demo.TestDataAbility",
  "icon": "$media:icon",
  "description": "$string:testdataability_description",
  "type": "data",
  "uri": "dataability://com.charjedu.ability_demo.TestDataAbility"
}
```

图 4-26　通过 DataAbility 模板生成 DataAbility 配置

默认添加了这个 Data Ability 的操作权限。Type 的类型为 data、uri：dataability 的本地资源访问地址。

生成的 DataAbility 的代码如代码示例 4-21 所示，代码默认定义了对数据的增、删、改、查方法，开发人员只需要在这些方法下添加自己的实现操作便可以了。

代码示例 4-21　根据 DataAbility 生成代码

```
public class TestDataAbility extends Ability {
    private static final HiLogLabel LABEL_LOG = new HiLogLabel(3, 0xD001100, "Demo");

    @Override
    public void onStart(Intent intent) {
        super.onStart(intent);
        HiLog.info(LABEL_LOG, "TestDataAbility onStart");
    }
```

```
    @Override
    public ResultSet query(Uri uri, String[] columns, DataAbilityPredicates predicates) {
        return null;
    }

    @Override
    public int insert(Uri uri, ValuesBucket value) {
        HiLog.info(LABEL_LOG, "TestDataAbility insert");
        return 999;
    }

    @Override
    public int delete(Uri uri, DataAbilityPredicates predicates) {
        return 0;
    }

    @Override
    public int update(Uri uri, ValuesBucket value, DataAbilityPredicates predicates) {
        return 0;
    }

    @Override
    public FileDescriptor openFile(Uri uri, String mode) {
        return null;
    }

    @Override
    public String[] getFileTypes(Uri uri, String mimeTypeFilter) {
        return new String[0];
    }

    @Override
    public PacMap call(String method, String arg, PacMap extras) {
        return null;
    }

    @Override
    public String getType(Uri uri) {
        return null;
    }
}
```

2. 设置 DataAbility 权限

默认情况下，ability 的配置中已经添加了 dataability 的操作权限，但是还是需要在全局手动把上面的权限添加到 reqPermissions 中，如图 4-27 所示。

```
"reqPermissions": [
  {
    "name": "com.charjedu.ability_demo.DataAbilityShellProvider.PROVIDER"
  },
```

图 4-27 dataability 的操作权限

在 build. gradle 中添加 complieOptions，将 annotaionEnabled 设置为 true，如图 4-28 所示。

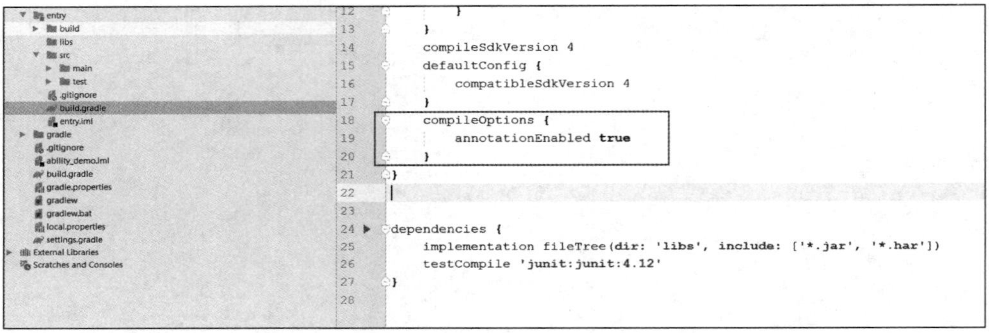

```
12          }
13      }
14      compileSdkVersion 4
15      defaultConfig {
16          compatibleSdkVersion 4
17      }
18      compileOptions {
19          annotationEnabled true
20      }
21  }
22
23
24  dependencies {
25      implementation fileTree(dir: 'libs', include: ['*.jar', '*.har'])
26      testCompile 'junit:junit:4.12'
27  }
28
```

图 4-28 将 annotaionEnabled 设置为 true

3. 通过 ORM 生成数据库

这里我们定义一个 OrmDatabase 的子类，通过注解的方式把这个抽象类定义成数据库对象，如图 4-29 所示。

```
import ohos.data.orm.OrmDatabase;
import ohos.data.orm.annotation.Database;

//创建数据库bookstore
//entities 是定义的表
//version 为数据库版本
@Database(entities = {User.class},version = 1)
public abstract class MyDataBase extends OrmDatabase {

}
```

图 4-29 ORM 生成数据库

entities：后面的对象定义了这个数据库的所有数据表；

version：数据库版本号。

在 dataability 中初始化数据库，如图 4-30 所示。

```
public class SimpleDataAbility extends Ability {
    private static final HiLogLabel LABEL_LOG = new HiLogLabel( type: 3,  domain: 0xD001100,  tag: "Demo");

    private OrmContext cxt = null;

    @Override
    public void onStart(Intent intent) {
        super.onStart(intent);
        HiLog.info(LABEL_LOG,  format: "SimpleDataAbility onStart");
        //数据库工具类
        DatabaseHelper databaseHelper = new DatabaseHelper( context: this);
        //数据库别名，数据库文件名，数据库类
        cxt = databaseHelper.getOrmContext( alias: "MyDataBase",  name: "MyDataBase.db", MyDataBase.class,  ...migrations: null);

    }
```

图 4-30　始化数据库

代码实现如代码示例 4-22 所示。

代码示例 4-22　初始化数据库

```
private OrmContext cxt = null;

@Override
public void onStart(Intent intent) {
    super.onStart(intent);
    HiLog.info(LABEL_LOG, "SimpleDataAbility onStart");
    //数据库工具类
    DatabaseHelper DatabaseHelper = new DatabaseHelper(this);
    //初始化数据库别名、数据库文件名、数据库类
    cxt = DatabaseHelper.getOrmContext("MyDatabase", "MyDatabase.db", MyDatabase.class,
null);

}
```

4. 通过 ORM 生成数据库表结构

定义 ORM 表数据结构，如图 4-31 所示。

```
@Entity(tableName = "User",indices = {
        @Index(value = {"userId"},name = "userId_index",unique = true)})
public class User extends OrmObject {

    @PrimaryKey(autoGenerate = true)
    private Integer userId;
    private String userName;
    private int age;
    private double balance;
```

图 4-31　定义 ORM 表数据结构

tableName：表名；

@Index：设置的索引信息。

为 User 类属性添加 getter 和 setter 方法,如代码示例 4-23 所示。

代码示例 4-23　创建 User 实体

```
import ohos.data.orm.OrmObject;
import ohos.data.orm.annotation.Entity;
import ohos.data.orm.annotation.Index;
import ohos.data.orm.annotation.PrimaryKey;

@Entity(tableName = "User",indices = {
        @Index(value = {"userId"},name = "userId_index",unique = true)})
public class User extends OrmObject {

    @PrimaryKey(autoGenerate = true)
    private Integer userId;
    private String userName;
    private int age;
    private double balance;

    @Override
    public String toString() {
        return "User{" +
"userId = " + userId +
", userName = '" + userName + '\'' +
", age = " + age +
", balance = " + balance +
                '}';
    }

    public Integer getUserId() {
        return userId;
    }

    public void setUserId(Integer userId) {
        this.userId = userId;
    }

    public String getUserName() {
        return userName;
    }

    public void setUserName(String userName) {
        this.userName = userName;
    }

    public int getAge() {
```

```
        return age;
    }

    public void setAge(int age) {
        this.age = age;
    }

    public double getBalance() {
        return balance;
    }

    public void setBalance(double balance) {
        this.balance = balance;
    }

}
```

4.4.3 DataAbility本地数据库数据操作

DataAbility操作SQLite中的数据，分两步执行：第一步，实现DataAbility中的增、删、改、查方法。第二步，在页面中通过DataAbilityHelper调用DataAbility。

1. 在DataAbility中实现数据增、删、改、查

首先，在SimpleDataAbility的Data Ability中初始化OrmContext对象，这里通过DatabaseHelper的getOrmContext方法，将数据库别名设置为MyDatabase，将数据库文件名称设置为MyDatabase.db，将数据库对象设置为MyDatabase.class，如代码示例4-24所示。

代码示例4-24 初始化OrmContext对象

```
private OrmContext cxt = null;

@Override
public void onStart(Intent intent) {
    super.onStart(intent);
    HiLog.info(LABEL_LOG, "SimpleDataAbility onStart");
    //数据库工具类
    DatabaseHelper DatabaseHelper = new DatabaseHelper(this);
    //初始化数据库别名、数据库文件名、数据库类
    cxt = DatabaseHelper.getOrmContext("MyDatabase", "MyDatabase.db", MyDatabase.class,
null);

}
```

1）Insert()

插入数据，其中 uri 为目标资源路径，ValuesBucket 为要新增的对象，如代码示例 4-25 所示。

代码示例 4-25　插入数据

```
@Override
public int insert(Uri uri, ValuesBucket value) {
    HiLog.info(LABEL_LOG, "SimpleDataAbility insert");
    if (cxt == null) {
        return 1;
    }

    //如果操作的表不是 User,则返回 - 1
    if(!uri.getDecodedPathList().get(1).equals("User")){
        return - 1;
    }

    User user = new User();
    user.setUserName(value.getString("userName"));
    user.setAge(value.getInteger("age"));
    user.setBalance(value.getInteger("balance"));

    //插入数据,如果插入不成功,则返回 - 1
    boolean isInserted = cxt.insert(user);
    if (!isInserted) {
        return - 1;
    }
    //插入成功后,刷新数据,如果刷新失败,则返回 - 1
    boolean flush = cxt.flush();
    if (!flush){
        return - 1;
    }

    //最后返回 user 的编号
    int i = Math.toIntExact(user.getRowId());
    return i;

}
```

2）Query()

查询数据，如代码示例 4-26 所示。

代码示例 4-26　查询数据

```
@Override
public ResultSet query(Uri uri, String[] columns, DataAbilityPredicates predicates) {
    if (cxt == null) {
        return null;
    }

    //如果操作的表不是 User,则返回 -1
    //String dataURL = "dataability:///com.charjedu.ability_demo.SimpleDataAbility/User";
    //添加了"/User"才可以
    if(!uri.getDecodedPathList().get(1).equals("User")){
        return null;
    }

    OrmPredicates ormPredicates = DataAbilityUtils.createOrmPredicates(predicates, User.class);
    return cxt.query(ormPredicates, columns);
}
```

3）Delete()

删除数据，如代码示例 4-27 所示。

代码示例 4-27　删除数据

```
@Override
public int delete(Uri uri, DataAbilityPredicates predicates) {
    //返回已经删除的条数
    if (cxt == null) {
        return -1;
    }
    //判断是不是 user 表
    if(!uri.getDecodedPathList().get(1).equals("User")){
        return -1;
    }
    OrmPredicates ormPredicates = DataAbilityUtils.createOrmPredicates(predicates, User.class);
    int delete = cxt.delete(ormPredicates);
    return delete;
}
```

4）Update()

更新数据，如代码示例 4-28 所示。

代码示例 4-28　更新数据

```
@Override
public int update(Uri uri, ValuesBucket value, DataAbilityPredicates predicates) {
```

```
    //返回已经更新的条数
    if (cxt == null) {
        return -1;
    }
    //判断是不是 user 表
    if(!uri.getDecodedPathList().get(1).equals("User")){
        return -1;
    }
    OrmPredicates ormPredicates = DataAbilityUtils.createOrmPredicates(predicates, User.class);
    int updateNum = cxt.update(ormPredicates,value);
    return updateNum;
}
```

2. 在页面中调用 DataAbility

在页面中构造查询的条件和参数，通过 DataAbilityHelper 调用 DataAbility 中的增、删、改、查方法。

1) query()

查询方法，其中 uri 为目标资源路径，columns 为想要查询的字段。开发者定义的查询条件可以通过 DataAbilityPredicates 来构建。查询用户表中 id 在 101～103 的用户，并把结果打印出来，如代码示例 4-29 所示。

代码示例 4-29　查询方法

```
DataAbilityHelper helper = DataAbilityHelper.creator(this);

//构造查询条件
DataAbilityPredicates predicates = new DataAbilityPredicates();
predicates.between("userId", 101, 103);

//进行查询
ResultSet resultSet = helper.query(uri,columns,predicates);

//处理结果
resultSet.goToFirstRow();
do{
    //在此处理 ResultSet 中的记录
}while(resultSet.goToNextRow());
```

2) insert()

新增方法，其中 uri 为目标资源路径，ValuesBucket 为要新增的对象。插入一条用户信息的方法如代码示例 4-30 所示。

代码示例 4-30　新增方法

```
DataAbilityHelper helper = DataAbilityHelper.creator(this);

//构造插入数据
ValuesBucket valuesBucket = new ValuesBucket();
valuesBucket.putString("name", "Tom");
valuesBucket.putInteger("age", 12);
helper.insert(uri, valuesBucket);
```

3) batchInsert()

批量插入方法和insert()类似。批量插入用户信息的代码,如代码示例4-31所示。

代码示例 4-31　批量插入方法

```
DataAbilityHelper helper = DataAbilityHelper.creator(this);

//构造插入数据
ValuesBucket[] values = new ValuesBucket[2];
value[0] = new ValuesBucket();
value[0].putString("name", "Tom");
value[0].putInteger("age", 12);
value[1] = new ValuesBucket();
value[1].putString("name", "Tom1");
value[1].putInteger("age", 16);
helper.batchInsert(uri, values);
```

4) delete()

删除方法,其中删除条件可以通过DataAbilityPredicates来构建。删除用户表中id在101～103的用户,如代码示例4-32所示。

代码示例 4-32　删除方法

```
DataAbilityHelper helper = DataAbilityHelper.creator(this);

//构造删除条件
DataAbilityPredicates predicates = new DataAbilityPredicates();
predicates.between("userId", 101,103);
helper.delete(uri,predicates);
```

5) update()

更新方法,更新数据由ValuesBucket传入,更新条件由DataAbilityPredicates来构建。更新id为102的用户,如代码示例4-33所示。

代码示例 4-33 更新方法

```
DataAbilityHelper helper = DataAbilityHelper.creator(this);

//构造更新条件
DataAbilityPredicates predicates = new DataAbilityPredicates();
predicates.equalTo("userId",102);

//构造更新数据
ValuesBucket valuesBucket = new ValuesBucket();
valuesBucket.putString("name", "Tom");
valuesBucket.putInteger("age", 12);
helper.update(uri, valuesBucket, predicates);
executeBatch()
```

6）executeBatch()

此方法用来执行批量操作。DataAbilityOperation 中提供了设置操作类型、数据和操作条件的方法，开发者可自行设置自己要执行的数据库操作。插入多条数据的代码如代码示例 4-34 所示。

代码示例 4-34 执行批量操作

```
DataAbilityHelper helper = DataAbilityHelper.creator(abilityObj, insertUri);

//构造批量操作
ValuesBucket value1 = initSingleValue();
DataAbilityOperation opt1 = DataAbilityOperation.newInsertBuilder(insertUri).
withValuesBucket(value1).build();
ValuesBucket value2 = initSingleValue2();
DataAbilityOperation opt2 = DataAbilityOperation.newInsertBuilder(insertUri).
withValuesBucket(value2).build();
ArrayList<DataAbilityOperation> operations = new ArrayList<DataAbilityOperation>();
operations.add(opt1);
operations.add(opt2);
DataAbilityResult[] result = helper.executeBatch(insertUri, operations);
```

4.4.4 跨设备访问 DataAbility

跨设备访问 DataAbility 是鸿蒙应用开发的一个特点，下面我们介绍如何在设备间进行数据的访问和操作。

1. URI 添加设备 ID

（1）跨设备场景下：dataability://device_id/com.charjedu.ability_demo.SimpleDataAbility。

(2) 本地设备下：dataability:///com.charjedu.ability_demo.SimpleDataAbility。

2. 跨设备访问需要的权限设置

开发者需要在 config.json 文件的 defPermissions 字段中自定义如下权限，如图 4-32 所示。

```
"reqPermissions": [
  {"name": "com.charjedu.ability_demo.DataAbilityShellProvider.PROVIDER"...},
  {"name": "ohos.permission.INTERNET"...},
  {"name": "ohos.permission.DISTRIBUTED_DATASYNC"...},
  {"name": "ohos.permission.READ_USER_STORAGE"...},
  {"name": "ohos.permission.servicebus.ACCESS_SERVICE"...},
  {"name": "ohos.permission.servicebus.BIND_SERVICE"...},
  {"name": "ohos.permission.DISTRIBUTED_DEVICE_STATE_CHANGE"...},
  {"name": "ohos.permission.GET_DISTRIBUTED_DEVICE_INFO"...},
  {"name": "ohos.permission.GET_BUNDLE_INFO"...}
],
"defPermissions": [
  {
    "name": "com.charjedu.ability_demo.DataAbilityShellProvider.PROVIDER",
    "grantMode": "system_grant"
  }
]
```

图 4-32 自定义权限

system_grant：安装后由系统自动授予。

同时需要在 MainAbility 中请求如下权限，如图 4-33 所示。

```java
public class MainAbility extends Ability {
    @Override
    public void onStart(Intent intent) {
        requestPermissionsFromUser(new String[]{
                "ohos.permission.DISTRIBUTED_DATASYNC",
                "ohos.permission.servicebus.BIND_SERVICE",
                "ohos.permission.servicebus.ACCESS_SERVICE"
        }, requestCode: 0);
        super.onStart(intent);
        super.setMainRoute(MainAbilitySlice.class.getName());

    }

}
```

图 4-33 在 MainAbility 中请求的权限

3. 允许其他应用访问的权限

Data Ability 的注册信息中一定要打开允许其他应用访问的权限，在 ability 的配置文件中添加"visible":true。否则跨设备访问的时候可能出现权限问题，如图 4-34 所示。

```
{
  "visible": true,
  "permissions": [
    "com.charjedu.ability_demo.DataAbilityShellProvider.PROVIDER"
  ],
  "name": "com.charjedu.ability_demo.da.SimpleDataAbility",
  "icon": "$media:icon",
  "description": "hap sample empty provider",
  "type": "data",
  "uri": "dataability://com.charjedu.ability_demo.SimpleDataAbility"
},
```

图 4-34 打开允许其他应用访问的权限

4.5 本章小结

本章通过 5 个章节，讲解了 3 种不同类型的 Ability 的用法：Page Ability 为开发者提供了构建页面的能力；Service Ability 为开发者提供了后端服务的开发能力；Data Ability 为开发人员提供了数据访问的统一入口。

第 5 章

鸿蒙 ACE JavaScript 应用框架

本章详细讲解鸿蒙 JavaScript 应用框架，通过本章的学习，读者可以全面了解鸿蒙 JavaScript 应用开发框架的语法特性和 API 的使用。

5.1 ACE JavaScript 框架介绍

鸿蒙基于 JavaScript 的轻应用 UI 框架是一种跨设备的高性能 UI 开发框架，如图 5-1 所示，支持声明式编程和跨设备多态 UI。

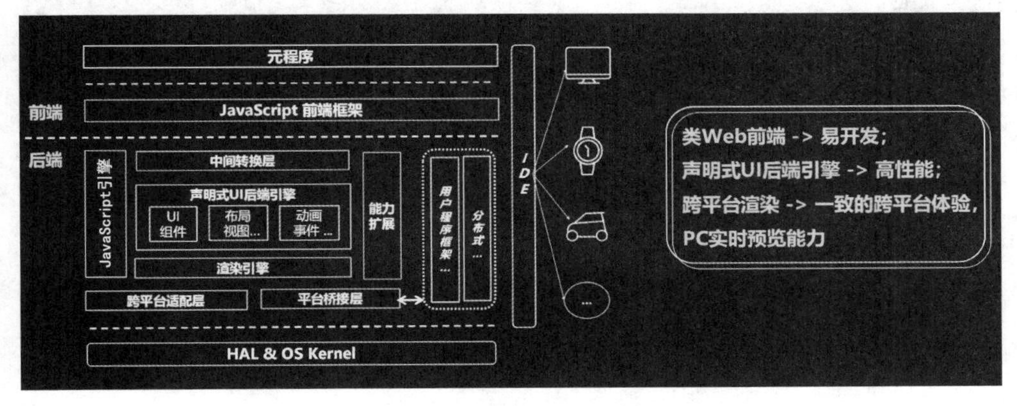

图 5-1 鸿蒙 ACE JavaScript

5.1.1 ACE JavaScript 框架特性

1. 声明式编程

JavaScript UI 框架采用类 HTML 和 CSS 声明式编程语言作为页面布局和页面样式的开发语言，页面业务逻辑则支持 ECMAScript 规范的 JavaScript 语言。JavaScript UI 框架提供的声明式编程，可以让开发者避免编写 UI 状态切换的代码，视图配置信息更加直观。

2. 跨设备

开发框架在架构上支持 UI 跨设备显示能力,运行时可自动映射到不同设备类型,开发者无感知,从而降低开发者多设备适配成本。

3. 高性能

开发框架包含了许多核心控件,如列表、图片和各类容器组件等,针对声明式语法进行了渲染流程的优化。

JavaScript 应用框架实现主要包含两部分。

(1) native 部分:使用 C++进行编写,实现框架主体,如图 5-2 所示。

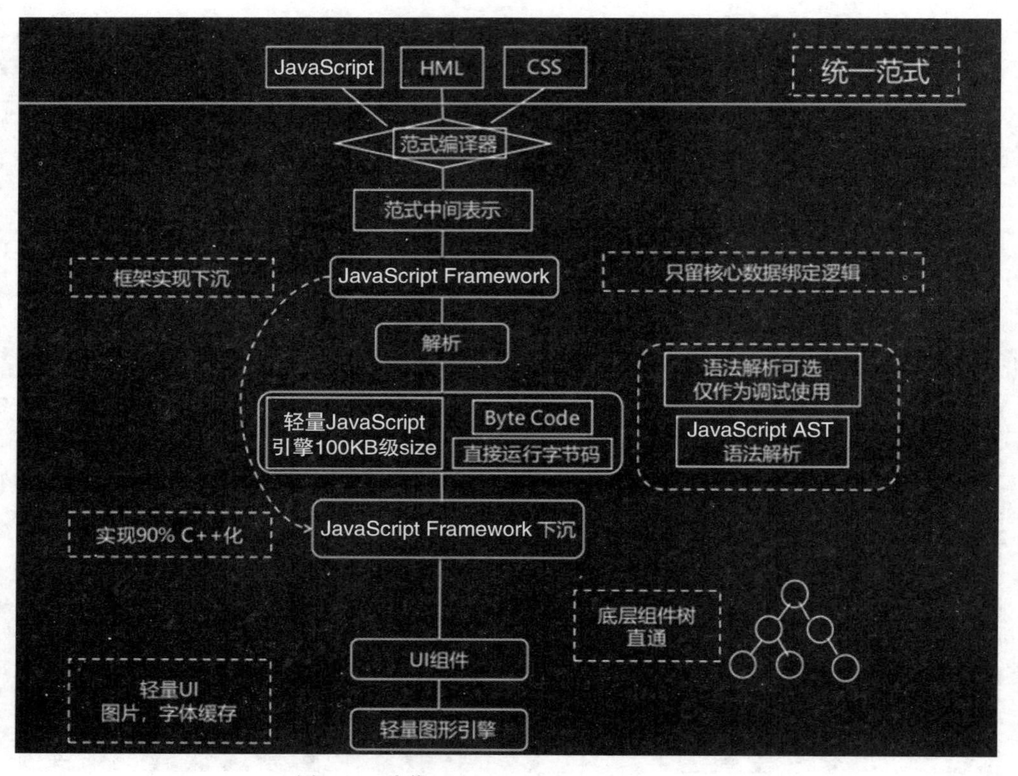

图 5-2 鸿蒙 ACE JavaScript Framework

(2) JavaScript 部分:提供 JavaScript 应用框架对用户 JavaScript 文件的运行时支持,并通过向引擎暴露一些全局方法和对象,支撑 JavaScript 运行时与 native 框架之间的交互。

5.1.2 ACE JavaScript 整体架构

ACE JavaScript UI 框架,如图 5-3 所示,包括应用层(Application)、前端框架层(Framework)、引擎层(Engine)和平台适配层(Porting Layer)。

JavaScript RunTime Core 在 IoT 设备上没有使用 V8,也没有使用 JSCore,而是选择了

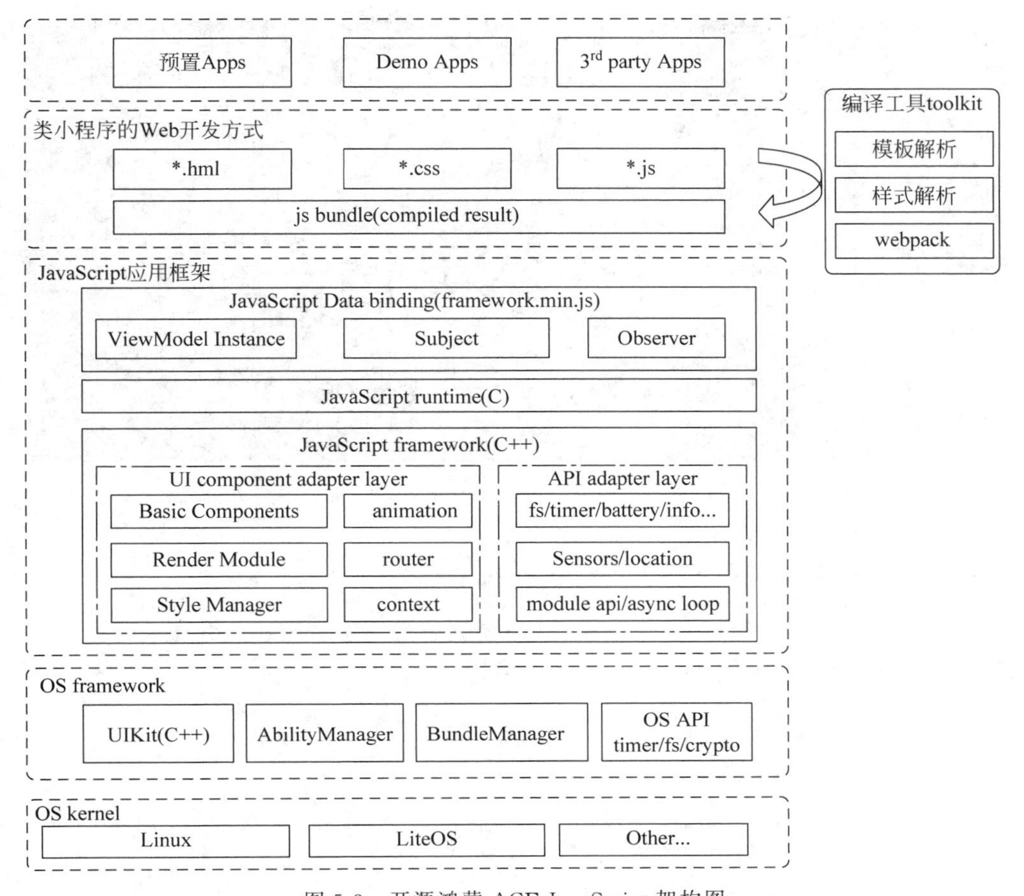

图 5-3 开源鸿蒙 ACE JavaScript 架构图

JerryScript。JerryScript 是用于物联网的超轻量 JavaScript 引擎。它能够在内存少于 64KB 的设备上执行 ECMAScript 5.1 源代码。

三星开源了 IoT.js 和 JerryScript。IoT.js 是一个使用 JavaScript 语言编写的物联网应用平台；JerryScript 是一个适用于嵌入式设备的小型 JavaScript 引擎。

三星创建这两个项目的目的是为了让 JavaScript 开发者能够构建物联网应用。物联网设备在 CPU 性能和内存空间上都有着严重的制约，因此，三星设计了 JerryScript 引擎，它能够运行在小于 64KB 内存上，且全部代码能够存储在不足 200KB 的只读存储(ROM)上。

OpenHarmony 是 HarmonyOS 的开源版，由华为捐赠给开放原子开源基金会(OpenAtom Foundation)开源。第一个开源版本支持在 128KB～128MB 设备上运行。代码仓库网址为 https://gitee.com/openharmony/ace_lite_jsfwk。

5.1.3 ACE JavaScript 运行流程

ACE JavaScript 运行原理图如图 5-4 所示。

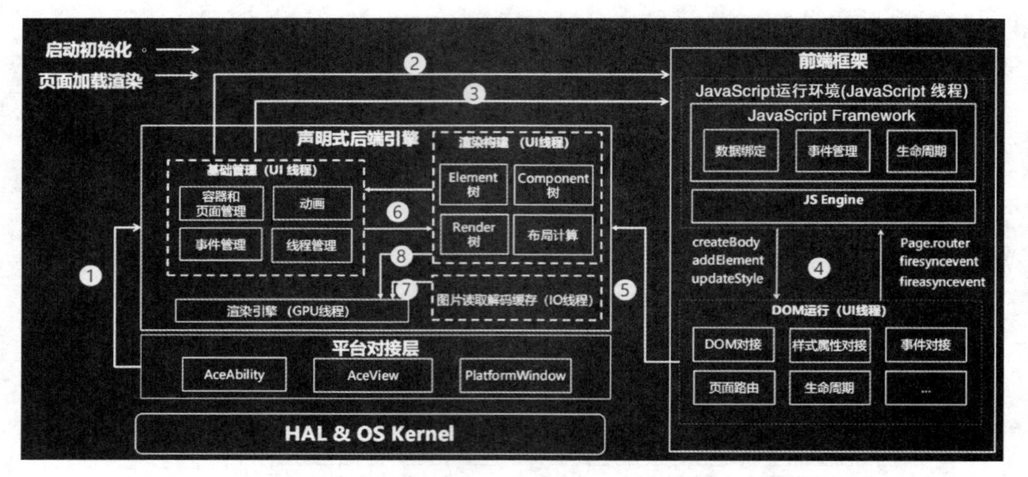

图 5-4　鸿蒙 ACE JavaScript 运行原理图

5.1.4　ACE JavaScript 数据绑定机制

为了实现单向数据绑定机制，JavaScript 应用框架使用 JavaScript 语言实现了一套简单的数据劫持框架，称为 RunTime-core，目录结构如图 5-5 所示。

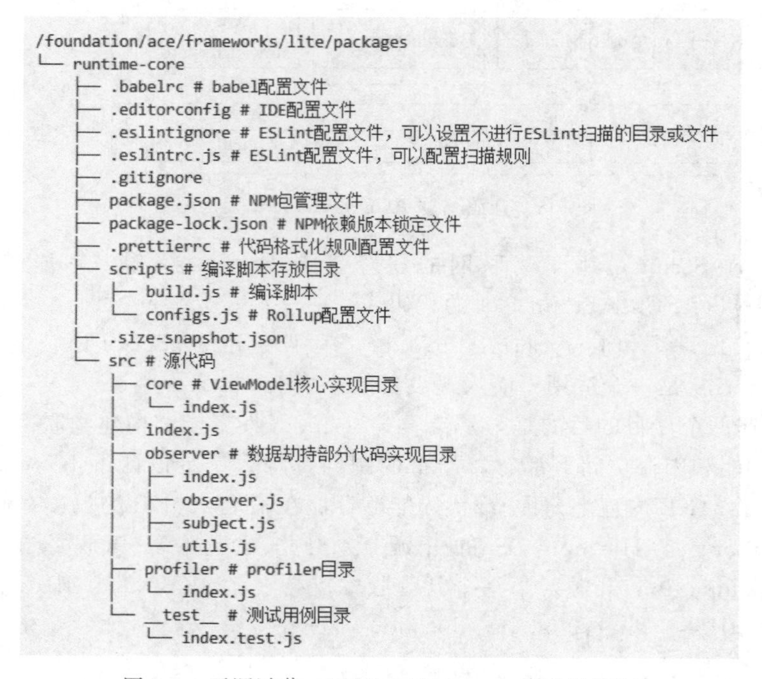

```
/foundation/ace/frameworks/lite/packages
└── runtime-core
    ├── .babelrc # babel配置文件
    ├── .editorconfig # IDE配置文件
    ├── .eslintignore # ESLint配置文件，可以设置不进行ESLint扫描的目录或文件
    ├── .eslintrc.js # ESLint配置文件，可以配置扫描规则
    ├── .gitignore
    ├── package.json # NPM包管理文件
    ├── package-lock.json # NPM依赖版本锁定文件
    ├── .prettierrc # 代码格式化规则配置文件
    ├── scripts # 编译脚本存放目录
    │   ├── build.js # 编译脚本
    │   └── configs.js # Rollup配置文件
    ├── .size-snapshot.json
    └── src # 源代码
        ├── core # ViewModel核心实现目录
        │   └── index.js
        ├── index.js
        ├── observer # 数据劫持部分代码实现目录
        │   ├── index.js
        │   ├── observer.js
        │   ├── subject.js
        │   └── utils.js
        ├── profiler # profiler目录
        │   └── index.js
        └── __test__ # 测试用例目录
            └── index.test.js
```

图 5-5　开源鸿蒙 ACE RunTime-core 代码结构图

鸿蒙 JavaScript RunTime-core 采用类似 Vue.js 框架设计模式，如图 5-6 所示，通过 Object.defineProperty 的 getter 和 setter，并结合观察者模式实现数据绑定。

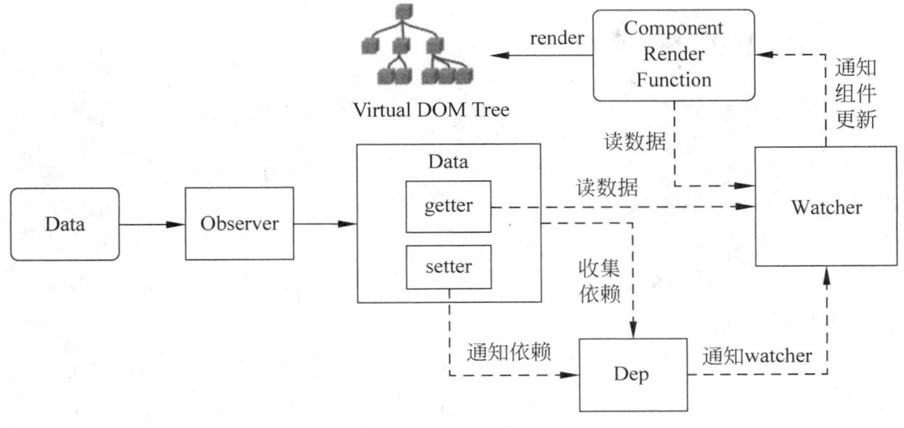

图 5-6　MVVM 模式图

5.2 ACE JavaScript 语法详细讲解

鸿蒙 HML 采用了目前流行的组件化开发思想，通过声明式的组件标记语言，使 UI 更加容易开发和复用。

5.2.1 HML 语法

HML(HarmonyOS Markup Language)鸿蒙标记语言如图 5-7 所示，是一套类 HTML 的标记语言，通过组件和事件构建出页面的内容。页面具备数据绑定、事件绑定、列表渲染、条件渲染和逻辑控制等高级能力。

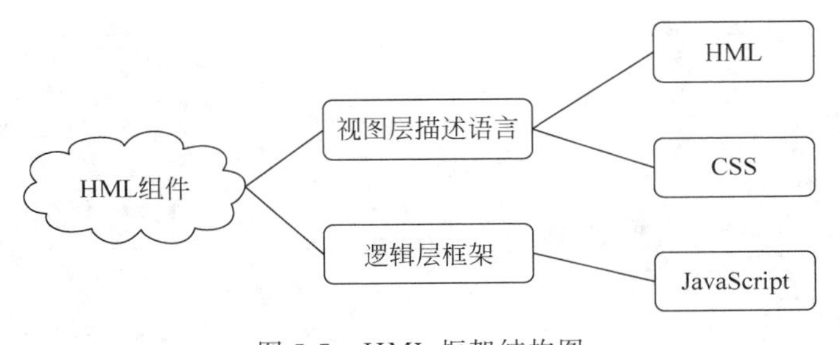

图 5-7　HML 框架结构图

下面我们介绍 HML 语法的特点，鸿蒙的 HML 采用了 MVVM(一种把视图与代码逻辑分离的界面层设计方法)设计模式，通过 Webpack(前端打包工具)进行 HML 文件的编译和打包。

1. HML 标记语言

HML 标记语言是鸿蒙 JavaScript 框架用来描述界面的标记语言，与 HTML 类似，如代码示例 5-1 所示。

代码示例 5-1 HML 标记语言

```
<div class = "item - container">
<text class = "item - title"> Image Show </text >
<div class = "item - content">
< image src = "/common/xxx.png" class = "image"></image >
</div >
</div >
```

2. 采用双向绑定机制

双向绑定机制是 MVVM 设计模式的核心部分，通过双向绑定机制彻底解决了界面与逻辑代码的耦合关系。鸿蒙的 ACE JavaScript 框架采用目前比较流行的双向绑定机制实现 UI 与代码逻辑分离，如代码示例 5-2 所示。

代码示例 5-2 双向绑定机制

```
< text > {{content}} </text >
//xxx.js
export default {
  data: {
 content: 'Hello World!',
  },
}
```

3. HML 界面事件绑定

在 HML 上通过事件绑定的回调函数接收一个事件对象参数，可以通过访问该事件对象获取事件信息，如代码示例 5-3 所示。

代码示例 5-3 界面事件响应

```
<!-- xxx.hml -->
< div >
<!-- 正常格式 -->
< div onclick = "clickfunc"></div >
<!-- 缩写 -->
< div @click = "clickfunc"></div >
</div >
//xxx.js
export default {
  data: {
    obj: '',
  },
```

```
clickfunc: function(e) {
    this.obj = 'Hello World';
   console.log(e);
  },
}
```

4. HML 列表渲染

对列表数据的渲染通过在 HML 标记上使用 for 指令，简化对界面标签元素的循环所带来的复杂操作。for 循环指令大大减少了界面开发的工作量，如代码示例 5-4 所示。

代码示例 5-4 界面列表渲染

```
<!-- xxx.hml -->
<div class = "array - container">
<!-- div 列表渲染 -->
<!-- 默认 $ item 代表数组中的元素, $ idx 代表数组中的元素索引 -->
<div for = "{{array}}" tid = "id">
<text>{{ $ idx}}.{{ $ item.name}}</text>
</div>
<!-- 自定义元素变量名称 -->
<div for = "{{value in array}}" tid = "id">
<text>{{ $ idx}}.{{value.name}}</text>
</div>
<!-- 自定义元素变量、索引名称 -->
<div for = "{{(index, value) in array}}" tid = "id">
<text>{{index}}.{{value.name}}</text>
</div>
</div>
//xxx.js
export default {
  data: {
    array: [
      {id: 1, name: 'jack', age: 18},
      {id: 2, name: 'tony', age: 18},
    ],
  },
}
```

tid 属性主要用来加速 for 循环的重渲染，旨在当列表中的数据有变更时，提高重新渲染的效率。tid 属性用来指定数组中每个元素的唯一标识，如果未指定，则数组中每个元素的索引为该元素的唯一 id。例如上述 tid="id"表示数组中的每个元素的 id 属性为该元素的唯一标识。for 循环支持的写法如下：

for="array"：其中 array 为数组对象，array 的元素变量默认为 $ item。

for="v in array"：其中 v 为自定义的元素变量，元素索引默认为 $ idx。

for="(i, v) in array"：其中元素索引为 i，元素变量为 v，遍历数组对象 array。

5. HML 条件渲染

条件渲染分为两种：if/elif/else 和 show。两种写法的区别在于：第一种写法里当 if 为 false 时，组件不会在 vdom 中构建，也不会渲染，而第二种写法里当 show 为 false 时，虽然也不渲染，但会在 vdom 中构建，如代码示例 5-5 所示。另外，当使用 if/elif/else 写法时，节点必须是兄弟节点，否则编译无法通过。

代码示例 5-5　界面条件渲染

```
<!-- xxx.hml -->
<div>
<text if = "{{show}}"> Hello - TV </text>
<text elif = "{{display}}"> Hello - Wearable </text>
<text else> Hello - World </text>
</div>
//xxx.js
export default {
  data: {
    show: false,
    display: true,
  },
}
```

优化渲染：show 方法。当 show 为真时，节点正常渲染；当 show 为假时，仅仅设置 display 样式为 none，如代码示例 5-6 所示。

代码示例 5-6　优化渲染

```
<!-- xxx.hml -->
<text show = "{{visible}}"> Hello World </text>
//xxx.js
export default {
  data: {
    visible: false,
  },
}
```

说明：禁止在同一个元素上同时设置 for 和 if 属性。

6. 逻辑控制块

<block>控制块使循环渲染和条件渲染变得更加灵活，block 在构建时不会被当作真实的节点进行编译。注意 block 标签只支持 for 和 if 属性，如代码示例 5-7 所示。

代码示例 5-7 <block>控制块

```
<!-- xxx.hml -->
<list>
<block for = "glasses">
<list - item type = "glasses">
<text>{{ $ item.name}}</text>
</list - item>
<block for = " $ item.kinds">
<list - item type = "kind">
<text>{{ $ item.color}}</text>
</list - item>
</block>
</block>
</list>
//xxx.js
export default {
  data: {
    glasses: [
      {name:'sunglasses', kinds:[{name:'XXX',color:'XXX'},{name:'XXX',color:'XXX'}]},
      {name:'nearsightedness mirror', kinds:[{name:'XXX',color:'XXX'}]},
    ],
  },
}
```

7. 模板引用

鸿蒙 ACE JavaScript 框架支持直接在 HML 中引用其他的 HML 模板文件，这种做法的目的是为了更好地复用视图。这里每个独立的 HML 都是一个自定义的组件，如代码示例 5-8 所示。

代码示例 5-8 模板引用

```
<!-- template.hml -->
<div class = "item">
<text> Name: {{name}}</text>
<text> Age: {{age}}</text>
</div>
<!-- index.hml -->
<element name = 'comp' src = '../../common/template.hml'></element>
<div>
<comp name = "Tony" age = "18"></comp>
</div>
```

5.2.2 CSS 语法

CSS 是描述 HML 页面结构的样式语言，所有组件均存在系统默认样式，也可在页面

CSS 样式文件中对组件、页面自定义不同的样式。

下面我们从样式的几个重要方面来了解鸿蒙应用中 CSS 的用法。

1. 鸿蒙的尺寸单位

需要注意的是这两种单位的使用场景：px 单位实际上是个弹性单位，帮助我们在不同分辨率下自动进行界面适配。

鸿蒙 ACE JavaScript 支持如下两种常见的尺寸单位。

(1) 逻辑像素 px。

- 默认屏幕具有的逻辑宽度为 720px，实际显示时会将页面布局缩放至屏幕实际宽度，如 100px 在实际宽度为 1440 物理像素的屏幕上，实际渲染为 200 物理像素(从 720px 向 1440 物理像素，所有尺寸放大 2 倍)。
- 当额外配置 autoDesignWidth 为 true 时，逻辑像素 px 将按照屏幕密度进行缩放，如 100px 在屏幕密度为 3 的设备上，实际渲染为 300 物理像素。当应用需要适配多种设备时，建议采用此方法。

(2) 百分比：表示该组件占父组件尺寸的百分比，如将组件的 width 设置为 50%，代表其宽度为父组件的 50%。

2. 样式引入方式

为了模块化管理和代码复用，CSS 样式文件支持使用@import 语句导入 css 文件。

每个页面目录下存在一个与布局 hml 文件同名的 css 文件，用来描述该 hml 页面中组件的样式，决定组件应该如何显示。

(1) 内部样式，支持使用 style、class 属性来控制组件的样式，如代码示例 5-9 所示。

代码示例 5-9　内部样式

```
<!-- index.hml -->
<div class = "container">
<text style = "color:red"> Hello World </text>
</div>
/* index.css */
.container {
  justify-content: center;
}
```

(2) 文件导入，合并外部样式文件。例如，在 common 目录中定义样式文件 style.css，并在 index.css 中进行导入，如代码示例 5-10 所示。

代码示例 5-10　外部样式

```
/* style.css */
.title {
  font-size: 50px;
}
```

```
/* index.css */
@import '../../common/style.css';
.container {
  justify-content: center;
}
```

3. 样式预编译

预编译提供了利用特有语法生成 css 的程序，可以提供变量、运算等功能，令开发者更便捷地定义组件样式，目前支持 less、sass 和 scss 的预编译。

说明：使用样式预编译时，需要将原 css 文件后缀改为 less、sass 或 scss，如将 index.css 改为 index.less、index.sass 或 index.scss。

当前文件使用样式预编译，例如将原 index.css 改为 index.less，如代码示例 5-11 所示。

代码示例 5-11　使用 less 预编译

```
/* index.less */
/* 定义变量 */
@colorBackground: #000000;
.container {
  background-color: @colorBackground; /* 使用当前 less 文件中定义的变量 */
}
```

引用预编译文件，例如 common 中存在 style.scss 文件，将原 index.css 改为 index.scss，并引入 style.scss，如代码示例 5-12 所示。

代码示例 5-12　使用 scss 预编译

```
/* style.scss */
/* 定义变量 */
$colorBackground: #000000;
/* 在 index.scss 中引用 */
/* index.scss */
/* 引入外部 scss 文件 */
@import '../../common/style.scss';
.container {
  background-color: $colorBackground; /* 使用 style.scss 中定义的变量 */
}
```

4. 媒体查询

媒体查询(Media Query)在移动设备上应用十分广泛，开发者经常需要根据设备的大致类型或者特定的特征和设备参数(例如屏幕分辨率)来修改应用的样式。为此媒体查询提供了如下功能：

(1) 针对设备和应用的属性信息,可以设计出相匹配的布局样式。

(2) 当屏幕发生动态改变时,例如分屏、横竖屏切换,应用页面布局同步更新。

说明:media(媒体)属性值默认为设备的真实尺寸大小、物理像素和真实的屏幕分辨率。勿与以 720px 为基准的项目配置宽度 px 混淆。

通用媒体特征如代码示例 5-13 所示。

代码示例 5-13　媒体查询

```
<!-- xxx.hml -->
<div>
<div class = "container">
<text class = "title"> Hello World </text>
</div>
</div>

/* xxx.css */
.container {
  width: 300px;
  height: 600px;
  background-color: #008000;
}
@media screen and (device-type: tv) {
  .container {
    width: 500px;
    height: 500px;
    background-color: #fa8072;
  }
}
@media screen and (device-type: tv) {
  .container {
    width: 300px;
    height: 300px;
    background-color: #008b8b;
  }
```

5. 自定义字体样式

font-face 用于定义字体样式。应用可以在 style 中定义 font-face 来指定相应的字体名和字体资源,然后在 font-family 样式中引用该字体。

说明:自定义字体可以是从项目中的字体文件或网络字体文件中加载的字体,字体格式支持 ttf 和 otf。

font-family：自定义字体的名称。

src：自定义字体的来源，支持如下类别：

- 项目中的字体文件：通过 URL 指定项目中的字体文件路径(只支持绝对路径)。
- 网络字体文件：通过 URL 指定网络字体的网址。
- 不支持设置多个 src。

这里通过一个简单案例介绍如何使用自定义样式，例如有以下的布局：

```
<div>
<text class = "demo - text">测试自定义字体</text>
</div>
```

ttf 文件通常放在 common 目录下。页面样式，如代码示例 5-14 所示。

代码示例 5-14 自定义样式

```
@font - face {
 font - family: HWfont;
 src: URL("/common/HWfont.ttf");
}
.demo - text {
font - family: HWfont;
}
```

6. 动画样式

组件普遍支持的动画样式可以在 style 或 css 中设置动态旋转、平移、缩放效果。

说明：@keyframes 的 from/to 不支持动态绑定。

对于不支持起始值或终止值缺省的情况，可以通过 from 和 to 显示指定起始和结束，如代码示例 5-15 所示。

代码示例 5-15 关键帧动画

```
@keyframes Go
{
from {
        background - color: #f76160;
        transform:translate(0px) rotate(0deg) scale(1.0);
}
    to {
        background - color: #09ba07;
        transform:translate(100px) rotate(180deg) scale(2.0);
```

```
    }
}
```

5.2.3 JavaScript 逻辑

鸿蒙 ACE JavaScript 框架通过 MVVM 设计模式实现界面 UI 与 JavaScript 逻辑的分离，这里我们来介绍一下 HML 逻辑层的用法。JavaScript 逻辑层支持 ES6 语法，但是需要通过引入 babel 进行处理。

1. ACE JavaScript 支持 ES6 模块化标准

由鸿蒙内置模块引入，所以不需要使用路径，内置模块通常以@system 开头，代码如下：

```
mport router from '@system.router';
```

自定义模块引入，通过相对路径引入，代码如下：

```
import utils from '../../common/utils.js';
```

2. $ refs 获取 DOM 元素

通过 $ refs 获取 DOM 元素，如代码示例 5-16 所示。

代码示例 5-16　通过 $ refs 获取 DOM 元素

```
<!-- index.hml -->
<div class = "container">
<image - animator class = "image - player" ref = "animator" images = "{{images}}" duration =
"1s" onclick = "handleClick"></image - animator>
</div>

//index.js
export default {
  data: {
    images: [
      { src: '/common/frame1.png' },
      { src: '/common/frame2.png' },
      { src: '/common/frame3.png' },
    ],
  },
  handleClick() {
const animator = this.$refs.animator;
//获取 ref 属性为 animator 的 DOM 元素
    const state = animator.getState();
    if (state === 'paused') {
```

```
        animator.resume();
      } else if (state === 'stopped') {
        animator.start();
      } else {
        animator.pause();
      }
    },
  };
```

3. 通过 $ element 方法获取 HML 元素

$ element 方法是鸿蒙 JavaScript 内置方法，用于获取 HML 元素，如代码示例 5-17 所示。

代码示例 5-17 获取 HML 元素

```
<!-- index.hml -->
<div class="container">
<image-animator class="image-player" id="animator"images="{{images}}" duration="1s"
onclick="handleClick"></image-animator>
</div>
//index.js
export default {
  data: {
    images: [
      { src: '/common/frame1.png' },
      { src: '/common/frame2.png' },
      { src: '/common/frame3.png' },
    ],
  },
  handleClick() {
//获取 id 属性为 animator 的 DOM 元素
const animator = this.$element('animator');
const state = animator.getState();
    if (state === 'paused') {
      animator.resume();
    } else if (state === 'stopped') {
      animator.start();
    } else {
      animator.pause();
    }
  },
};
```

5.2.4 多语言支持

鸿蒙支持多语言，通过在文件组织中指定的 i18n 文件夹内放置每个语言地区下的资源定义文件即可，资源文件命名为“语言-地区.json”格式，例如英文(美国)的资源文件命名为en-US.json。当开发框架无法在应用中找到系统语言的资源文件时，默认使用en-US.json中的资源内容。

资源文件用于存放应用在多种语言场景下的资源内容，开发框架使用JSON文件保存资源定义。

由于不同语言针对单复数有不同的匹配规则，在资源文件中的使用zero、one、two、few、many、other定义不同单复数场景下的词条内容。例如中文不区分单复数仅存在other场景，英文存在one、other场景，阿拉伯语存在上述6种场景。

以en-US.json和ar-AE.json为例，资源文件内容格式，如代码示例5-18所示。

代码示例5-18　多语言定义

```
{
"strings": {
"hello": "Hello world!",
"object": "Object parameter substitution - {name}",
"array": "Array type parameter substitution - {0}",
"symbol": "@#$%^&*()_+-={}[]\\|:;\"'<>,./?",
"people": {
"one": "one person",
"other": "{count} people"
        }
    },

"files": {
"image": "image/en_picture.PNG"
    }
}
```

其他语言的配置文件，代码如下：

```
{
"strings": {
"plurals": {
"zero": "أحد لا",
"one": "وحده",
"two": "اثنان",
"few": "اشخاص ستة",
"many": "شخص خمسون",
"other": "شخص مائة"
        }
    }
}
```

上面介绍了如何定义多语言资源配置文件的方法，下面介绍如何引用资源。

注意：在应用开发的页面中使用多语言的语法，包含简单格式化和单复数格式化两种，都可以在 hml 或 js 中使用。

简单格式化方法是在应用中使用 $ t 方法引用资源，$ t 既可以在 hml 中使用，也可以在 js 中使用。系统将根据当前语言环境和指定的资源路径（通过 $ t 的 path 参数设置），显示对应语言的资源文件中的内容，如代码示例 5-19 所示。

代码示例 5-19　简单格式化方法

```
<!-- xxx.hml -->
<div>
<!-- 不使用占位符,text 中显示"Hello world!" -->
<text>{{ $t('strings.hello') }}</text>
<!-- 具名占位符格式,运行时将占位符{name}替换为"Hello world" -->
<text>{{ $t('strings.object', { name: 'Hello world' }) }}</text>
<!-- 数字占位符格式,运行时将占位符{0}替换为"Hello world" -->
<text>{{ $t('strings.array', ['Hello world']) }}</text>
<!-- 先在 js 中获取资源内容,再在 text 中显示"Hello world" -->
<text>{{ hello }}</text>
<!-- 先在 js 中获取资源内容,并将占位符{name}替换为"Hello world",再在 text 中显示"Object
parameter substitution - Hello world" -->
<text>{{ replaceObject }}</text>
<!-- 先在 js 中获取资源内容,并将占位符{0}替换为"Hello world",再在 text 中显示"Array type
parameter substitution - Hello world" -->
<text>{{ replaceArray }}</text>

<!-- 获取图片路径 -->
<image src="{{ $t('files.image') }}" class="image"></image>
<!-- 先在 js 中获取图片路径,再在 image 中显示图片 -->
<image src="{{ replaceSrc }}" class="image"></image>
</div>


//xxx.js
//下面为在 js 文件中的使用方法
export default {
  data: {
    hello: '',
    replaceObject: '',
    replaceArray: '',
    replaceSrc: '',
  },
  onInit() {
```

```
    //简单格式化
    this.hello = this.$t('strings.hello');
      this.replaceObject = this.$t('strings.object', { name: 'Hello world' });
      this.replaceArray = this.$t('strings.array', ['Hello world']);
  this.replaceSrc = this.$t('files.image');
    },
  }
```

单复数格式化示例,如代码示例 5-20 所示。

代码示例 5-20　单复数格式化

```
<!-- xxx.hml -->
<div>
<!-- 传递数值为 0 时: "0 people" 阿拉伯语中此处匹配 key 为 zero 的词条 -->
<text>{{ $tc('strings.plurals', 0) }}</text>
<!-- 传递数值为 1 时: "one person" 阿拉伯语中此处匹配 key 为 one 的词条 -->
<text>{{ $tc('strings.plurals', 1) }}</text>
<!-- 传递数值为 2 时: "2 people" 阿拉伯语中此处匹配 key 为 two 的词条 -->
<text>{{ $tc('strings.plurals', 2) }}</text>
<!-- 传递数值为 6 时: "6 people" 阿拉伯语中此处匹配 key 为 few 的词条 -->
<text>{{ $tc('strings.plurals', 6) }}</text>
<!-- 传递数值为 50 时: "50 people" 阿拉伯语中此处匹配 key 为 many 的词条 -->
<text>{{ $tc('strings.plurals', 50) }}</text>
<!-- 传递数值为 100 时: "100 people" 阿拉伯语中此处匹配 key 为 other 的词条 -->
<text>{{ $tc('strings.plurals', 100) }}</text>
</div>
```

5.3　ACE JavaScript 布局

鸿蒙目前支持 Flexbox 与 Grid 两种布局方式,Flexbox 是默认容器组件的布局方式。

5.3.1　FlexBox 布局

FlexBox 是一种当页面需要适应不同的屏幕大小及设备类型时确保元素拥有恰当的行为的布局方式。引入弹性盒布局模型的目的是提供一种更加有效的方式来对一个容器中的子元素进行排列、对齐和分配空白空间。

FlexBox 布局中有两个重要的概念:Flex 容器和 Flex 项目,如图 5-8 所示。

CSS 的 Grid 和 FlexBox 相结合将是解决布局的最佳方案。虽然浏览器对 CSS Grid 和 FlexBox 的属性未完全支持,但对于实现布局而言,这已是一种非常完美的结合。

鸿蒙操作系统的布局方案结合这两种布局方式,有效地解决了多屏响应的问题。

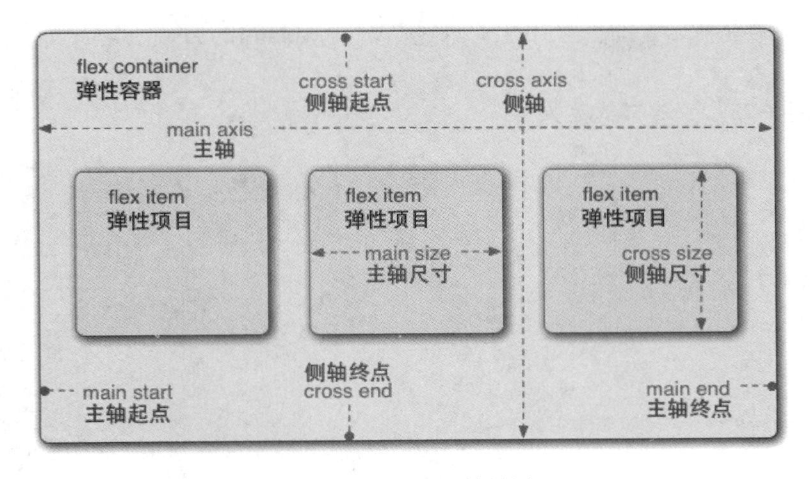

图 5-8　FlexBox 效果图

5.3.2　Grid 布局

网格布局(Grid)如图 5-9 所示,是最强大的 CSS 布局方案。它将网页划分成一个个网格,可以任意组合不同的网格,从而可以设计出各种各样的布局。以前,只能通过复杂的 CSS 框架达到的效果,现在已经在浏览器内置了。

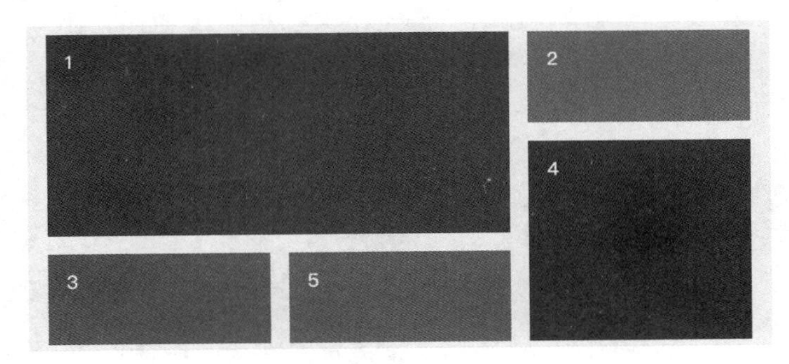

图 5-9　Grid 效果图

可以通过 Flex 或者 Grid 进行相应布局设计,达到鸿蒙多屏流转响应的目的。这里我们来看一个 Grid 布局的案例,元素布局如代码示例 5-21 所示。

代码示例 5-21　Grid 布局

```
<div class = "common grid - parent">
<div class = "grid - child grid - left - top"></div>
<div class = "grid - child grid - left - bottom"></div>
<div class = "grid - child grid - right - top"></div>
<div class = "grid - child grid - right - bottom"></div>
</div>
```

下面，通过设置将元素的父级用代码(display: grid;)转化为网格布局，如代码示例 5-22 所示。

代码示例 5-22 grid 样式定义

```
.common {
  width: 400px;
  height: 400px;
  background-color: #ffffff;
  align-items: center;
  justify-content: center;
  margin: 24px;
}
.grid-parent {
  display: grid;
  grid-template-columns: 50% 50%;
  grid-columns-gap: 24px;
  grid-rows-gap: 24px;
  grid-template-rows: 50% 50%;
}
.grid-child {
  width: 100%;
  height: 100%;
  border-radius: 8px;
}
.grid-left-top {
  grid-row-start: 0;
  grid-column-start: 0;
  grid-row-end: 0;
  grid-column-end: 0;
  background-color: #3f56ea;
}
.grid-left-bottom {
  grid-row-start: 1;
  grid-column-start: 0;
  grid-row-end: 1;
  grid-column-end: 0;
  background-color: #00aaee;
}
.grid-right-top {
  grid-row-start: 0;
  grid-column-start: 1;
  grid-row-end: 0;
  grid-column-end: 1;
  background-color: #00bfc9;
}
```

```
.grid-right-bottom {
  grid-row-start: 1;
  grid-column-start: 1;
  grid-row-end: 1;
  grid-column-end: 1;
  background-color: #47cc47;
}
```

CSS Grid 可轻松构建复杂的 Web 设计。它的工作原理是将元素转换为具有行和列的网格容器,以便将子元素放置在网格中所需的位置。Grid 案例展示如图 5-10 所示。

图 5-10 Grid 案例图

5.4 ACE JavaScript 内置组件

组件(Component)结构如图 5-11 所示,是构建页面的核心,每个组件通过对数据和方法的简单封装,实现独立的可视、可交互功能单元。组件之间相互独立,随取随用,也可以在需求相同的地方重复使用。开发者还可以通过组件间合理的搭配定义满足业务需求的新组件,从而减少开发量。

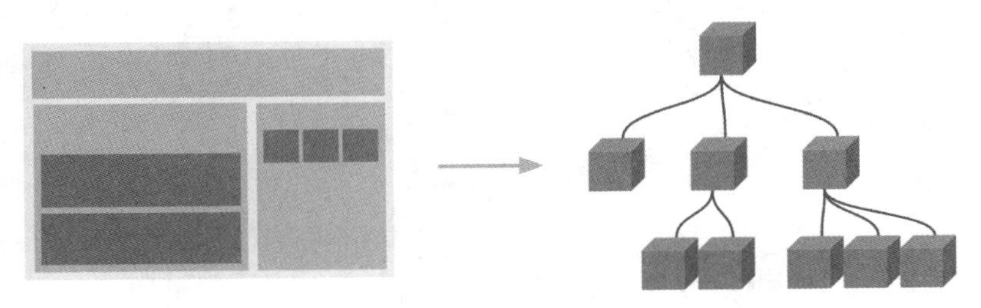

图 5-11 组件结构图

内置组件分类：根据组件的功能，目前鸿蒙的版本可以将组件分为以下四大类，如图 5-12 所示。

组件类型	主要组件
基础组件	text、image、progress、rating、span、marquee、image-animator、divider、search、menu、chart
容器组件	div、list、list-item、stack、swiper、tabs、tab-bar、tab-content、popup、list-item-group、refresh、dialog
媒体组件	video
画布组件	canvas

图 5-12 鸿蒙内置组件分类图

5.4.1 基础组件

下面介绍几种常见的基础组件，这些组件的名称和 HTML 中的组件名称相同，但是其实已经不再是 DOM 对象了，变成了鸿蒙操作系统内置的组件。

1. DIV 组件

基础容器，用作页面结构的根节点或将内容进行分组。和页面中使用 div 标签类似，支持所有鸿蒙设备，如图 5-13 所示。

手机	平板	智慧屏	智能穿戴
支持	支持	支持	支持

图 5-13 DIV 组件对设备支持情况

DIV 组件在 HML 中的定义如代码示例 5-23 所示。

代码示例 5-23 DIV 组件定义

```
< div class = "container">
< div class = "flex - box">
< div class = "flex - item color - primary"></div>
< div class = "flex - item color - warning"></div>
< div class = "flex - item color - success"></div>
</div>
</div>
```

2. Image 组件

图片组件，用来渲染并展示图片。本组件需要访问网络，需要申请 ohos. permission. INTERNET 权限(如果使用云端路径)。

首先，需要在配置文件 config. json 下的 module 下添加 reqPermissions，如代码示例 5-24 所示。

代码示例 5-24　申请权限

```
"module": {
"abilities": [],
"reqPermissions": [
        {
"name": "ohos.permission.INTERNET"
        }
 ]
}
```

现在就可以在 HML 中使用 URL 的方式引用网络图片网址了。

```
<image
src="http://blog.51itcto.com/wp-content/uploads/2020/11/1062-220x150.jpeg">
</image>
```

3. chart 组件

图表组件,用于呈现线性图、柱状图、量规图界面。下面分别介绍 chart 组件的几种图表的用法。

线性图如图 5-14 所示,使用方法如代码示例 5-25 所示。

代码示例 5-25　线性图

```
<!-- xxx.hml -->
<div class="container">
<stack class="chart-region">
<image class="chart-background" src="common/background.png"></image>
<chart class="chart-data" type="line" ref="linechart" options="{{lineOps}}" datasets
 ="{{lineData}}"></chart>
</stack>
<button value="Add data" onclick="addData"></button>
</div>
/* xxx.css */
.container {
  flex-direction: column;
  justify-content: center;
  align-items: center;
}
.chart-region {
  height: 400px;
  width: 700px;
}
.chart-background {
  object-fit: fill;
}
```

```
.chart - data {
  width: 700px;
  height: 600px;
}
//xxx.js
export default {
  data: {
    lineData: [
      {
        strokeColor: '#0081ff',
        fillColor: '#cce5ff',
        data: [763, 550, 551, 554, 731, 654, 525, 696, 595, 628, 791, 505, 613, 575, 475,
553, 491, 680, 657, 716],
        gradient: true,
      }
    ],
    lineOps: {
      xAxis: {
        min: 0,
        max: 20,
        display: false,
      },
      yAxis: {
        min: 0,
        max: 1000,
        display: false
      },
      series: {
        lineStyle: {
          width: "5px",
          smooth: true,
        },
        headPoint: {
          shape: "circle",
          size: 20,
          strokeWidth: 5,
          fillColor: '#ffffff',
          strokeColor: '#007aff',
          display: true
        },
        loop: {
          margin: 2,
          gradient: true,
        }
      }
    },
```

```
    },
    addData() {
      this.$refs.linechart.append({
        serial: 0,
        data: [Math.floor(Math.random() * 400) + 400]
      })
    }
}
```

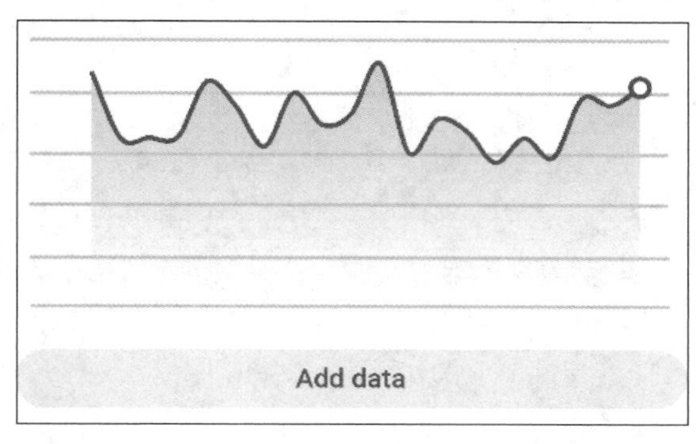

图 5-14　线性图

柱状图如图 5-15 所示，使用方法如代码示例 5-26 所示。

代码示例 5-26　柱状图

```
<!-- xxx.hml -->
<div class="container">
<stack class="data-region">
<image class="data-background" src="common/background.png"></image>
<chart class="data-bar" type="bar" id="bar-chart" options="{{barOps}}" datasets=
"{{barData}}"></chart>
</stack>
</div>
/* xxx.css */
.container {
  flex-direction: column;
  justify-content: center;
  align-items: center;
}
.data-region {
  height: 400px;
  width: 700px;
}
.data-background {
  object-fit: fill;
}
```

```
.data-bar {
  width: 700px;
  height: 400px;
}
//xxx.js
export default {
  data: {
    barData: [
      {
        fillColor: '#f07826',
        data: [763, 550, 551, 554, 731, 654, 525, 696, 595, 628],
      },
      {
        fillColor: '#cce5ff',
        data: [535, 776, 615, 444, 694, 785, 677, 609, 562, 410],
      },
      {
        fillColor: '#ff88bb',
        data: [673, 500, 574, 483, 702, 583, 437, 506, 693, 657],
      },
    ],
    barOps: {
      xAxis: {
        min: 0,
        max: 20,
        display: false,
        axisTick: 10
      },
      yAxis: {
        min: 0,
        max: 1000,
        display: false
      },
    },
  }
}
```

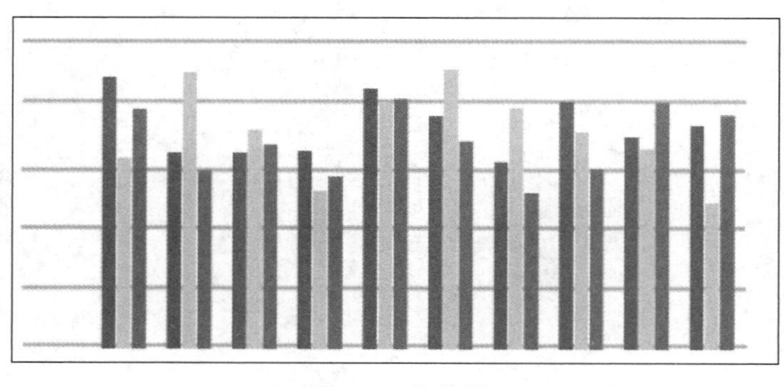

图 5-15　柱状图

量规图如图 5-16 所示，使用方法如代码示例 5-27 所示。

代码示例 5-27 量规图

```
<!-- xxx.hml -->
<div class = "container">
<div class = "gauge - region">
<chart class = "data - gauge" type = "gauge" percent = "50"></chart>
</div>
</div>
/* xxx.css */
.container {
  flex - direction: column;
  justify - content: center;
  align - items: center;
}
.gauge - region {
  height: 400px;
  width: 400px;
}
.data - gauge {
  colors: #83f115, #fd3636, #3bf8ff;
  weights: 4, 2, 1;
}
```

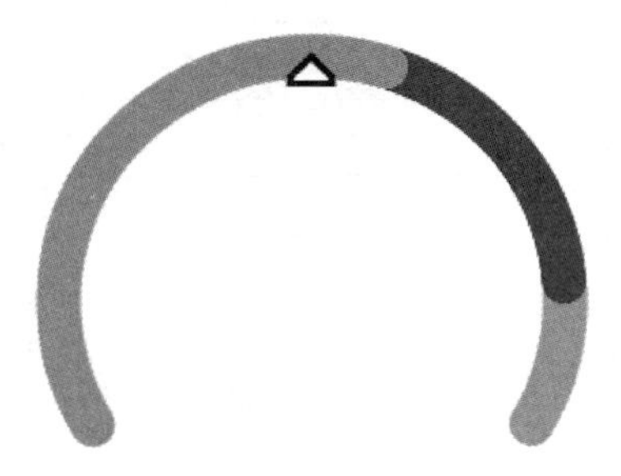

图 5-16 量规图

5.4.2 媒体组件

视频播放组件，本组件需要访问网络，因此需要申请 ohos.permission.INTERNET 权限(如果使用云端路径)，用法如代码示例 5-28 所示。

代码示例 5-28 Video 组件

```
<video ref = "avplayer" controls = "false"
autoplay = "true"
            style = "width: 100 % ;height: 100 % ;
```

```
object - fit: cover;"
          onclick = "resumePlayer"
src = "/res/video/driver.mp4"
          onFinish = "pausePlayer">
</video>
```

5.4.3 画布组件

画布组件用于自定义绘制图形。组件的设备支持情况如图 5-17 所示。

手机	平板	智慧屏	智能穿戴
支持	支持	支持	支持

图 5-17 canvas 对设备的支持情况

注意：canvas 对象的获取，只能在页面生命周期 onShow()中获取 Context。

canvas 画布的方法如表 5-1 所示，需要注意的是 getContext 方法不支持在 onInit 和 onReady 中进行调用，只能在 onShow 方法中调用。

表 5-1 canvas 方法

名　称	参　数	描　述
getContext	string	不支持在 onInit 和 onReady 中进行调用

下面介绍一下 canvas 的用法：

```
<canvas id = "board" class = "board"></canvas>
```

在 onShow 方法中获取 Context 对象：

```
var heroCanvas  =  this. $ element("heroCanvas")
this.heroCxt  =  heroCanvas.getContext("2d")
```

下面通过一个自由涂鸦画板的例子，了解 canvas 的使用，canvas 实现涂鸦效果，通过在 canvas 上绑定 ontouchend、ontouchstart、ontouchmove 这 3 个监听事件，这些监听事件的对象可以获取监听的 touch 事件的坐标信息，实现涂鸦功能，如代码示例 5-29 所示。

代码示例 5-29 涂鸦画板

```
<canvas id = "board" ontouchend = "paintEnd" ontouchstart = "painStart" ontouchmove = "paint"
class = "board"></canvas>
```

```
#逻辑实现
export default {
    data: {
        cxt: {}
    },
    onInit() {
    },
    onShow() {
        this.cxt = this.$element("board").getContext("2d");
    },
    painStart(e) {
        this.cxt.beginPath();
        this.cxt.strokeStyle = "white"
        this.cxt.lineWidth = 10
        this.cxt.lineCap = "round"
        this.cxt.lineJoin = "round"
        //绘制起点
        this.cxt.moveTo(e.touches[0].localX,e.touches[0].localY)
    },
    paint(e) {
        console.error(e.touches[0].localX);
        this.cxt.lineTo(e.touches[0].localX,e.touches[0].localY);
        this.cxt.stroke();
    },
    paintEnd() {
        this.cxt.closePath();
    },
}
```

5.5　自定义组件

自定义组件是用户根据业务需求,将已有的组件组合,封装成的新组件,可以在工程中多次调用,从而提高代码的可读性。自定义组件通过 element 引入宿主页面,使用方法如代码示例 5-30 所示。

代码示例 5-30　自定义组件的引入方式

```
<element name='comp' src='../../common/component/comp.hml'></element>
<div>
<comp prop1='xxxx' @child1="bindParentVmMethod"></comp>
</div>
```

(1) name 属性指自定义组件名称(非必填),组件名称对大小写不敏感,默认使用小写。src 属性指自定义组件 hml 文件路径(必填),若没有设置 name 属性,则默认使用 hml 文件名作为组件名。

(2) 事件绑定：自定义组件中绑定子组件事件使用(on|@)child1 语法，子组件通过 this. $emit('child1', { params: '传递参数' })触发事件并进行传值，父组件执行 bindParentVmMethod 方法并接收子组件传递的参数。

注意：子组件中使用驼峰命名法命名的事件，在父组件中进行绑定时需要使用短横线分隔命名形式，例如：@children-event 表示绑定子组件的 childrenEvent 事件，如 @children-event="bindParentVmMethod"。

自定义组件中的内置对象说明，如图 5-18 所示。

属性	类型	描述
data	Object/Function	页面的数据模型，类型是对象或者函数，如果类型是函数，则返回值必须是对象。属性名不能以$或_开头，不要使用保留字for、if、show、tid。 data与private和public不能重合使用
props	Array/Object	props用于组件之间的通信，可以通过<tag xxxx='value'>方式传递给组件；props名称必须用小写，不能以$或_开头，不要使用保留字for、if、show、tid。目前props的数据类型不支持Function
computed	Object	用于在读取或设置时进行预先处理，计算属性的结果会被缓存。计算属性名不能以$或_开头，不要使用保留字

图 5-18　自定义组件的内置对象说明

5.5.1　自定义组件的定义

注意：自定义组件与 page 中的组件结构是一样的，包含 hml、css、js 文件。

下面具体介绍创建组件的步骤。

步骤 1：在 common 目录常见公开组件，一个目录就是一个组件，每个组件分为三部分，index. hml 文件、index. css 文件和 index. js 文件，如图 5-19 所示。

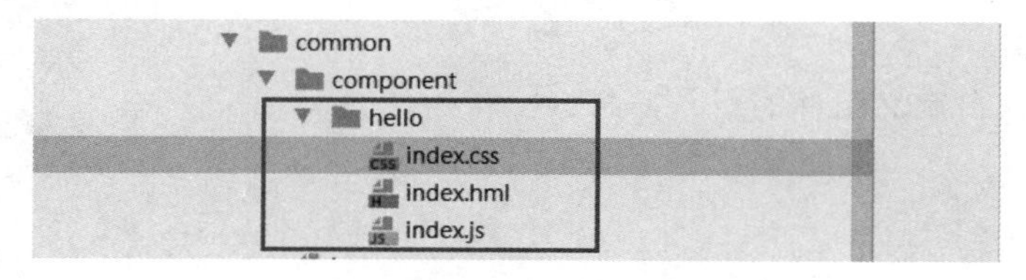

图 5-19　自定义组件

步骤 2：编写好自定义组件后，可通过 element 在需要引用的地方引入，如代码示例 5-31 所示。

代码示例 5-31 自定义组件的引入

```
<!-- 通过 element 引用自定义组件：name 组件名 -->
<element name = "hello"
        src = "../../common/component/hello/index.hml">
</element>

<div class = "container">
<hello></hello>                    //组件声明
</div>
```

5.5.2 自定义组件事件与交互

自定义组件创建好后，可以向一个组件添加输入和输出属性，如图 5-20 所示，这样就可以复用组件的逻辑了。

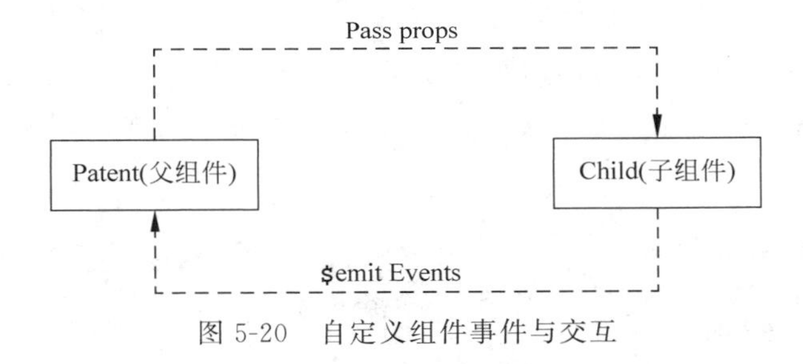

图 5-20 自定义组件事件与交互

自定义组件可以通过 props 声明属性，父组件通过设置属性向子组件传递参数。camelCase(驼峰命名法)的 prop 名，在外部父组件传递参数时需要使用 kebab-case（短横线分隔命名)形式，即当属性 compProp 在父组件引用时需要转换为 comp-prop。向自定义组件添加 props，通过父组件向下传递参数的示例如代码示例 5-32 所示。

代码示例 5-32 props 属性

```
<!-- comp.hml -->
<div class = "item">
<text class = "title - style">{{compProp}}</text>
</div>
//comp.js
 export default {
props: ['compProp'],
 }
<!-- xxx.hml -->
<element name = 'comp' src = '../../common/component/comp/comp.hml'></element>
<div class = "container">
<comp comp - prop = "{{title}}"></comp>
</div>
```

子组件可以通过固定值 default 设置默认值,当父组件没有设置该属性时,将使用其默认值。此情况下 props 属性必须为对象形式,不能使用数组形式,示例如代码示例 5-33 所示。

代码示例 5-33 props 默认值

```
<!-- comp.hml -->
<div class = "item">
<text class = "title - style">{{title}}</text>
</div>
//comp.js
 export default {
props: {
      title: {
        default: 'title',
      },
    },
 }
```

本示例中加入了一个 text 组件用于显示标题,标题的内容是一个自定义属性,用于显示用户设置的标题内容,当用户没有设置时显示默认值 title。在引用该组件时添加该属性的设置,代码如下:

```
<!-- xxx.hml -->
<element name = 'comp' src = '../../common/component/comp/comp.hml'></element>
<div class = "container">
<comp title = "自定义组件"></comp>
</div>
```

父子组件之间数据的传递是单向的,只能从父组件向子组件传递,子组件不能直接修改父组件传递下来的值,可以将 props 传入的值用 data 接收后作为默认值,再对 data 的值进行修改,如代码示例 5-34 所示。

代码示例 5-34 props 的单向传递

```
//comp.js
 export default {
props: ['defaultCount'],
   data() {
     return {
       count: this.defaultCount,
     };
   },
   onClick() {
     this.count = this.count + 1;
   },
 }
```

1. $ watch 监控数据改变

如果需要观察组件中属性的变化，可以通过 $ watch 方法增加属性变化回调。使用方法如代码示例 5-35 所示。

代码示例 5-35 $ watch

```
//comp.js
export default {
  props: ['title'],
  onInit() {
    this.$watch('title', 'onPropertyChange');
  },
  onPropertyChange(newV, oldV) {
    console.info('title 属性变化 ' + newV + '' + oldV);
  },
}
```

2. computed 计算属性

自定义组件中经常需要在读取或设置某个属性时进行预先处理，以便提高开发效率，此种情况需要使用 computed 字段。computed 字段中可通过设置属性的 getter 和 setter 方法在属性读写的时候进行触发，使用方式如代码示例 5-36 所示。

代码示例 5-36 computed 计算属性

```
//comp.js
export default {
  props: ['title'],
  data() {
    return {
      objTitle: this.title,
      time: 'Today',
    };
  },
  computed: {
    message() {
      return this.time + '' + this.objTitle;
    },
    notice: {
      get() {
        return this.time;
      },
      set(newValue) {
        this.time = newValue;
      },
    },
  },
```

```
    onClick() {
      console.info('get click event ' + this.message);
      this.notice = 'Tomorrow';
    },
  }
```

这里声明的第一个计算属性 message 默认只有 getter 函数，message 的值会取决于 objTitle 的值的变化。getter 函数只能读取而不能改变值的大小，当需要向计算属性赋值的时候可以提供一个 setter 函数，如示例中的 notice。

3. 组件的输出属性

输出属性定义在组件的声明标签上，格式如下：

```
<hello title = "hello world" @out = "callbackfunc"></hello>
```

注意：这里 title 是组件的输入属性，@out 是输出属性名，@out 的值是父组件中定义的方法，可以通过这个函数从组件内将数据传出去。

子组件也可以通过绑定的事件向上传递参数，在自定义事件中添加传递参数的代码如代码示例 5-37 所示。

代码示例 5-37　组件的输出属性

```
<!-- comp.hml -->
<div class = "item">
<text class = "text - style" onclick = "childClicked">单击这里查看隐藏文本</text>
<text class = "text - style" if = "{{showObj}}"> hello world</text>
</div>
//comp.js
 export default {
   childClicked () {
this.$emit('eventType1', {text: '收到子组件参数'});
      this.showObj = !this.showObj;
   },
 }
```

子组件向上传递参数 text，父组件接收时通过 e.detail 获取参数，如代码示例 5-38 所示。

代码示例 5-38　获取参数

```
<!-- xxx.hml -->
<div class = "container">
<text >父组件: {{text}}</text>
```

```
<comp @event-type1="textClicked"></comp>
</div>
//xxx.js
 export default {
   data: {
     text: '开始',
   },
   textClicked (e) {
     this.text = e.detail.text;
   },
 }
```

5.6 页面路由

鸿蒙 ACE JavaScript 框架的路由提供两种方式。

JavaScript 框架中创建一个 Ability，这个 Ability 管理一个 JavaScript Component 页面组件，一个 Component 下有多个页面组件，可实现这些页面组件间的跳转，还可以通过 JavaScript 的@system.router 模块进行导航。

我们可以创建多个 Ability，多个 Ability 间的路由跳转，可以通过 Java 中的 PA 进行跳转。

5.6.1 单页面路由

页面之间跳转是由系统接口通过 Ability 进行跳转的，所以它不是简单的页面 hash 路由，而是通过 FA 的方式进行路由跳转的。

注意：页面路由需要在页面渲染完成之后才能调用，在 onInit 和 onReady 生命周期中页面还处于渲染阶段，禁止调用页面路由方法。

下面具体介绍页面路由的使用步骤。

导入模块的代码如下：

```
import router from '@system.router';
```

1. 路由方法

router.push(OBJECT)跳转到应用内的指定页面，如代码示例 5-39 所示。

代码示例 5-39 router.push

```
//在当前页面中
router.push({
```

```
    uri: 'pages/routerpage2/routerpage2',
    params: {
        data1: 'message',
        data2: {
            data3: [123,456,789],
            data4: {
                data5: 'message'
            },
        },
    },
});
//在 routerpage2 页面中
console.info('showData1:' + this.data1);
console.info('showData3:' + this.data2.data3);
```

2. router.replace(OBJECT)

用应用内的某个页面替换当前页面，并销毁被替换的页面，如代码示例 5-40 所示。

代码示例 5-40 router.replace

```
//在当前页面中
router.replace({
  uri: 'pages/detail/detail',
  params: {
      data1: 'message',
  },
});
//在 detail 页面中
console.info('showData1:' + this.data1);
```

3. router.back(OBJECT)

返回上一页面或指定的页面，如代码示例 5-41 所示。

代码示例 5-41 router.back

```
//index 页面
router.push({
  uri: 'pages/detail/detail',
});

//detail 页面
router.push({
  uri: 'pages/mall/mall',
});

//mall 页面通过 back,将返回 detail 页面
```

```
router.back();
//detail 页面通过 back,将返回 index 页面
router.back();
//通过 back,返回 detail 页面
router.back({path:'pages/detail/detail'});
```

4. router.clear()

清空页面栈中的所有历史页面,仅保留当前页面作为栈顶页面,代码如下:

```
router.clear();
```

5. router.getLength()

获取当前页面栈内的页面数量,代码如下:

```
var size = router.getLength();
console.log('pages stack size = ' + size);
```

6. router.getState()

获取当前页面的状态信息,代码如下:

```
var page = router.getState();
console.log('current index = ' + page.index);
console.log('current name = ' + page.name);
console.log('current path = ' + page.path);
```

5.6.2 多页面路由

我们可以创建多个 Ability,多个 Ability 间的路由跳转,可以通过 Java 中的 PA 进行跳转。

5.7 应用 JavaScript 接口

通过原生系统接口,可以让界面层 JavaScript 获得访问系统的原生功能,例如:系统资源信息、分布式调度、设备管理等。本章详细介绍原生接口的使用方法和使用场景。

5.7.1 弹框

系统弹框有两种形式,一种是 Toast,另一种是 Dialog。下面具体介绍弹框的使用步骤。

步骤 1:导入模块。其代码如下:

```
import prompt from '@system.prompt';
```

步骤 2：使用 prompt.showToast(OBJECT)显示文本弹窗。其代码如下：

```
prompt.showToast({
  message: 'Message Info',
  duration: 2000,
});
```

prompt.showDialog(OBJECT) 在页面内显示对话框，如代码示例 5-42 所示。

代码示例 5-42　prompt.showDialog()

```
prompt.showDialog({
  title: 'Title Info',
  message: 'Message Info',
  buttons: [
    {
      text: 'button',
      color: '#666666',
    },
  ],
success: function(data) {
    console.log('dialog success callback,click button : ' + data.index);
 },
  cancel: function() {
    console.log('dialog cancel callback');
  },
});
```

5.7.2　网络访问

如果需要访问外部数据，则可以通过系统接口查询网络信息。

注意：需要开启权限 ohos.permission.INTERNET。默认支持 https，如果要支持 http，则需要在 config.json 里增加 network 标签，并设置属性标识 "cleartextTraffic"：true。

下面具体介绍网络访问的使用步骤。

步骤 1：导入模块。其代码如下：

```
import fetch from '@system.fetch';
```

步骤 2：配置权限，在 config.json 中添加权限，如代码示例 5-43 所示。

代码示例 5-43　配置权限

```
"module": {
"abilities": [],
"reqPermissions": [
      {
"name": "ohos.permission.INTERNET"
      }
]
}
```

步骤 3：配置 http 支持，如代码示例 5-44 所示。

代码示例 5-44　配置 http 支持

```
"deviceConfig": {
"default": {
"network": {
"usesCleartext": true
      }
    }
  },
```

步骤 4：调用 fetch.fetch(OBJECT)，通过网络获取数据，如代码示例 5-45 所示。

代码示例 5-45　fetch 方法

```
fetch.fetch({
  URL: 'http://www.path.com',
  responseType: 'text',
  success: function(response) {
    console.log('response code:' + response.code);
    console.log('response data:' + response.data);
  },
  fail: function(data, code) {
    console.log('fail callback');
  },
});
```

5.7.3　分布式迁移

分布式迁移提供了一个主动迁移接口及一系列页面生命周期回调，以支持将应用迁移到其他设备。如果迁移到的设备上已经运行该 FA，则生命周期 onNewRequest 将被回调。API 的设备支持情况，如表 5-2 所示。

表 5-2　API 的设备支持情况

API	手　机	平　板	智　慧　屏	智能穿戴
FeatureAbility. continueAbility()	支持	支持	支持	支持
onStartContinuation()	支持	支持	支持	支持
onSaveData(OBJECT)	支持	支持	支持	支持
onRestoreData(OBJECT)	支持	支持	支持	支持
onCompleteContinuation(code)	支持	支持	支持	支持

下面具体介绍分布式迁移的使用步骤。

步骤 1：申请分布式迁移权限，在 config. json 的"reqPermissions"添加以下权限申请。

```
ohos.permission.DISTRIBUTED_DATASYNC
```

步骤 2：通过 FeatureAbility 发起迁移。

FeatureAbility. continueAbility()为主动进行 FA 迁移的入口。

- 参数：无；
- 返回值：json 字符串，指示是否成功，字段如表 5-3 所示。

表 5-3　continueAbility 返回值

参数名	类型	非空	说　明
code	number	是	0：发起迁移成功 非 0：发起迁移失败，原因见 data
data	Object	是	成功：返回 null 失败：携带错误信息，类型为 String

下面通过伪代码示例展示了一个备忘录 FA 从发起到完成迁移的过程。发起迁移 FA 如代码示例 5-46 所示。

代码示例 5-46　发起迁移

```
const injectRef = Object.getPrototypeOf(global) || global
injectRef.regeneratorRunTime = require('@babel/RunTime/regenerator')
import prompt from '@system.prompt'

export default {
  data: {
    continueAbilityData: {
      remoteData1: 'self define continue data for distribute',
      remoteData2: {
        item1: 0,
        item2: true,
        item3: 'inner string'
      },
      remoteData3: [1, 2, 3]
```

```
        }
    },

    //shareData 的数据会在 onSaveData 触发时与 saveData 一起传送到迁移目标 FA,并绑定到其
//shareData 数据段上
    //shareData 的数据可以直接使用 this 访问.eg:this.remoteShareData1
    shareData: {
      remoteShareData1: 'share data for distribute',
      remoteShareData2: {
        item1: 0,
        item2: false,
        item3: 'inner string'
      },
      remoteShareData3: [4, 5, 6]
    },
    tryContinueAbility: async function() {
      //应用进行迁移
      let result = await FeatureAbility.continueAbility();
      console.info("result:" + JSON.stringify(result));
    },
    onStartContinuation: function onStartContinuation() {
      //判断当前的状态是否适合迁移
      console.info("trigger onStartContinuation");
      return true;
    },
    onCompleteContinuation: function onCompleteContinuation(code) {
      //迁移操作完成,code 返回结果
      console.info("trigger onCompleteContinuation: code = " + code);
    },
    onSaveData: function onSaveData(saveData) {
      //将数据保存到 savedData 中进行迁移
      var data = this.continueAbilityData;
      Object.assign(saveData, data)
    }
}
```

迁移到 FA 后，会调用 onRestoreData 方法，恢复缓存的数据 restoreData，代码如下：

```
onRestoreData: function onRestoreData(restoreData) {
  //收到迁移数据,恢复
  this.continueAbilityData = restoreData;
}
```

下面看一看这些迁移过程中的回调方法的用法和作用。

1. onStartContinuation()

FA 发起迁移时的回调,在此回调中应用可以根据当前状态决定是否迁移。

- 参数:无;
- 返回值:Boolean 类型,true 表示允许进行迁移,false 表示不允许迁移。

2. onSaveData(OBJECT)

保存状态数据的回调,开发者需要向参数对象中填入需迁移到目标设备上的数据。参数如表 5-4 所示,无返回值。

表 5-4 saveData 参数

参数名	类型	必填	说明
savedData	Object	是	出参,可以往其中填入可被序列化的自定义数据

说明:shareData 中的数据在迁移时,会自动迁移到目标设备上。

3. onRestoreData(OBJECT)

恢复发起迁移时 onSaveData 方法所保存的数据的回调。参数如表 5-5 所示,无返回值。

表 5-5 onRestoreData 参数

参数名	类型	必填	说明
restoreData	Object	是	用于恢复应用状态的对象,其中的数据及结构由 onSaveData 决定

4. onCompleteContinuation(code)

迁移完成的回调,在调用端被触发,表示应用迁移到目标设备上的结果。参数如表 5-6 所示,无返回值。

表 5-6 onCompleteContinuation 返回参数说明

参数名	类型	必填	说明
code	number	是	迁移完成的结果。0:成功;-1:失败

注意:在 HarmonyOS 中,分布式任务调度平台对搭建 HarmonyOS 的多设备所构建的"超级虚拟设备终端"提供统一的组件管理能力,为应用定义统一的能力基础、接口形式、数据结构、服务描述语言、屏蔽硬件的差异、支持远程启动、远程调用、业务无缝迁移等分布式任务。

在后面章节中会详细讲解分布式任务调度的一些其他方法。

5.8 系统 JavaScript 接口

系统能力为 JavaScript 调用操作系统底层数据提供接口，通过系统提供的 API 让 JavaScript 可以通过这些接口获取地理位置、传感器、系统应用等相关信息。

5.8.1 消息通知

消息通知提供系统底层的通知能力，调用起来也非常简单，首先导入@system.notification，代码如下：

```
import notification from '@system.notification';
```

notification.show(OBJECT)用于显示通知，如代码示例 5-47 所示。

代码示例 5-47　消息通知

```
notification.show({
  contentTitle: 'title info',
  contentText: 'text',
  clickAction: {
    bundleName: 'com.huawei.testapp',
    abilityName: 'notificationDemo',
    uri: '/path/to/notification',
  },
});
```

5.8.2 地理位置

获取地理位置信息，同样需要用到系统底层能力，API 对设备的支持情况如表 5-7 所示。

表 5-7　API 对设备的支持情况

API	手　机	平　板	智　慧　屏	智能穿戴
geolocation.getLocation	支持	支持	支持	支持
geolocation.getLocationType	支持	支持	支持	支持
geolocation.subscribe	支持	支持	支持	支持
geolocation.unsubscribe	支持	支持	支持	支持
geolocation.getSupportedCoordTypes	支持	支持	支持	支持

注意：地理位置的获取，需要申请权限 ohos.permission.LOCATION。

步骤 1：导入地理位置模块。其代码如下：

```
import geolocation from '@system.geolocation';
```

步骤 2：使用 geolocation. getLocation(OBJECT)获取设备的地理位置，参数如表 5-8 所示。

表 5-8 方法参数

参数名	类型	必填	说明
timeout	number	否	超时时间，单位为 ms，默认值为 30000ms。 设置超时是为了防止出现权限被系统拒绝、定位信号弱或者定位设置不当，而导致请求阻塞的情况。超时后会使用 fail 回调函数。 取值范围为 32 位正整数。如果设置值小于或等于 0，则系统按默认值处理
coordType	string	否	坐标系的类型，可通过 getSupportedCoordTypes 获取可选值，缺省值为 wgs84
success	Function	否	接口调用成功的回调函数
fail	Function	否	接口调用失败的回调函数
complete	Function	否	接口调用结束的回调函数

success 返回值，如表 5-9 所示。

表 5-9 success 返回值

参 数 名	类 型	说 明
longitude	number	设备位置信息：经度
latitude	number	设备位置信息：纬度
altitude	number	设备位置信息：海拔
accuracy	number	设备位置信息：精确度
time	number	设备位置信息：时间

fail 返回错误代码，如表 5-10 所示。

表 5-10 fail 返回值

错 误 码	说 明
601	获取定位权限失败，失败原因：用户拒绝
602	权限未声明
800	超时
801	系统位置开关未打开
802	该次调用结果未返回，前接口又被重新调用，该次调用失败并返回错误码

获取地理位置信息，如代码示例5-48所示。

代码示例5-48 地理位置信息

```
geolocation.getLocation({
  success: function(data) {
    console.log('success get location data. latitude:' + data.latitude);
  },
  fail: function(data, code) {
    console.log('fail to get location. code:' + code + ', data:' + data);
  },
});
```

1. geolocation.getLocationType(OBJECT)

获取当前设备支持的定位类型。Object参数如表5-11所示，如代码示例4-49所示。

表5-11 Object参数

参数名	类型	必填	说明
success	Function	否	接口调用成功的回调函数
fail	Function	否	接口调用失败的回调函数
complete	Function	否	接口调用结束的回调函数

success返回值，如表5-12所示。

表5-12 success返回值

参数名	类型	说明
types	Array<string>	可选的定位类型['gps', 'network']

代码示例5-49 获取当前设备支持的定位类型

```
geolocation.getLocationType({
  success: function(data) {
    console.log('success get location type:' + data.types[0]);
  },
  fail: function(data, code) {
    console.log('fail to get location. code:' + code + ', data:' + data);
  },
});
```

2. geolocation.subscribe(OBJECT)

订阅设备的地理位置信息。如果多次调用，则只有最后一次的调用生效。Object参数如表5-13所示，调用方法如代码示例5-50所示。

表 5-13 Object 参数

参数名	类型	必填	说明
coordType	string	否	坐标系的类型，可通过 getSupportedCoordTypes 获取可选值，默认值为 wgs84
success	Function	是	位置信息发生变化的回调函数
fail	Function	否	接口调用失败的回调函数

代码示例 5-50 订阅设备的地理位置信息

```
geolocation.subscribe({
  success: function(data) {
    console.log('get location. latitude:' + data.latitude);
  },
  fail: function(data, code) {
    console.log('fail to get location. code:' + code + ', data:' + data);
  },
});
```

geolocation.unsubscribe()取消订阅设备的地理位置信息，代码如下：

```
geolocation.unsubscribe();
```

success 返回值，如表 5-14 所示。

表 5-14 success 返回值

参数名	类型	说明
longitude	number	设备位置信息：经度
latitude	number	设备位置信息：纬度
altitude	number	设备位置信息：海拔
accuracy	number	设备位置信息：精确度
time	number	设备位置信息：时间

fail 返回错误代码，如表 5-15 所示。

表 5-15 fail 返回值

错误码	说明
601	获取定位权限失败，失败原因：用户拒绝
602	权限未声明
801	系统位置开关未打开

3. geolocation.getSupportedCoordTypes()

获取设备支持的坐标系类型。

■ 返回值;
■ 字符串数组,表示坐标系类型,如[wgs84, gcj02]。

```
var types = geolocation.getSupportedCoordTypes();
```

5.8.3　设备信息

设备信息的获取,对于开发鸿蒙应用来讲,非常重要,这里首先需要导入@system.device模块,注意,获取device信息时需要在onShow方法中调用。

```
import device from '@system.device';
```

获取设备方法,如代码示例5-51所示。

代码示例5-51　设备信息

```
device.getInfo({
  success: function(data) {
    console.log('success get device info brand:' + data.brand);
  },
  fail: function(data, code) {
    console.log('fail get device info code:'+ code + ', data: ' + data);
  },
});
```

5.8.4　应用管理

对于已经安装的HAP信息,可以通过导入@system.package模块来获取相关应用的信息,这里需要申请权限才可以使用,代码如下:

```
import pkg from '@system.package';
```

注意: 地理位置的获取,需要申请权限ohos.permission.GET_BUNDLE_INFO。

应用管理方法使用如代码示例5-52所示。

代码示例5-52　应用管理

```
pkg.hasInstalled({
  bundleName: 'com.example.bundlename',
  success: function(data) {
```

```
    console.log('package has installed: ' + data);
  },
  fail: function(data, code) {
    console.log('query package fail, code: ' + code + ', data: ' + data);
  },
});
```

5.8.5 媒体查询

鸿蒙 JavaScript 应用框架提供了在 CSS 的媒体查询，同时也可以通过系统接口的方式获取媒体查询信息，这样方便开发者在代码中通过媒体查询进行不同类型媒体的操作。这里首先需要导入@system.mediaquery 模块，代码如下：

```
import mediaquery from '@system.mediaquery';
```

媒体查询提供匹配媒体的方法及监听回调的方法如下。

1. mediaquery.matchMedia(condition)

根据媒体查询条件，创建 MediaQueryList 对象，代码如下：

```
var mMediaQueryList = mediaquery.matchMedia('(max-width: 466)');
```

2. MediaQueryList.addListener(OBJECT)

向 MediaQueryList 添加回调函数，回调函数应在 onShow 生命周期之前添加，即需要在 onInit 或 onReady 生命周期里添加。

媒体查询需要注意 this 指向的问题，如果需要访问当前实例，则应尽量使用箭头函数。实现方法如代码示例 5-53 所示。

代码示例 5-53 媒体查询

```
import mediaquery from '@system.mediaquery';
export default {
  onReady() {
    var mMediaQueryList = mediaquery.matchMedia('(max-width: 466)');
    function maxWidthMatch(e) {
      if (e.matches) {
        //do something
        //这里无法访问外部 this
      }
    }
mMediaQueryList.addListener(maxWidthMatch);  //这样写无法访问外部 this
mMediaQueryList.addListener(e =>{
      //这里访问外部 this
```

```
})
  },
}
```

3. MediaQueryList.removeListener(OBJECT)

移除 MediaQueryList 中的回调函数，代码如下：

```
query.removeListener(minWidthMatch);
```

5.8.6　振动

通过系统接口提供的振动控制，API 的设备支持情况如表 5-16 所示。注意地理位置的获取，需要申请权限 ohos.permission.VIBRATE。

表 5-16　API 的设备支持情况

API	手　机	平　板	智　慧　屏	智能穿戴
vibrator.vibrate	支持	支持	不支持	支持

首先需要导入@system.vibrator 模块，代码如下：

```
import vibrator from '@system.vibrator';
```

vibrator.vibrate(OBJECT)OBJECT 参数如表 5-17 所示。

表 5-17　Object 参数

参　数　名	类　型	必　填	说　明
mode	string	否	振动的模式，其中 long 表示长振动，short 表示短振动，默认值为 long

振动调用的方法如下：

```
vibrator.vibrate({
mode: 'short',
});
```

5.8.7　应用配置

获取应用当前的语言和地区信息可通过@system.configuration 模块获取，代码如下：

```
import configuration from '@system.configuration';
```

configuration. getLocale()获取应用当前的语言和地区。默认与系统的语言和地区同步。返回值对象,如表 5-18 所示。

表 5-18 方法返回值

参数名	类型	说明
language	string	语言。例如:zh
countryOrRegion	string	国家或地区。例如:CN
dir	string	文字布局方向。取值范围: ltr:从左到右; rtl:从右到左

获取应用配置信息,如代码示例 5-54 所示。

代码示例 5-54 获取应用配置信息

```
const localeInfo = configuration.getLocale();
console.info(localeInfo.language);
```

5.9 多实例接口

ACE JavaScript UI 框架支持多 Ability 实例管理,不同 Ability 实例可绑定不同窗口实例,并能指定不同 JavaScript Component 入口,运行互不影响。不同 Ability 实例的实例名互不相同,在编写应用时,开发人员需调用接口设置实例名。

说明:多实例接口应用通过 AceAbility 类中 setInstanceName()接口设置该实例的实例名称,需要在 super. onStart(Intent)前调用此接口。注意:多实例应用的 module. js 字段中有多个实例项,使用时需选择当前实例对应的项。

实例名称与应用配置文件(config. json)中 module. js 数组下对象的 name 的值对应,若上述字段的值为{实例名称},则需要在应用 Ability 实例的 onStart 方法中调用此接口并将实例名设置为{实例名称}。若用户未修改实例名,而使用了默认值 default,则无须调用此接口。

通过 DevEco Studio 在 src/main/js 目录创建 JavaScript Component,如图 5-21 所示。

可以创建多个 Page Ability,通过 setInstanceName 绑定所创建的 JavaScript Component 的名字。这样可以在一个应用中有多个 Page Ability,通过这个方式,可以实现轻应用的开发,如代码示例 5-55 所示。

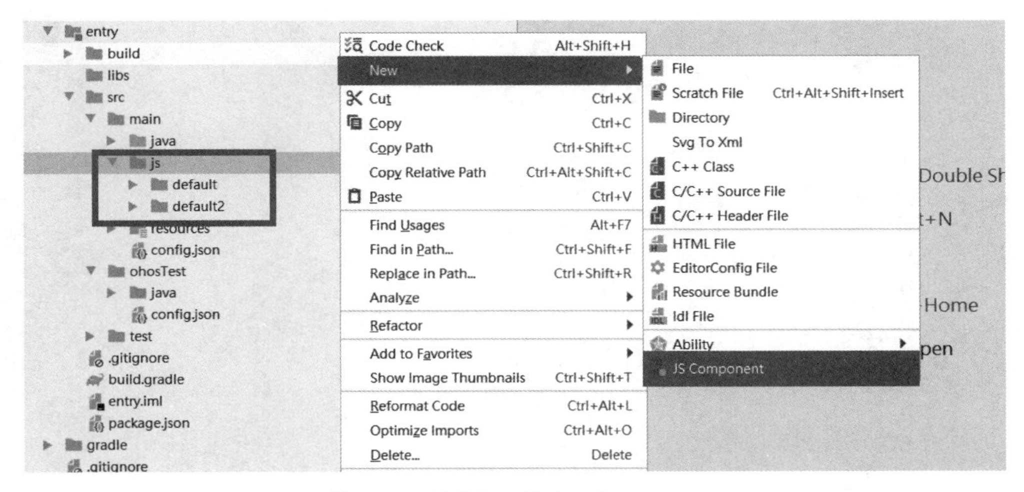

图 5-21　创建 JavaScript Component

代码示例 5-55　通过 setInstanceName 绑定所创建的 JavaScript Component 的名字

```
public class MainAbility extends AceAbility {
    @Override
    public void onStart(Intent intent) {
      //config.json 配置文件中 ability.js.name 的标签值
  setInstanceName("JSComponentName");
super.onStart(intent);
    }
}
```

通过 setInstanceName 绑定已创建好的 JavaScript Component，同时可以创建多个 Page Ability。通过多个 Page Ability 绑定不同的 JavaScript Component 可以实现把一个较大的项目，分为多个小的安装项目，这将在后面的游戏场景中用到。

5.10　本章小结

本章是鸿蒙应用开发的基础知识篇，鸿蒙 ACE JavaScript 框架提供了面向 IoT 设备开发的基础框架。本章通过 9 节，分别介绍了 ACE JavaScript 应用框架的语法，以及接口调用。ACE JavaScript 框架适用于富设备，如手机、TV、手表等，同时也适用于安装了开源鸿蒙系统的 IoT 设备开发，如轻设备、智能终端等。ACE JavaScript 的部分功能目前在开源鸿蒙操作系统的 IoT 设备是无法使用的。

第三篇　分布式开发篇

学习目标

通过本篇的学习，你将学习到鸿蒙的分布式调度、分布式数据服务，以及分布式文件系统的使用，并对鸿蒙的高级用法有更多了解。

主要内容如下：

- 鸿蒙分布式任务调度；
- 鸿蒙分布式数据服务；
- 鸿蒙的分布式文件服务。

如何学习本篇：

- 学习本章的读者可结合每节中的案例反复编码练习；
- 可以通过学习分布式内容，把之前非分布式功能改造为鸿蒙分布式支持的多功能。

第 6 章

鸿蒙分布式任务调度

6.1 分布式任务调度

在 HarmonyOS 中，分布式任务调度平台对搭载 HarmonyOS 的多设备所构筑的“超级虚拟终端”提供统一的组件管理能力，为应用定义统一的能力基线、接口形式、数据结构、服务描述语言等，并且屏蔽硬件差异；HarmonyOS 支持远程启动、远程调用、业务无缝迁移等分布式任务。

6.1.1 分布式任务调度介绍

(1) 分布式任务调度平台，如图 6-1 所示，在底层实现 Ability(分布式任务调度的基本组件)跨设备的启动/关闭、连接及断开连接、迁移等能力，实现跨设备的组件管理。

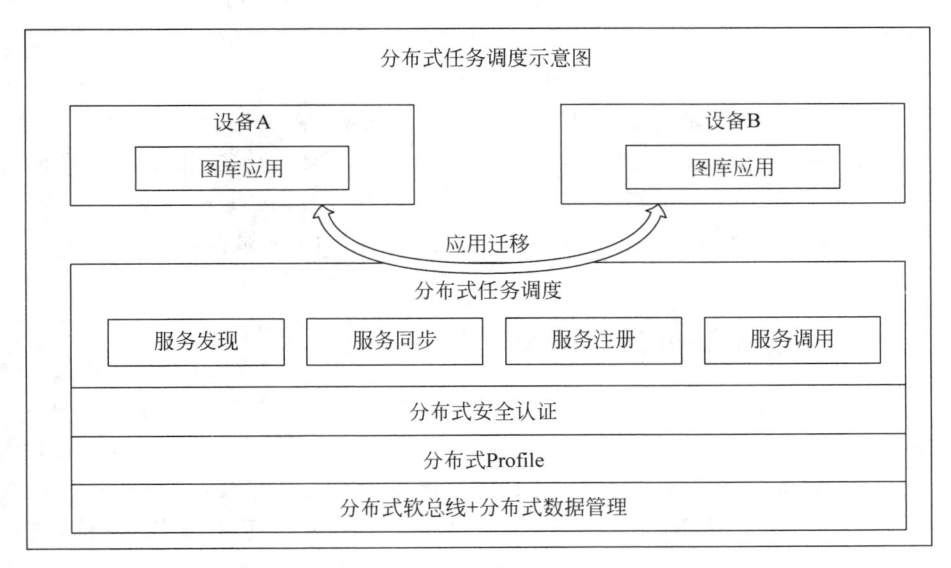

图 6-1 分布式服务调度

（2）启动和关闭：向开发者提供管理远程 Ability 的能力，即支持启动 Page 模板的 Ability，以及启动、关闭 Service 和 Data 模板的 Ability。

（3）连接和断开连接：向开发者提供跨设备控制服务（Service 和 Data 模板的 Ability）的能力，开发者可以通过与远程服务连接及断开连接实现获取或注销跨设备管理服务的对象，以便达到和本地一致的服务调度。

（4）迁移能力：向开发者提供跨设备业务的无缝迁移能力，开发者可以通过调用 Page 模板 Ability 的迁移接口，将本地业务无缝迁移到指定设备中，从而打通设备间壁垒。

6.1.2 分布式任务调度约束与限制

使用鸿蒙分布式任务调度功能，需要注意以下几点。

（1）开发者需要在 Intent 中设置支持分布式的标记（例如：Intent. FLAG_ABILITYSLICE_MULTI_DEVICE 表示该应用支持分布式调度），否则将无法获得分布式能力。权限弹出框如图 6-2 所示。

是否允许"MyApplication"多设备协同访问？
设备间数据交换。
禁止 始终允许

图 6-2 权限弹出窗口

（2）开发者可以向 config. json 中的 reqPermissions 字段添加多设备协同访问的权限申请：第三方应用使用｛"name"："ohos. permission. DISTRIBUTED_DATASYNC"｝。

（3）PA（Particle Ability，Service 和 Data 模板的 Ability）的调用支持连接及断开连接、启动及关闭这四类行为，在进行调度时：

- 开发者必须在 Intent 中指定 PA 对应的 bundleName 和 abilityName；
- 当开发者需要跨设备启动、关闭或连接 PA 时，需要在 Intent 中指定对端设备的 deviceId。开发者可通过如设备管理类 DeviceManager 提供的 getDeviceList 获取指定条件下匿名化处理的设备列表，实现对指定设备 PA 的启动/关闭及连接管理。

（4）FA（Feature Ability，Page 模板的 Ability）的调用支持启动和迁移行为，在进行调度时：

- 需要开发者在 Intent 中指定对端设备的 deviceId、bundleName 和 abilityName；
- FA 的迁移实现相同 bundleName 和 abilityName 的 FA 跨设备迁移，因此需要指定迁移设备的 deviceId。

6.1.3 分布式调度场景介绍

开发者在应用中集成分布式调度能力，如图 6-3 所示，通过调用指定能力的分布式接口，实现跨设备能力调度。根据 Ability 模板及意图的不同，分布式任务调度向开发者提供以下 6 种能力：启动远程 FA、启动远程 PA、关闭远程 PA、连接远程 PA、断开连接远程 PA 和 FA 跨设备迁移。下面以设备 A（本地设备）和设备 B（远程设备）为例，进行场景介绍。

设备 A 启动设备 B 的 FA：在设备 A 上通过本地应用提供的启动按钮，启动设备 B 上对

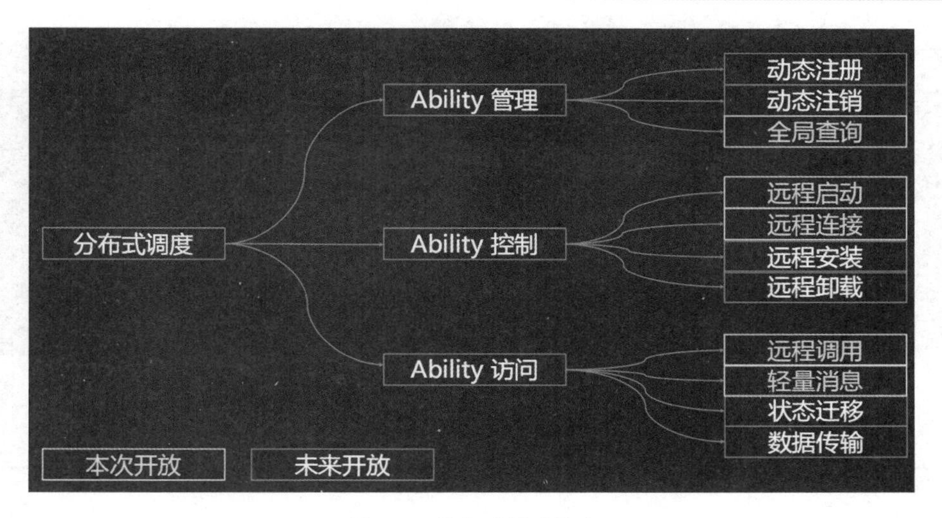

图 6-3 分布式调度能力

应的 FA。例如,设备 A 控制设备 B 打开相册,只需开发者在启动 FA 时指定打开相册的意图。

设备 A 启动设备 B 的 PA:在设备 A 上通过本地应用提供的启动按钮,启动设备 B 上指定的 PA。例如:开发者在启动远程服务时通过意图指定音乐播放服务,即可实现设备 A 启动设备 B 音乐播放的能力。

设备 A 关闭设备 B 的 PA:在设备 A 上通过本地应用提供的关闭按钮,关闭设备 B 上指定的 PA。类似启动的过程,开发者在关闭远程服务时通过意图指定音乐播放服务,即可实现关闭设备 B 上该服务的能力。

设备 A 连接设备 B 的 PA:在设备 A 上通过本地应用提供的连接按钮,连接设备 B 上指定的 PA。连接后,通过其他相关功能按钮实现控制对端 PA 的能力。通过连接关系,开发者可以实现跨设备的同步服务调度,实现如大型计算任务互助等价值场景。

设备 A 与设备 B 的 PA 断开连接:在设备 A 上通过本地应用提供断开连接的按钮,将之前已连接的 PA 断开连接。

设备 A 的 FA 迁移至设备 B:设备 A 通过本地应用提供的迁移按钮,将设备 A 的业务无缝迁移到设备 B 中。通过业务迁移能力,打通设备 A 和设备 B 间的壁垒,实现如文档跨设备编辑、视频从客厅到房间跨设备接续播放等场景。

6.1.4 分布式调度接口说明

分布式调度平台提供的连接和断开连接 PA、启动远程 FA、启动和关闭 PA 及迁移 FA 的能力,是实现更多价值性场景的基础。

1. 连接远程 PA

connectAbility(Intent intent, IAbilityConnection conn)接口提供连接指定设备上 PA 的能力,如图 6-4 所示,Intent 中指定待连接 PA 的设备 deviceId、bundleName 和 abilityName。当连接成功后,通过在 conn 定义的 onAbilityConnectDone 回调中获取对端 PA 的服务代

理,两者的连接关系则由 conn 维护,具体的参数定义如表 6-1 所示。

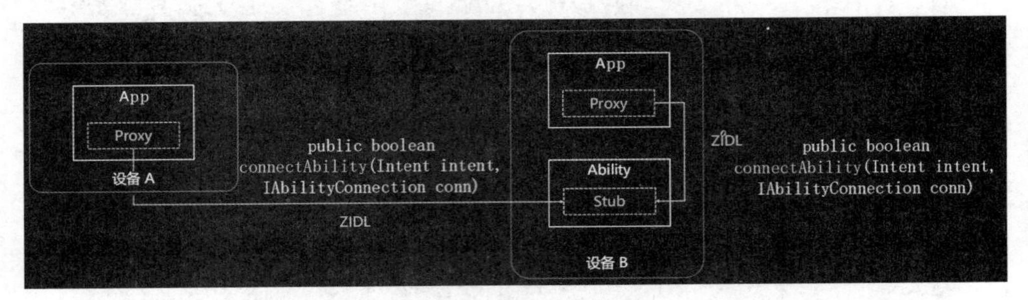

图 6-4 提供连接到指定设备上 PA 的能力

表 6-1 connectAbility 参数表

参数名	类 型	说 明
intent	ohos. aafwk. content. Intent	开发者需要在 intent 对应的 Operation 中指定待连接 PA 的设备 deviceId、bundleName 和 abilityName
conn	ohos. aafwk. ability. IAbilityConnection	当连接成功或失败时,作为连接关系的回调接口。该接口提供连接完成和断开连接完成时的处理逻辑,开发者可根据具体的场景进行定义

2. 启动远程 FA/PA

startAbility(Intent intent)接口提供启动指定设备上 FA 和 PA 的能力,如图 6-5 所示。Intent 中指定待启动 FA/PA 的设备 deviceId、bundleName 和 abilityName,具体参数定义如表 6-2 所示。

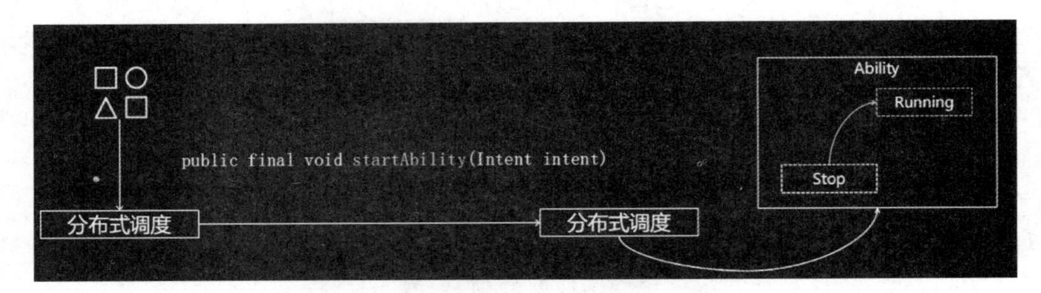

图 6-5 startAbility 调度

表 6-2 startAbility 参数表

参数名	类 型	说 明
intent	ohos. aafwk. content. Intent	当开发者需要调用该接口启动远程 PA 时,需要指定待启动 PA 的设备 deviceId、bundleName 和 abilityName。若不指定设备 deviceId,则无法跨设备调用 PA。 类似地,在启动 FA 时,也需要开发者指定启动 FA 的设备 deviceId、bundleName 和 abilityName

分布式调度平台还会提供与上述功能相对应的断开远程 PA 的连接和关闭远程 PA 的接口，相关的参数与连接、启动的接口类似。

断开远程 PA 连接：disconnectAbility(IAbilityConnection conn)。

关闭远程 PA：stopAbility(Intent intent)。

3. 迁移 FA

continueAbility(String deviceId)接口提供将本地 FA 迁移到指定设备上的能力，需要开发者在调用时指定目标设备的 deviceId，具体参数定义如表 6-3 所示。

表 6-3　deviceId 说明

参数名	类型	说　明
deviceId	String	当开发者需要调用该接口将本地 FA 迁移时，需要指定目标设备的 deviceId

Ability 和 AbilitySlice 类均需要实现 IAbilityContinuation 及其方法，这样才可以实现 FA 迁移。

6.2 实现跨设备打开 FA

打开或者关闭远程设备的 FA，需要满足几个条件。

(1) 使用同一个账号登录多台设备，如图 6-6 所示。

图 6-6　使用同一个账号登录多台设备

(2) 需要同时在多台设备上安装相同应用,如图 6-7 所示。

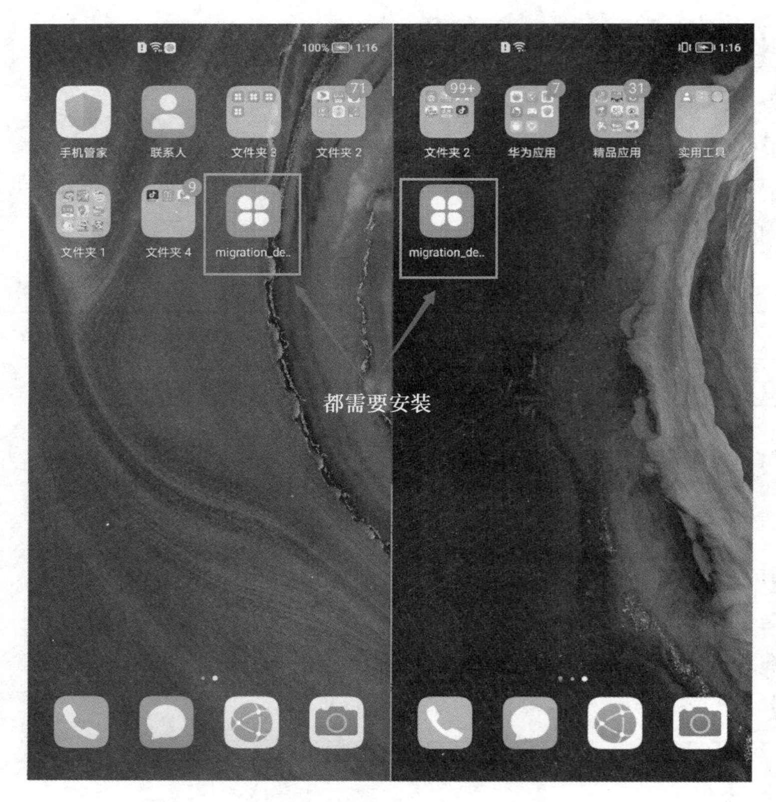

图 6-7 需要同时在多台设备上安装相同应用

(3) 开启权限: 涉及对在线组网设备的查询,该项能力需要开发者在对应的 config.json 中声明获取设备列表及设备信息的权限,代码如下:

```
{
"reqPermissions": [
{ "name": "ohos.permission.DISTRIBUTED_DATASYNC"}
        { "name": "ohos.permission.DISTRIBUTED_DEVICE_STATE_CHANGE"},
        { "name": "ohos.permission.GET_DISTRIBUTED_DEVICE_INFO" },
        { "name": "ohos.permission.GET_BUNDLE_INFO"}
    ]
}
```

使用分布式能力要求开发者在 Ability 对应的 config.json 中声明多设备协同访问的权限: 第三方应用使用{"name": "ohos.permission.DISTRIBUTED_DATASYNC"},如图 6-8 所示。

(4) 获取在线设备信息,如代码示例 6-1 所示。

```
public class MainAbility extends Ability {
    @Override
    public void onStart(Intent intent) {
        requestPermissionsFromUser(new String[]{
                "ohos.permission.DISTRIBUTED_DATASYNC",
                "ohos.permission.servicebus.BIND_SERVICE",
                "ohos.permission.servicebus.ACCESS_SERVICE"
        }, requestCode: 0);
        super.onStart(intent);
        super.setMainRoute(MainAbilitySlice.class.getName());
    }
}
```

图 6-8 声明多设备协同访问的权限

代码示例 6-1 获取在线设备信息

```
//调用 DeviceManager 的 getDeviceList 接口,通过 FLAG_GET_ONLINE_DEVICE 标记获得在线设备列表
List<DeviceInfo> onlineDevices = DeviceManager.getDeviceList(DeviceInfo.FLAG_GET_ONLINE_
DEVICE);
//判断组网设备是否为空
if (onlineDevices.isEmpty()) {
    return null;
}
int numDevices = onlineDevices.size();
ArrayList<String> deviceIds = new ArrayList<>(numDevices);
ArrayList<String> deviceNames = new ArrayList<>(numDevices);
onlineDevices.forEach((device) ->{
    deviceIds.add(device.getDeviceId());
    deviceNames.add(device.getDeviceName());
});
//以选择首个设备作为目标设备为例
//开发者也可按照具体场景,通过别的方式进行设备选择
String selectDeviceId = deviceIds.get(0);
return selectDeviceId;
```

(5)根据 deviceId 连接打开远程设备 FA,如代码示例 6-2 所示。

代码示例 6-2 打开远程设备 FA

```
Intent intent1 = new Intent();
Operation operation = new Intent.OperationBuilder()
        .withDeviceId(device_id)
        .withBundleName("com.charjedu.migration_demo")
        .withAbilityName("com.charjedu.migration_demo.NewsAbility")
        .withFlags(Intent.FLAG_ABILITYSLICE_MULTI_DEVICE)
        .build();
intent1.setOperation(operation);
```

```
//通过 AbilitySlice 包含的 startAbility 接口实现跨设备启动 FA
startAbility(intent1);
```

启动与关闭的行为类似，开发者只需要在 Intent 中指定待调度 PA 的 deviceId、bundleName 和 abilityName，并以 operation 的形式封装到 Intent 内。通过 AbilitySlice（Ability）包含的 startAbility()和 stopAbility()接口即可实现相应功能。

6.3 实现跨设备 FA 迁移

FA 迁移可以打通设备间的壁垒，有助于不同能力的设备进行互助。

跨设备迁移（下文简称“迁移”）支持将 Page 在同一用户的不同设备间迁移，以便支持用户对无缝切换的诉求。以 Page 从设备 A 迁移到设备 B 为例，迁移动作主要步骤如下：

（1）设备 A 上的 Page 请求迁移。

（2）HarmonyOS 处理迁移任务，并回调设备 A 上 Page 的保存数据方法，用于保存迁移必需的数据。

（3）HarmonyOS 在设备 B 上启动同一个 Page，并回调其恢复数据方法。

注意：一个应用可能包含多个 Page，仅需要在支持迁移的 Page 中通过以下方法实现 IAbilityContinuation 接口。同时，此 Page 所包含的所有 AbilitySlice 也需要实现此接口。

---

1. 实现 IAbilityContinuation 接口

onStartContinuation()：Page 请求迁移后，系统首先回调此方法，开发者可以在此回调中决策当前是否可以执行迁移，例如，弹框让用户确认是否开始迁移。

onSaveData()：如果 onStartContinuation()返回值为 true，则系统回调此方法，开发者在此回调中保存必须传递到另外设备上以便恢复 Page 状态的数据。

onRestoreData()：源侧设备上 Page 完成保存数据后，系统在目标侧设备上回调此方法，开发者在此回调中接收用于恢复 Page 状态的数据。注意，在目标侧设备上的 Page 会重新启动其生命周期，无论其启动模式如何配置，且系统回调此方法的时机在 onStart()之前。

onCompleteContinuation()：目标侧设备上恢复数据一旦完成，系统就会在源侧设备上回调 Page 的此方法，以便通知应用迁移流程已结束。开发者可以在此检查迁移结果是否成功，并在此处理迁移结束的动作，例如，应用可以在迁移完成后终止自身生命周期。

onRemoteTerminated()：如果开发者使用 continueAbilityReversibly()而不是 continueAbility()，则此后可以在源侧设备上使用 reverseContinueAbility()进行回迁。这种场景下，相当于同一个 Page（的两个实例）同时在两个设备上运行，迁移完成后，如果目标侧设备上 Page 因任何原因终止，则源侧 Page 通过此回调接收终止通知。

2. 请求迁移

实现 IAbilityContinuation 的 Page 可以在其生命周期内，调用 continueAbility()或

continueAbilityReversibly()请求迁移。两者的区别是,通过后者发起的迁移此后可以进行回迁,如代码示例6-3所示。

代码示例 6-3　请求迁移

```
try {
    continueAbility();
} catch (IllegalStateException e) {
    //Maybe another continuation in progress.
    ...
}
```

以 Page 从设备 A 迁移到设备 B 为例,如图 6-9 所示,详细的流程如下:

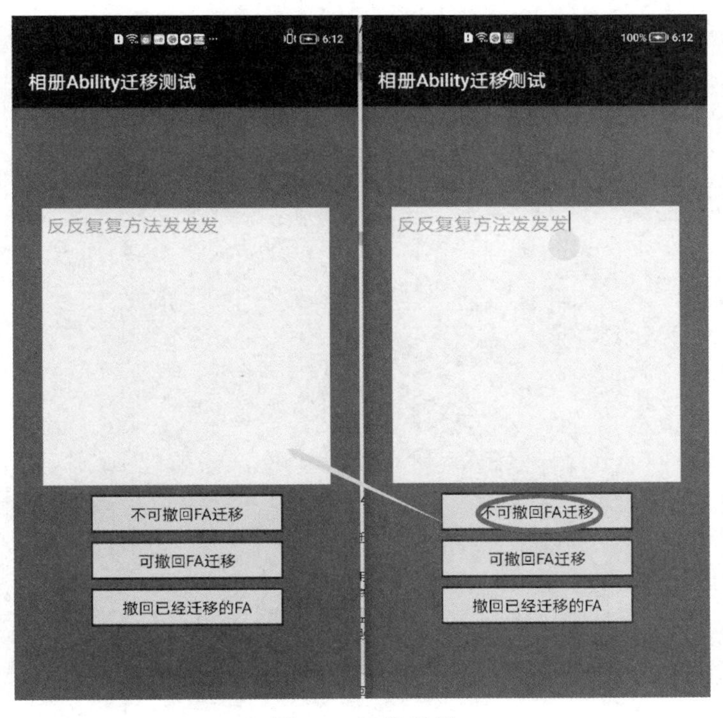

图 6-9　迁移案例

(1) 设备 A 上的 Page 请求迁移。

(2) 系统回调设备 A 上 Page 及其 AbilitySlice 栈中所有 AbilitySlice 实例的 IAbilityContinuation. onStartContinuation()方法,以确认当前是否可以立即迁移。

(3) 如果可以立即迁移,则系统回调设备 A 上 Page 及其 AbilitySlice 栈中所有 AbilitySlice 实例的 IAbilityContinuation. onSaveData()方法,以便保存迁移后恢复状态所必需的数据。

(4) 如果保存数据成功,则系统在设备 B 上启动同一个 Page,并恢复 AbilitySlice 栈,然后回调 IAbilityContinuation.onRestoreData()方法，传递此前保存的数据，此后设备 B 上此 Page 从 onStart()开始其生命周期回调。

(5) 系统回调设备 A 上 Page 及其 AbilitySlice 栈中所有 AbilitySlice 实例的 IAbilityContinuation.onCompleteContinuation()方法，通知数据恢复成功与否。

Page 从设备 A 迁移到设备 B，如代码示例 6-4 所示。

代码示例 6-4　Page 从设备 A 迁移到设备 B

```
private TextField textField;
private String content;
@Override
public void onStart(Intent intent) {
    super.onStart(intent);
    super.setUIContent(ResourceTable.Layout_ability_album);

    Button btn1 = (Button)findComponentById(ResourceTable.Id_btn1);
    Button btn2 = (Button)findComponentById(ResourceTable.Id_btn2);
    Button btn3 = (Button)findComponentById(ResourceTable.Id_btn3);
    textField = (TextField) findComponentById(ResourceTable.Id_txt_m) ;
    textField.setText(content);

    btn1.setClickedListener(component -> {
        String device_id = DeviceUtils.getDeviceId();
continueAbility(device_id);
    });

    btn2.setClickedListener(component -> {
        String device_id = DeviceUtils.getDeviceId();
 continueAbilityReversibly(device_id);
    });

    btn3.setClickedListener(component -> {
        if (getContinuationState() == ContinuationState.REMOTE_RUNNING){
reverseContinueAbility();
        }
    });
}

@Override
public void onActive() {
    super.onActive();
}

@Override
public void onForeground(Intent intent) {
```

```
    super.onForeground(intent);
}

@Override
public boolean onStartContinuation() {
    //是否可以迁移,true 表示迁移,false 表示不能迁移
    return true;
}

@Override
public boolean onSaveData(IntentParams intentParams) {
    //迁移之前,通过 intentParams 保存需要恢复的数据
  intentParams.setParam("data",textField.getText());
    return true;
}

@Override
public boolean onRestoreData(IntentParams intentParams) {
    //迁移完成后,通过 intentParmas 恢复数据
  content = intentParams.getParam("data").toString();
    return true;
}

@Override
public void onCompleteContinuation(int i) {

}
```

6.4 实现跨设备可撤回 FA 迁移

使用 continueAbilityReversibly()请求迁移并完成后，源侧设备上已迁移的 Page 可以发起回迁，以便使用户活动重新回到此设备。

continueAbility(deviceid)迁移应用后，是不可以撤回的，如果希望迁移后能够撤回应用，则可以通过 continueAbilityReversibly(deviceid)方法实现，以这种方法迁移后，可以通过 reverseContinueAbility 撤回应用，代码如下：

```
String deviceid = DeviceUtils.getDeviceId();
if (deviceid != null){
    continueAbilityReversibly(deviceid);
}
```

撤回迁移调用,代码如下:

```
reverseContinueAbility();
```

以 Page 从设备 A 迁移到设备 B 后请求回迁为例,详细的流程如下:

(1) 设备 A 上的 Page 请求回迁。

(2) 系统回调设备 B 上 Page 及其 AbilitySlice 栈中所有 AbilitySlice 实例的 IAbilityContinuation. onStartContinuation()方法,以确认当前是否可以立即迁移。

(3) 如果可以立即迁移,则系统回调设备 B 上 Page 及其 AbilitySlice 栈中所有 AbilitySlice 实例的 IAbilityContinuation. onSaveData()方法,以便保存回迁后用于恢复状态所必需的数据。

(4) 如果保存数据成功,则系统将设备 A 上的 Page 恢复 AbilitySlice 栈,然后回调 IAbilityContinuation. onRestoreData()方法,传递此前保存的数据。

(5) 如果数据恢复成功,则系统终止设备 B 上 Page 的生命周期。

第7章

鸿蒙分布式数据服务

7.1 分布式数据服务介绍

分布式数据服务(Distributed Data Service,DDS)如图7-1所示,为应用程序提供不同设备间数据库数据分布式的能力。通过调用分布式数据接口,应用程序将数据保存到分布式数据库中。通过结合账号、应用和数据库三元组,分布式数据服务对属于不同的应用的数据进行隔离,保证不同应用之间的数据不能通过分布式数据服务互相访问。在通过可信认证的设备间,分布式数据服务支持应用数据相互同步,为用户提供在多种终端设备上一致的数据访问体验。

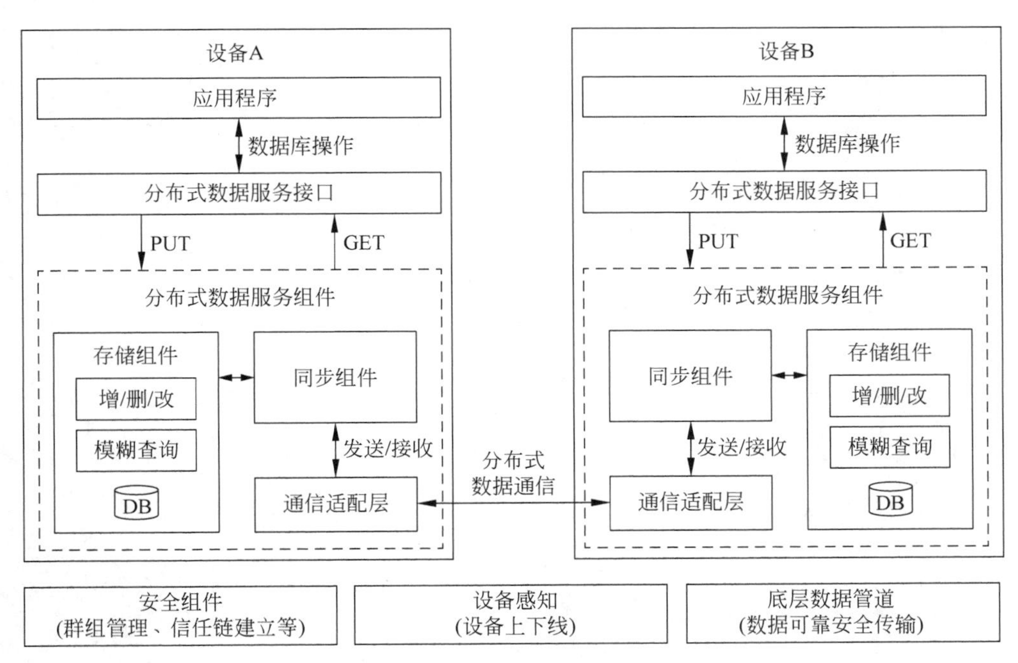

图7-1 分布式数据服务接口

应用程序通过调用分布式数据服务接口实现分布式数据库创建、访问、订阅功能，服务接口通过操作服务组件提供的能力，将数据存储至存储组件，存储组件调用同步组件实现数据同步，同步组件使用通信适配层将数据同步至远端设备，远端设备通过同步组件接收数据，并更新至本端存储组件，通过服务接口提供给应用程序使用。

分布式数据服务的数据模型仅支持KV数据模型，不支持外键、触发器等关系型数据库中的技术点。

分布式数据服务支持的KV数据模型规格有以下几种。

(1) 设备协同数据库，Key最大支持896B，Value最大支持4MB。

(2) 单版本数据库，Key最大支持1KB，Value最大支持4MB。

(3) 每个应用程序最多支持同时打开16个KvStore。

由于支持的存储类型不完全相同等原因，分布式数据服务无法完全代替业务沙箱内数据库数据的存储功能，开发人员需要确定要进行分布式同步的数据，并把这些数据保存到分布式数据服务中。

分布式数据服务当前不支持应用程序自定义冲突解决策略。

分布式数据服务当前流控机制针对KvStore的接口1s最大访问1000次，1min最大访问10000次。KvManager的接口1s最大访问50次，1min最大访问500次。

7.2 分布式数据库权限设置

要使用分布式数据库，首先需要申请ohos.permission.DISTRIBUTED_DATASYNC权限，代码如下：

```
"reqPermissions": [
      {
"name": "ohos.permission.DISTRIBUTED_DATASYNC"
      }
   ]
```

除了需要在config.json中申请权限外，还需要在MainAbility.java中主动申请权限，如代码示例7-1所示。

代码示例7-1 主动申请权限

```
public class MainAbility extends Ability {

    @Override
    public void onStart(Intent intent) {
  requestPermissionsFromUser(new String[]{
"ohos.permission.DISTRIBUTED_DATASYNC"
        },0);
        super.onStart(intent);
```

```
        super.setMainRoute(MainAbilitySlice.class.getName());
    }


}
```

7.3 分布式数据库的基本操作

下面通过以下几个步骤来测试分布式数据库的操作。

步骤1：导入的包名。其代码如下：

```
import ohos.data.distributed.common.KvManager;
import ohos.data.distributed.common.KvManagerConfig;
import ohos.data.distributed.common.KvManagerFactory;
```

步骤2：根据配置构造分布式数据库管理类实例。这里首先需要根据应用上下文创建KvManagerConfig对象，然后创建分布式数据库管理器实例，代码如下：

```
KvManagerConfig config = new KvManagerConfig(this);
KvManager kvManager = KvManagerFactory.getInstance().createKvManager(config);
```

步骤3：获取/创建单版本分布式数据库。其代码如下：

```
String storeID = "test";
Options options = new Options();
options.setCreateIfMissing(true).setEncrypt(false).setKvStoreType(KvStoreType.SINGLE_
VERSION);
SingleKvStore singleKvStore = kvManager.getKvStore(options, storeID);
```

步骤4：将数据写入单版本分布式数据库。其代码如下：

```
String key = "website";
String value = "51itcto.com: 我的博客";
singleKvStore.putString(key, value);
```

步骤5：查询单版本分布式数据库数据。其代码如下：

```
String data = singleKvStore.getString("test"); //test 为 key
```

步骤6：删除单版本分布式数据库数据。其代码如下：

```
singleKvStore.delete(key);
singleKvStore.delete("test"); //test 是 key
```

步骤 7：关闭单版本分布式数据库。其代码如下：

```
kvManager.closeKvStore(singleKvStore);
```

步骤 8：删除单版本分布式数据库。其代码如下：

```
kvManager.deleteKvStore(storeID);
```

7.4 订阅分布式数据变化

订阅分布式数据变化。首先需要在客户端实现 KvStoreObserver 接口，然后构造并注册 KvStoreObserver 实例。实现方法如代码示例 7-2 所示。

代码示例 7-2 把观察者和数据库绑定

```
//创建观察者类
class KvStoreObserverClient implements KvStoreObserver {
    @Override
    public void onChange(ChangeNotification notification) {
        List<Entry> insertEntries = notification.getInsertEntries();
    }
}

//把观察者和数据库绑定
KvStoreObserver kvStoreObserverClient = new KvStoreObserverClient();
singleKvStore.subscribe(SubscribeType.SUBSCRIBE_TYPE_ALL, kvStoreObserverClient);
```

7.5 手动同步分布式数据库

手动同步分布式数据库分为两步：第一步获取已连接的设备列表；第二步选择同步方式进行数据同步。

以下为单版本分布式数据库进行数据同步的代码示例，其中同步方式为 PUSH_ONLY，如代码示例 7-3 所示。

代码示例 7-3 单版本分布式数据库进行数据同步

```
List<DeviceInfo> deviceInfoList = kvManager.getConnectedDevicesInfo(DeviceFilterStrategy.
NO_FILTER);
List<String> deviceIdList = new ArrayList<>();
for (DeviceInfo deviceInfo : deviceInfoList) {
    deviceIdList.add(deviceInfo.getId());
}
singleKvStore.sync(deviceIdList, SyncMode.PUSH_ONLY);
```

7.6 分布式数据库的谓词查询

分布式数据库是 Key-Value 数据库，如果将复杂的对象类型的值保存到 Value 中，如：{"id":1,"Name":"张飒"}，这时需要查询 Value 中 Name="张飒"的数据，这样就可以通过谓词进行查询了。

这里把 3 个对象的数据保存到 Value 中，如图 7-2 所示。

```
singleKvStore.putString( s: "key_1", s1: "{\"id\":1,\"name\":\"zhangsan\",\"age\":18}"); ❶
singleKvStore.putString( s: "key_2", s1: "{\"id\":2,\"name\":\"lisi\",\"age\":28}");   ❷
singleKvStore.putString( s: "key_3", s1: "{\"id\":3,\"name\":\"wangwu\",\"age\":38}"); ❸
```

图 7-2　把 3 个对象的数据保存到 Value 中

上面的数据如何存储到 Key-Value 数据库中呢？

首先需要创建一个 Schema 的数据库结构，这个 Schema 结构用于定义 Value 的数据库结构。数据库的创建基于这个 Schema 进行创建。

FieldNode 为用来定义 Schema 的字段，可以设置这个字段的属性、默认值等。

schema.getRootFieldNode().appendChild(f1)；通过 appendChild 把 FieldNode 追加到 Schema 中。

下面我们定义了一个 id、name、age 的表结构数据 Schema，同时将 Schema 的索引设置为 id，如代码示例 7-4 所示。

代码示例 7-4　表结构数据 Schema

```
FieldNode f1 = new FieldNode("id");
f1.setType(FieldValueType.INTEGER);
f1.setNullable(false);
f1.setDefault(0);
FieldNode f2 = new FieldNode("name");
f2.setType(FieldValueType.STRING);
f2.setDefault("");
f2.setNullable(false);
FieldNode f3 = new FieldNode("age");
f3.setType(FieldValueType.INTEGER);
f3.setDefault(0);
f3.setNullable(false);


Schema schema = new Schema();
List<String> indexList = new ArrayList<>();
indexList.add("$.id");
schema.setIndexes(indexList);
```

```
schema.getRootFieldNode().appendChild(f1);
schema.getRootFieldNode().appendChild(f2);
schema.getRootFieldNode().appendChild(f3);

schema.setSchemaMode(SchemaMode.COMPATIBLE);

config = new KvManagerConfig(this);
kvManager = KvManagerFactory.getInstance().createKvManager(config);
Options options = new Options();
options.setSchema(schema);
options.setCreateIfMissing(true).setAutoSync(true).setEncrypt(false).setKvStoreType
(KvStoreType.SINGLE_VERSION);

singleKvStore = kvManager.getKvStore(options, "test1");
class KvStoreObserverClient implements KvStoreObserver {
    @Override
    public void onChange(ChangeNotification notification) {

    }
}

//把观察者和数据库绑定
KvStoreObserver kvStoreObserverClient = new KvStoreObserverClient();
singleKvStore.subscribe(SubscribeType.SUBSCRIBE_TYPE_ALL, kvStoreObserverClient);
```

接下来,我们就可以通过谓词查询上面保存的数据了,这里通过 Query 的 equalTo 方法,查询 value 中 name 等于 lisi 的数据,如代码示例 7-5 所示。

代码示例 7-5 通过谓词查询

```
Query query = Query.select();
query.equalTo("$.name","lisi");
List<Entry> entries = singleKvStore.getEntries(query);
if(entries.size() > 0) {
    new ToastDialog(DatabaseAbilitySlice.this)
            .setText(entries.get(0).getValue().getString())
            .show();
}
```

第 8 章

鸿蒙分布式文件服务

8.1 分布式文件系统介绍

分布式文件服务能够为用户设备中的应用程序提供多设备之间的文件共享能力，支持相同账号下同一应用文件的跨设备访问，应用程序可以不感知文件所在的存储设备，能够在多个设备之间无缝获取文件。

8.1.1 分布式文件系统基本概念

1. 分布式文件

分布式文件是指依赖于分布式文件系统，分散存储在多个用户设备上的文件，应用间的分布式文件目录互相隔离，不同应用的文件不能互相访问。

2. 文件元数据

文件元数据是用于描述文件特征的数据，包含文件名、文件大小、创建时间、访问时间、修改时间等信息。

8.1.2 分布式文件系统运作机制

分布式文件服务采用无中心节点的设计，每个设备都存储一份全量的文件元数据和本设备上产生的分布式文件，元数据在多台设备间互相同步，当应用需要访问分布式文件时，分布式文件服务首先查询本设备上的文件元数据，获取文件所在的存储设备，然后对存储设备上的分布式文件服务发起文件访问请求，将文件内容读取到本地，如图 8-1 所示。

8.1.3 分布式文件系统约束与限制

(1) 应用程序如需使用分布式文件服务完整功能，则需要申请 ohos.permission.DISTRIBUTED_DATASYNC 权限。

(2) 多个设备需要打开蓝牙，以便连接同一 WLAN 局域网，并且登录相同华为账号才能实现文件的分布式共享。

(3) 存在多设备并发写的场景下，为了保证文件独享，开发者需要对文件进行加锁

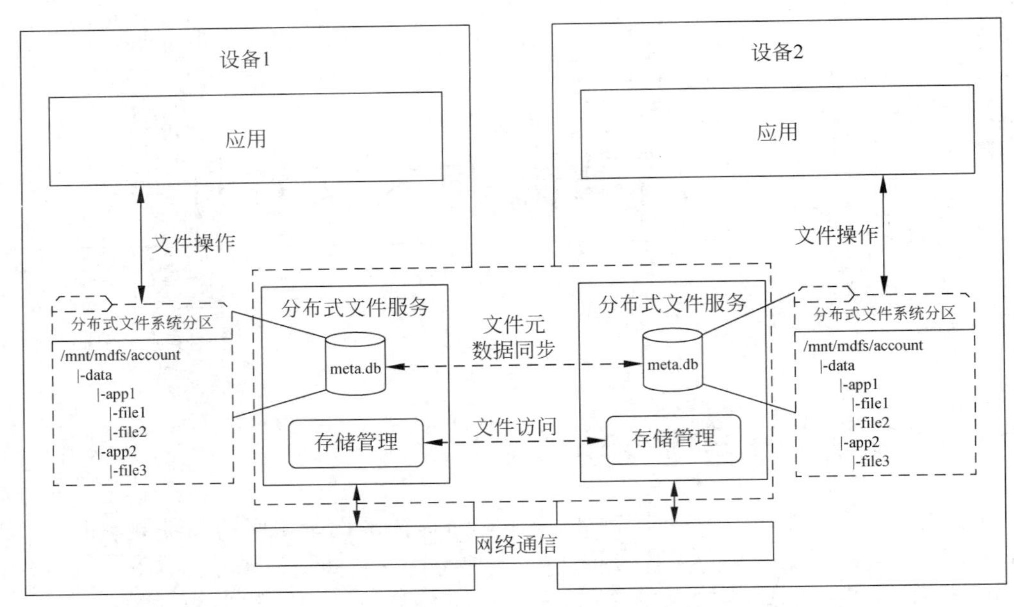

图 8-1　分布式文件服务运作示意图

保护。

(4) 应用访问分布式文件时,如果文件所在设备离线,则文件不能访问此文件。

(5) 非持锁情况下,如果并发写冲突,则后一次会覆盖前一次。

(6) 网络情况差时,如果访问存储在远端的分布式文件,则可能长时间不返回或返回失败,应用需要考虑对这种场景的处理。

(7) 当两台设备有同名文件时,同步元数据时会产生冲突,分布式文件服务会根据时间戳将文件按创建的先后顺序重命名,为避免此场景,建议应用在文件名上进行设备区分,例如,deviceId+时间戳。

8.2 分布式文件系统操作

应用可以通过分布式文件服务实现多个设备间的文件共享,设备 1 上的应用 A 创建了分布式文件 a,设备 2 上的应用 A 能够通过分布式文件服务读写设备 1 上的文件 a。

分布式文件兼容 POSIX 文件操作接口,应用使用 Context. getDistributedDir()接口获取目录后,可以直接使用 libc 或 JDK 访问分布式文件,如表 8-1 所示。

表 8-1　分布式文件服务 API 功能介绍

接　口　名	描　　述
Context. getDistributedDir()	获取文件的分布式目录

应用可以通过 Context. getDistributedDir()接口获取属于自己的分布式目录，然后通过 libc 或 JDK 接口，在该目录下创建、删除、读写文件或目录。

设备 1 上的应用 A 创建文件 hello. txt，并写入内容"Hello World"，代码如下：

```
Context context;
...                                    //context 初始化
File distDir = context.getDistributedDir();
String filePath = distDir + File.separator + "hello.txt";
FileWriter fileWriter = new FileWriter(filePath,true);
fileWriter.write("Hello World");
fileWriter.close();
```

设备 2 上的应用 A 通过 Context. getDistributedDir()接口获取分布式目录。设备 2 上的应用 A 读取文件 hello. txt，代码如下：

```
FileReader fileReader = new FileReader(filePath);
char[] buffer = new char[1024];
fileReader.read(buffer);
fileReader.close();
System.out.println(buffer);
```

第四篇　应用实战篇

学习目标

本篇通过3个应用开发案例，让读者熟悉鸿蒙应用开发技巧。通过本篇的学习，读者将可以基于鸿蒙ACE框架开发跨设备的应用程序。

主要内容如下：

- 鸿蒙智能手表应用开发(Java版)；
- 鸿蒙手机+TV游戏开发案例(Java+JavaScript版)；
- 鸿蒙手机+TV应用开发案例(Java+JavaScript版)；
- 鸿蒙的应用签名与发布。

如何学习本篇：

- 学习本章的读者可结合每小节中的案例反复编码练习。

第 9 章

智慧手表应用开发案例(Java 版)

本章完成一个运行在鸿蒙智能手表上的名为天气预报的 App。App 界面比较简单,默认通过设备定位获取当前位置及所在地的天气信息,也可以通过单击表盘下面的语音按钮,通过语音搜索其他地区的天气的信息。

9.1 天气预报 App 介绍

主页面的效果图如图 9-1 所示,表盘的中间部分显示热门的地区名称,表盘头部显示的是当前所选的天气信息。底部是一个语音按钮,可以通过语音搜索其他地区的天气情况,搜索的结果显示在表盘的头部位置。

默认通过地理位置定位获取位置所在地区的天气信息,并显示在表盘的头部位置,单击表盘中间的城市名称,表盘头部的天气信息会随之变化。

图 9-1 天气预报主页面显示效果图

9.2 天气预报 App 技术点

本案例中使用的技术点如下。

(1) ListContainer 组件:显示热门地区列表,单击地区名称,便可获取单击地区的详细天气信息。

(2) EventHandler 线程间通信:事件循环器,循环处理从该 EventRunner 所创建的新线程的事件队列中获取 InnerEvent 事件或者 Runnable 任务。InnerEvent 是 EventHandler 投递的事件。这里通过 EventHandler 来处理搜索事件。

(3) JSON 数据解析:通过第三方 JSON 包,解析网络传输的 JSON 数据。

(4) 网络数据加载:通过封装获取网络数据的方法类,获取第三方接口的 Restful 风格接口数据。

(5) 地理位置定位：通过获取设备的位置信息，进行默认天气查询，如图 9-2 所示。

图 9-2　获取设备的位置信息

(6) 语音搜索：如果需要查找其他地区的天气信息，则可以通过语音搜索的方式查询。

9.3　天气预报 App 界面实现

App 界面采用垂直方向布局，从上到下采用水平方向布局组件摆放天气图标和温度；Text 文本组件占一行；中间的城市列表采用 ListContainer 组件；底部的语音按钮采用 Image 组件，如图 9-3 所示。

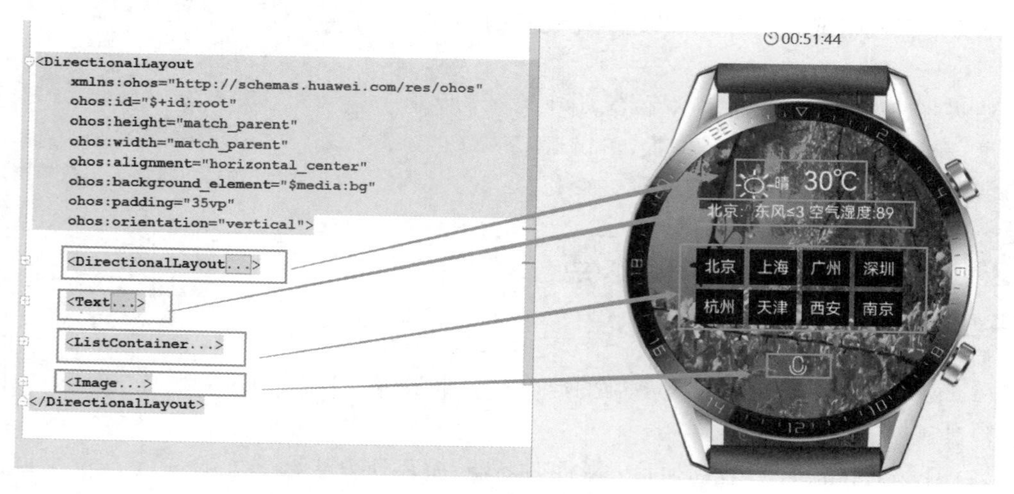

图 9-3　App 界面分解图

App 界面采用垂直方向布局，height 和 width 适配父容器，中间的子元素的排列方式 alignment 为水平居中对齐，背景通过 background_element 设置为 media 目录下的图片 bg.jpg，内填充为 35vp，如代码示例 9-1 所示。

代码示例 9-1

```
<DirectionalLayout
    xmlns:ohos = "http://schemas.huawei.com/res/ohos"
    ohos:id = "$ + id:rootlayout"
    ohos:height = "match_parent"
    ohos:width = "match_parent"
    ohos:alignment = "horizontal_center"
    ohos:background_element = "$ media:bg"
    ohos:padding = "35vp"
    ohos:orientation = "vertical">


</DirectionalLayout>
```

水平方向布局组件摆放天气图标和温度如代码示例 9-2 所示。水平方向布局组件的高度自适应子组件的高度，宽度适应父容器。

代码示例 9-2

```
<DirectionalLayout
    ohos:height = "match_content"
    ohos:width = "match_parent"
    ohos:alignment = "horizontal_center"
    ohos:orientation = "horizontal">

<Image
        ohos:height = "30vp"
        ohos:width = "30vp"
        ohos:scale_mode = "inside"
        ohos:image_src = "$ graphic:qing"
        ohos:top_padding = "5vp"/>

<Text
        ohos:text = "晴"
        ohos:height = "match_content"
        ohos:width = "match_content"
        ohos:right_margin = "10vp"
        ohos:text_color = "#fff"
        ohos:text_size = "12fp"/>

<Text
        ohos:text = "30℃"
        ohos:height = "match_content"
        ohos:width = "match_content"
        ohos:right_margin = "8vp"
        ohos:text_color = "#fff"
        ohos:text_size = "20fp"/>
</DirectionalLayout>
```

Text 文本组件占一行，如代码示例 9-3 所示，左右边距为 10vp，并将文字大小设置为 12fp。

代码示例 9-3

```
<Text
    ohos:id = "$ + id:tips"
    ohos:text = "北京：东风≤3 空气湿度:89"
    ohos:height = "match_content"
    ohos:width = "match_content"
    ohos:layout_alignment = "horizontal_center"
    ohos:right_margin = "10vp"
    ohos:left_margin = "10vp"
    ohos:text_color = "#fff"
    ohos:text_size = "12fp"/>
```

中间为城市列表，采用 ListContainer 组件，如代码示例 9-4 所示。上边距设置为 20vp，高度适应子组件的高度，宽度适应父容器的宽度。

代码示例 9-4

```
<ListContainer
    ohos:id = "$ + id:list"
    ohos:top_margin = "20vp"
    ohos:height = "match_content"
    ohos:width = "match_parent"/>
```

ListContainer 组件包含两个子组件，一个水平方向布局组件和一个文本组件。

水平方向组件，一行 4 个文本组件，这里单独定义一个布局文件，用于后面通过代码进行控制，如代码示例 9-5 所示。

代码示例 9-5　/resources/base/layout/list_item_layout. xml

```
<?xml version = "1.0" encoding = "utf - 8"?>
<DirectionalLayout
    xmlns:ohos = "http://schemas.huawei.com/res/ohos"
    ohos:id = "$ + id:li"
    ohos:height = "match_content"
    ohos:width = "match_parent"
    ohos:alignment = "horizontal_center"
    ohos:orientation = "horizontal" />
```

文本组件，这里也是独立创建的布局文件，后面需要通过代码进行处理，动态地将数据加入上面的布局组件中，如代码示例 9-6 所示。

代码示例 9-6　/resources/base/layout/item_layout. xml

```
<?xml version = "1.0" encoding = "utf - 8"?>
<Text
```

```
    xmlns:ohos = "http://schemas.huawei.com/res/ohos"
    ohos:id = "$ + id:city1"
    ohos:height = "match_content"
    ohos:width = "match_content"
    ohos:background_element = "#000"
    ohos:margin = "2vp"
    ohos:padding = "6vp"
    ohos:text_color = "#fff"
    ohos:text_size = "12vp"/>
```

底部的语音按钮采用 Image 组件，如代码示例 9-7 所示。

代码示例 9-7

```
<Image
    ohos:id = "$ + id:sound_search"
    ohos:top_margin = "20vp"
    ohos:image_src = "$graphic:yy"
    ohos:layout_alignment = "bottom|center"
    ohos:scale_mode = "inside"
    ohos:height = "20vp"
    ohos:width = "20vp" />
```

上面的天气图标通过 iconfont 网站获取 svg 图标文件，如图 9-4 所示，再通过 IDE 将 svg 格式转为 xml 格式，最后添加到 graphic 目录中。

图 9-4　下载免费的 svg 图标文件

9.4 天气预报 App 核心代码

通过上面的步骤完成了 App 布局，接下来实现 App 中的逻辑。

9.4.1 配置 App 中所需的权限

首先需要为项目添加权限，因为需要用到设备的位置信息、网络方面的信息，所以需要提前配置好。配置 config.json 权限，代码如下：

```
"reqPermissions": [
  {
"name": "ohos.permission.LOCATION_IN_BACKGROUND"
  },
  {
"name": "ohos.permission.LOCATION"
  },
  {
"name": "ohos.permission.INTERNET"
  },
  {
"name": "ohos.permission.GET_NETWORK_INFO"
  },
  {
"name": "ohos.permission.SET_NETWORK_INFO"
  }
]
```

同时需要在 MainAbility 中请求获取这些权限，如图 9-5 所示。

```
public class MainAbility extends Ability {
    @Override
    public void onStart(Intent intent) {
        super.onStart(intent);
        super.setMainRoute(MainAbilitySlice.class.getName());
        requestPermissionsFromUser(new String[]{
                "ohos.permission.INTERNET",
                "ohos.permission.LOCATION_IN_BACKGROUND",
                "ohos.permission.LOCATION",
        }, requestCode: 0);
    }
}
```

图 9-5 在 MainAbility 中请求获取这些权限

9.4.2 创建 ListContainer 数据类

这里一行显示 4 个城市信息，所以需要两个数据类：一个是城市类 CityModel，如代码示例 9-8 所示，另一个是每行显示的数据类 DataModel，数据类是一组 CityModel 的集合，如代码示例 9-9 所示。

代码示例 9-8 城市信息

```
public class CityModel {
    public String cityName;
    public String cityCode;


    public CityModel(String cityName, String cityCode) {
        this.cityName = cityName;
        this.cityCode = cityCode;
    }


}
```

DataModel 数据类是 ListContainer 每行显示的数据信息，只有一个属性 cityModels，保存的是一组城市数据。

代码示例 9-9 ListContainer 每行显示的数据信息

```
public class DataModel {

    //每行显示多条数据
    public ArrayList < CityModel > cityModels;

    public DataModel(ArrayList < CityModel > cityModels) {
        this.cityModels = cityModels;
    }
}
```

9.4.3 创建 ListContainer 数据提供类

ListContainer 每一行可以显示不同的数据，因此需要适配不同的数据结构，使其都能添加到 ListContainer 上。

创建 ListItemProvider.java，继承自 BaseItemProvider，如代码示例 9-10 所示。

代码示例 9-10 创建 ListContainer 数据提供类

```
public class ListItemProvider extends BaseItemProvider {

    private AbilitySlice mSlice;
    //每行显示多个
    private ArrayList < DataModel > dataMos = new ArrayList <>();
```

```
    private OnItemClickListener listener;

    public ListItemProvider(AbilitySlice mSlice, OnItemClickListener listener) {
        this.mSlice = mSlice;
        this.listener = listener;
    }

    public void setData(ArrayList<CityModel> cityModels){
        dataMos.clear();
        int i = 0;
        ArrayList<CityModel> tempList = new ArrayList<>();
        for (CityModel mo: cityModels){
            if (i == 4){
                i = 0;
                dataMos.add(new DataModel(tempList));
                System.out.println("i:" + i);
                tempList = new ArrayList<>();
            }
            tempList.add(mo);
            i++;
        }
        dataMos.add(new DataModel(tempList));
        System.out.println("size:" + dataMos.size());
        //刷新列表数据
        notifyDataChanged();
    }

    @Override
    public int getCount() {
        return dataMos.size();
    }

    @Override
    public Object getItem(int i) {
        return null;
    }

    @Override
    public long getItemId(int i) {
        return 0;
    }

    @Override
    public Component getComponent(int i, Component component, ComponentContainer
componentContainer) {
        Component list = LayoutScatter.getInstance(mSlice)
```

```
            .parse(ResourceTable.Layout_list_item_layout,null,false);
        if (!(list instanceof  ComponentContainer)){
            return null;
        }
        ComponentContainer rootLayout = (ComponentContainer)list;
        DataModel dataModel = dataMos.get(i);
        for (CityModel mo : dataModel.cityModels){
            System.out.println(" ------ :" + mo.cityName);
            Text item = (Text)LayoutScatter.getInstance(mSlice)
                    .parse(ResourceTable.Layout_item_layout,null,false);
            item.setText(mo.cityName);
            rootLayout.addComponent(item);
            item.setClickedListener(new Component.ClickedListener() {
                @Override
                public void onClick(Component component) {
                    listener.onItemClick(mo,i);
                }
            });
        }
        return list;
    }

    public interface OnItemClickListener {
        void onItemClick(CityModel mo, int position);
    }
}
```

1. ListItemProvider 3 个成员属性

ListItemProvider 为 ListContainer 提供数据适配,该类需要 3 个成员属性,代码如下:

```
private AbilitySlice mSlice;
    //每行显示多个
    private ArrayList<DataModel> dataModelArrayList = new ArrayList<>();

    private OnItemClickListener listener;
```

(1) AbilitySlice mSlice:页面 Slice 类。

(2) ArrayList<DataModel> dataModelArrayList:每行显示的数据集合,这些数据通过类中的方法初始化。

(3) OnItemClickListener listener:单击每个 Item 数据绑定的监听事件,OnItemClickListener 是一个接口,定义了一个 onItemClick 方法,该方法需要在组件外实现,有两个参数:城市数据信息和所在的列表中的位置编号,代码如下:

```
public interface OnItemClickListener {
    void onItemClick(CityModel mo, int position);
}
```

2. ListItemProvider 类构造方法

这里需要传入页面类和单击事件响应的方法接口，代码如下：

```
public ListItemProvider(AbilitySlice mSlice, OnItemClickListener listener) {
    this.mSlice = mSlice;
    this.listener = listener;
}
```

3. 初始化每行的数据集合

该方法为每行设置 4 个城市信息，通过外部传入一组城市详细信息，按 4 个一组添加到 dataModelArrayList 属性中，如代码示例 9-11 所示。

代码示例 9-11　每行设置 4 个城市信息

```
public void setData(ArrayList < CityModel > cityModels){
dataModelArrayList.clear();
    int i = 0;
    ArrayList < CityModel > tempList =  new ArrayList <>();
    for (CityModel mo: cityModels){
        if (i == 4){
            i = 0;
dataModelArrayList.add(new DataModel(tempList));
            System.out.println("i:" + i);
            tempList  = new ArrayList <>();
        }
        tempList.add(mo);
        i++;
    }
dataModelArrayList.add(new DataModel(tempList));
    System.out.println("size:" + dataModelArrayList.size());
    //刷新列表数据
    notifyDataChanged();
}
```

4. 实现 BaseItemProvider 接口中的方法

getCount 方法返回 ListContainer 需要显示的行数，这里是上面所设置的 dataModelArrayList 中的城市列表的 Size。

getComponent 方法给 ListContainer 绑定每一行的组件，i 表示 getCount 中的数据下标，通过 i 获取一种 DataModel 中的城市列表，遍历并赋值给预先设置的子容器 Layout_list_item_layout 和 Layout_item_layout。

LayoutScatter.getInstance(mSlice)方法获取指定XML布局中定义的组件，Layout_list_item_layou：水平布局组件，Layout_item_layou：文本组件。

遍历每组城市数据，动态地将组件添加到布局组件中，并返回设置好的布局组件list，如代码示例9-12所示。

代码示例9-12 实现BaseItemProvider接口中的方法

```java
@Override
public int getCount() {
    return dataModelArrayList.size();
}

@Override
public Object getItem(int i) {
    return null;
}

@Override
public long getItemId(int i) {
    return 0;
}

@Override
public Component getComponent(int i, Component component, ComponentContainer componentContainer) {
    Component list = LayoutScatter.getInstance(mSlice)
            .parse(ResourceTable.Layout_list_item_layout,null,false);
    if (!(list instanceof  ComponentContainer)){
        return null;
    }
    ComponentContainer rootLayout = (ComponentContainer)list;
    DataModel dataModel = dataModelArrayList.get(i);
    for (CityModel mo : dataModel.cityModels){
        System.out.println("------:" + mo.cityName);
        Text item = (Text)LayoutScatter.getInstance(mSlice)
                .parse(ResourceTable.Layout_item_layout,null,false);
        item.setText(mo.cityName);
        rootLayout.addComponent(item);
        item.setClickedListener(new Component.ClickedListener() {
            @Override
            public void onClick(Component component) {
                listener.onItemClick(mo,i);
            }
        });
    }
   return list;
}
```

9.4.4 绑定 ListContainer 数据提供类

在 MainAbilitySlice 中引入 ListItemProvider 类,代码如下:

```
private ListContainer listContainer;
private ListItemProvider listItemProvider;
private ArrayList<CityModel> cities = new ArrayList<>();
```

这里模拟几条城市详细信息,通过 ListItemProvider 类的 setData 方法将详细信息添加到每行显示的列表集合中,再通过 BindListContainer 方法绑定 ListContainer,如代码示例 9-13 所示。

代码示例 9-13 模拟城市数据

```
private void initCity() {
    cities.add(new CityModel("北京", "110000"));
    cities.add(new CityModel("上海", "310115"));
    cities.add(new CityModel("广州", "440100"));
    cities.add(new CityModel("深圳", "440300"));
    cities.add(new CityModel("杭州", "330100"));
    cities.add(new CityModel("天津", "120100"));
    cities.add(new CityModel("西安", "610100"));
    cities.add(new CityModel("南京", "320100"));
    listItemProvider.setData(cities);
}

private void BindListContainer() {
    listContainer = (ListContainer) findComponentById(ResourceTable.Id_list);
    listItemProvider = new ListItemProvider(this, this);
    listContainer.setItemProvider(listItemProvider);
}
```

在 onStart 方法中调用上面的两种方法,如图 9-6 所示。

```
@Override
public void onStart(Intent intent) {
    super.onStart(intent);
    super.setUIContent(ResourceTable.Layout_ability_main);
    BindListContainer();
    initCity();
}
```

图 9-6 在 onStart 方法中调用上面的两种方法

9.4.5 处理 ListContainer 单击事件处理

MainAbilitySlice 类实现 OnItemClickListener 接口,如图 9-7 所示。

```
public class MainAbilitySlice extends AbilitySlice implements ListItemProvider.OnItemClickListener {

    private ListContainer listContainer;
    private ListItemProvider listItemProvider;
    private ArrayList<CityModel> cities = new ArrayList<>();
    private Text txtTips;
    private Image imgSound;
```

图 9-7 ListContainer 中的每项单击事件

实现接口方法,代码如下:

```
@Override
public void onItemClick(CityModel mo, int position) {
    loadData(mo); //获取远程数据,该方法将在 9.4.8 节实现
}
```

单击每个列表中的城市,获取该城市的数据信息,绑定到界面并显示。下面我们实现数据的远程获取和数据绑定。

9.4.6 多线程处理事件和网络请求

鸿蒙开发禁止在主线程处理耗时的操作,这里我们通过 EventHandler 实现多线程处理事件和获取数据。

下面创建一个用来切换线程的帮助类,如代码示例 9-14 所示。

代码示例 9-14 创建一个用来切换线程的帮助类

```
public class MyEventHandler {

    //在主线程执行
    public static void runMainUI(Runnable runnable){
        //切换到主线程
        EventRunner runner = EventRunner.getMainEventRunner();
        //获取主线程中的 runner
        EventHandler eventHandler = new EventHandler(runner);
        eventHandler.postSyncTask(runnable);
    }

    //在子线程执行任务
    public static void runBackGround(Runnable runnable){
        //创建一个子线程
        EventRunner runner = EventRunner.create(true);
        EventHandler eventHandler = new EventHandler(runner);
        eventHandler.postTask(runnable,0, EventHandler.Priority.IMMEDIATE);
    }
}
```

9.4.7 格式化 JSON 数据

在获取网络数据之前，首先应掌握如何把 JSON 字符串转换成 Java 对象。

Jackson 当前使用比较广泛，用来序列化和反序列化 JSON 的 Java 开源框架。Jackson 社区相对比较活跃，更新速度也比较快，从 GitHub 中的统计来看，Jackson 是目前最流行的 JSON 解析器之一。

Gradle 配置，如图 9-8 所示，在项目目录下的 build. gradle 文件中添加以下配置：

```
implementation group: 'com.fasterxml.jackson.core', name: 'jackson-databind', version: '2.12.1'
implementation group: 'com.fasterxml.jackson.core', name: 'jackson-annotations', version: '2.12.1'
implementation group: 'com.fasterxml.jackson.core', name: 'jackson-core', version: '2.12.1'
```

```
dependencies {
    implementation fileTree(dir: 'libs', include: ['*.jar', '*.har'])
    testCompile 'junit:junit:4.12'
    implementation group: 'com.fasterxml.jackson.core', name: 'jackson-databind', version: '2.12.1'
    implementation group: 'com.fasterxml.jackson.core', name: 'jackson-annotations', version: '2.12.1'
    implementation group: 'com.fasterxml.jackson.core', name: 'jackson-core', version: '2.12.1'

}
```

图 9-8 Gradle 配置

接下来，使用 Jackson 把下面的 JSON 数据转换成 Java 对象，这里需要创建一个和 JSON 相同格式的 Java 类，如图 9-9 所示。

```
{
    "status": "1",
    "count": "1",
    "info": "OK",
    "infocode": "10000",
    "lives": [
        {
            "province": "北京",
            "city": "北京市",
            "adcode": "110000",
            "weather": "晴",
            "temperature": "6",
            "winddirection": "西南",
            "windpower": "≤3",
            "humidity": "21",
            "reporttime": "2021-02-18 15:30:22"
        }
    ]
}
```

图 9-9 JSON 文件

Weather 类,如代码示例 9-15 所示。

代码示例 9-15 Weather 类,把 JSON 转换成类对象

```
class Weather{
    public String status;
    public String count;
    public String info;
    public String infocode;
    public List<Lives> lives;
    static class Lives {
        public String province;
        public String city;
        public String adcode;
        public String weather;
        public String temperature;
        public String winddirection;
        public String windpower;
        public String humidity;
        public String reporttime;
    }
}
```

使用 ObjectMapper 方法把 JSON 字符串转换成 Java 对象,如代码示例 9-16 所示。

代码示例 9-16 ObjectMapper 方法把 JSON 字符串转换成 Java 对象

```
ObjectMapper mapper = new ObjectMapper();
try {
    Weather weather = mapper.readValue(jsonStr,Weather.class);
 //......
} catch (JsonProcessingException e) {
    e.printStackTrace();
}
```

9.4.8 封装网络访问类获取网络数据

单击每行中的城市名称,获取该城市的天气数据信息,通过城市的编号调用第三方数据接口,并返回单击城市的天气信息。

在本案例中,我们使用高德地图的天气接口获取数据。

```
https://restapi.amap.com/v3/weather/weatherInfo?city=110000&key=申请的 key
```

这里我们使用封装类 MyNet,通过类的 get 方法获取接口的数据,如代码示例 9-17 所示。

代码示例 9-17　封装 MyNet 类请求远程数据

```
Map < String, String > params = new HashMap <>();
params.put("city", co.cityCode);
params.put("key", "申请的 key 字符串");
MyNet myNet = new MyNet();
myNet.get("https://restapi.amap.com/v3/weather/weatherInfo", params, new IMyNet.NetListener() {
    @Override
    public void onSuccess(String jsonStr) {
        ObjectMapper mapper = new ObjectMapper();
        try {
            Weather weather = mapper.readValue(jsonStr,Weather.class);
            bindWeatherView(weather);
        } catch (JsonProcessingException e) {
            e.printStackTrace();
        }

    }

    @Override
    public void onError(String error) {

    }
});
```

获取数据后，绑定到页面，如代码示例 9-18 所示。

代码示例 9-18　获取数据后，绑定到页面

```
    private void bindWeatherView(Weather weather) {
        //北京：东风≤3,空气湿度:89
         String txtTips = weather.lives.get(0).city + ":" + weather.lives.get(0).
winddirection + "风" + weather.lives.get(0).windpower + "空气湿度:" + weather.lives.get(0).
humidity;
        Text tips = (Text) findComponentById(ResourceTable.Id_tips);
        tips.setText(txtTips);

        Text temp = (Text) findComponentById(ResourceTable.Id_temperature);
        temp.setText(weather.lives.get(0).temperature + "℃");

        Text weth = (Text) findComponentById(ResourceTable.Id_weather);
        weth.setText(weather.lives.get(0).weather);

        Image wicon = (Image)findComponentById(ResourceTable.Id_wicon);
        if (weather.lives.get(0).weather.contains("雪")){
            wicon.setImageAndDecodeBounds(ResourceTable.Media_xue);
        }
```

```
    }
}
```

接下来创建 MyNet 类,定义一个接口,定义 get 方法和回调接口 NetListener,如代码示例 9-19 所示。

代码示例 9-19 创建接口

```
public interface IMyNet {
    void get(String URL, Map < String,String > params,NetListener listener);
    interface NetListener {
        void onSuccess(String jsonStr);
        void onError(String error);
    }
}
```

实现 IMyNet 接口,实现 get 方法,如代码示例 9-20 所示。

代码示例 9-20 实现远程访问接口

```
public class MyNet implements IMyNet{

    private NetManager netManager;

    public MyNet() {
        netManager = NetManager.getInstance(null);
    }

    @Override
    public void get(String URL, Map < String, String > params,NetListener listener) {
       String finalURL = NetUtil.buildParams(URL, params);
        MyEventHandler.runBackGround(new Runnable() {
            @Override
            public void run() {
                doGet(finalURL,listener);
            }
        });
    }

    private void doGet(String finalURL, NetListener listener) {
        NetHandle netHandle = netManager.getDefaultNet();
        HttpURLConnection connection = null;
        InputStream inputStream = null;
        ByteArrayOutputStream bos = null;

        try {
```

```
                URL URL = new URL(finalURL);
                URLConnection URLConnection = netHandle.openConnection(URL, Proxy.NO_PROXY);
                if (URLConnection instanceof HttpURLConnection){
                    connection = (HttpURLConnection) URLConnection;
                }
                connection.setRequestMethod("GET");
                connection.connect();
                if (connection.getResponseCode() == 200) {
                    inputStream = connection.getInputStream();
                    bos = new ByteArrayOutputStream();
                    int readLen;
                    Byte[] Bytes = new Byte[1024];
                    while ((readLen = inputStream.read(Bytes)) != -1){
                        bos.write(Bytes,0,readLen);
                    }
                    String result = bos.toString();
                    MyEventHandler.runMainUI(new Runnable() {
                        @Override
                        public void run() {
                            listener.onSuccess(result);
                        }
                    });
                }else {
                    listener.onError("请求错误: " + connection.getResponseCode());
                }
            }catch (Exception e){
                listener.onError("请求失败: " + e.toString());
            }finally {
                if (connection != null){
                    connection.disconnect();
                    NetUtil.close(inputStream);
                    NetUtil.close(bos);
                }
            }
        }
    }
```

创建一个用来处理 URL 解析的工具类，如代码示例 9-21 所示。

代码示例 9-21　URL 解析工具类

```
public class NetUtil {

    public static String buildParams(String URL, Map<String,String> params){
        if(params == null){
            return URL;
```

```
        }
        StringBuilder stringBuilder = new StringBuilder(URL);
        boolean isFirst = true;
        for (String key:params.keySet()){
            String value = params.get(key);
            if (key != null &&value!= null){
                if (isFirst){
                    isFirst = false;
                    stringBuilder.append("?");
                }else {
                    stringBuilder.append("&");
                }
            }
            stringBuilder.append(key)
                    .append(" = ")
                    .append(value);
        }
        return stringBuilder.toString();
    }

    public static void close(Closeable closeable){
        if (closeable!= null){
            try {
                closeable.close();
            }catch (Exception e){
                e.printStackTrace();
            }
        }
    }
}
```

9.4.9　通过设备地理定位获取默认天气

App启动后，默认通过设备地理位置定位并获取坐标信息，然后把这个坐标信息转换成位置信息。

这里需要使用两个类：Locator类和GeoConvert类。

实例化Locator对象，所有与基础定位能力相关的功能API都由Locator提供。应用需要自行实现系统定义好的回调接口，并将其实例化，如代码示例9-22所示。

代码示例9-22　获取地理位置信息

```
private void loadLocation() {
    RequestParam requestParam = new RequestParam(RequestParam.SCENE_DAILY_LIFE_SERVICE);
```

```
        Locator locator = new Locator(MainAbilitySlice.this);
        locator.startLocating(requestParam, locatorCallback);
    }
```

实例化 LocatorCallback 对象，用于向系统提供位置上报的途径。

系统在定位成功并确定设备的实时位置时，会通过 onLocationReport 接口上报给应用。应用程序可以在 onLocationReport 接口的实现中完成自己的业务逻辑，如代码示例 9-23 所示。

代码示例 9-23 获取经度和纬度，再通过 GeoConvert 类进行位置信息转换

```
MyLocatorCallback locatorCallback = new MyLocatorCallback();

//获取地理位置成功回调类
private class MyLocatorCallback implements LocatorCallback {

    //返回成功的地理位置信息
    @Override
    public void onLocationReport(Location location) {

        double latitude = location.getLatitude();
        double longitude = location.getLongitude();

        //坐标转化地理位置信息
        GeoConvert geoConvert = new GeoConvert();

        try {
            List<GeoAddress> addrs = geoConvert.getAddressFromLocation(latitude,longitude, 1);

        /**
         * getLocale()  zh_CN
         * getAdministrativeArea()xx 省
         * getLocality()  xx 市
         * getSubLocality() xx 区
         * getSubAdministrativeArea() xx 街道
         * getPlaceName() xx 单位
         * getRoadName() xx 路
         * getSubRoadName() 122 号
         * getDescriptionsSize() 0 下面的 index = 0
         * getDescriptions(0)  xx 省 xx 市 xx 区 xxx 单位
         */

        } catch (IOException e) {
            System.out.println(e.getStackTrace());
```

```
            }
        }

        @Override
        public void onStatusChanged(int type) {
            System.out.println("onStatusChanged" + type);
        }

        @Override
        public void onErrorReport(int type) {
            System.out.println("onErrorReport");
        }
    }
```

在 onLocationReport 返回的坐标信息中，获取经度和纬度，再通过 GeoConvert 类进行位置信息转换。

实例化 GeoConvert 对象，所有与（逆）地理编码转化能力相关的功能 API 都由 GeoConvert 提供。坐标转化地理位置信息的代码如下所示：

```
geoConvert.getAddressFromLocation(纬度值, 经度值, 1);
```

9.4.10　通过语音查询天气

在使用语音识别 API 时，将与 ASR 相关的类添加至工程，如代码示例 9-24 所示。

代码示例 9-24　使用语音识别 API

```
//提供 ASR 引擎执行时所需要传入的参数类
import ohos.ai.asr.AsrIntent;
//错误码的定义类
import ohos.ai.asr.util.AsrError;
//加载语音识别 Listener
import ohos.ai.asr.AsrListener;
//提供调用 ASR 引擎服务接口的类
import ohos.ai.asr.AsrClient;
//ASR 回调结果中的关键字封装类
import ohos.ai.asr.util.AsrResultKey;
```

创建一个 AsrClient 对象。context 为应用上下文信息，应为 ohos.aafwk.ability.Ability 或 ohos.aafwk.ability.AbilitySlice 的实例或子类实例。

这里我们创建了一个 soundSearch 方法，如代码示例 9-25 所示，通过鸿蒙的 AI 系统接口 AsrClient 类监听语言，并返回识别的信息，通过识别的城市信息进行接口搜索。

代码示例 9-25　语音查询实现

```
private void soundSearch() {
    AsrClient asrClient = AsrClient.createAsrClient(MainAbilitySlice.this).orElse(null);
```

```
AsrIntent asrIntent = new AsrIntent();
//设置后置的端点检测(VAD)时间
asrIntent.setVadEndWaitMs(2000);
//设置前置的端点检测(VAD)时间
asrIntent.setVadFrontWaitMs(4800);
//设置语音识别的超时时间
asrIntent.setTimeoutThresholdMs(20000);
asrClient.startListening(asrIntent);
//需要将 buffer 替换为真实的声频数据
Byte[] buffer = new Byte[]{0, 1, 0, 10, 1};
//对于长度大于 1280 的声频,需要多次调用 writePcm 分段传输
asrClient.writePcm(buffer, 1280);
asrClient.init(asrIntent, new AsrListener() {
    @Override
    public void onInit(PacMap pacMap) {
        System.out.println("----------- onInit ---------");
    }

    @Override
    public void onBeginningOfSpeech() {
        System.out.println("---- onBeginningOfSpeech ----");
    }

    @Override
    public void onRmsChanged(float v) {
        System.out.println("---- onRmsChanged -----");
    }

    @Override
    public void onBufferReceived(Byte[] Bytes) {
        System.out.println("------ onBufferReceived -----");
    }

    @Override
    public void onEndOfSpeech() {
        System.out.println("------ onEndOfSpeech ---");
    }

    @Override
    public void onError(int i) {
        System.out.println("------- onError --------" + i);
    }

    @Override
    public void onResults(PacMap pacMap) {
        System.out.println("----- onResults --------");
```

```
            }

            @Override
            public void onIntermediateResults(PacMap pacMap) {

            }

            @Override
            public void onEnd() {
                System.out.println("------onEnd--------");
            }

            @Override
            public void onEvent(int i, PacMap pacMap) {

            }

            @Override
            public void onAudioStart() {
                System.out.println("-----onAudioStart----");
            }

            @Override
            public void onAudioEnd() {
                System.out.println("-----onAudioEnd-----");
            }
        });
    //开始监听语音
        asrClient.startListening(asrIntent);
    }
```

AsrListener 中的 onResults(PacMap results)方法返回的结果封装在 JSON 格式的文件中,需要解析才能得到,结果说明如表 9-1 所示。

表 9-1　AsrListener 中的 onResults(PacMap results)方法返回的结果

返回结果	结果类型	结果说明

{
"result":[{
"confidence":0,
"ori_word":"你 好 ",
"pinyin":"NI3 HAO3 ",
"word":"你好。"
}]
}
	JSON	识别结果

续表

返回结果	结果类型	结果说明
{ "confidence":xxx }	Double	识别结果的置信度
{ "word":"xxx" }	String	识别结果的文本内容

示例结果(JSON)如代码示例 9-26 所示。

代码示例 9-26 语音返回的 JSON 结果

```
{
"engine_type":"local_engine",
"result":[
{
"confidence":0,
"ori_word":"你 好 ",
"pinyin":"NI3 HAO3 ",
"word":"你好."
}],
"result_type":"lvcsr",
"scenario_type":5
}
```

9.5 本章小结

本章通过一个智能手表应用案例,演示了如何通过鸿蒙 ACE Java UI 框架开发一个网络应用程序的整个流程,读者可以跟随本章的案例代码实现该案例中的功能,同时也可以在此基础上进行功能创新。

第 10 章

多设备游戏开发案例（JavaScript 版）

本章将开发一款基于鸿蒙多设备的五子棋游戏：实现 AI 人机对战、多设备联机对战。游戏过程中可以存盘并将游戏流转到附近其他手机或者 TV 设备，实现游戏界面在不同设备的适配，以及大屏和手机显示不同的内容。

10.1 五子棋游戏功能介绍

鸿蒙多设备五子棋游戏是在传统的五子棋游戏的基础之上加入鸿蒙元素，鸿蒙五子棋游戏的主要功能如下：

(1) 选择 AI 人机对战，或者选择与其他设备玩家联机对战。

(2) 游戏可以在多个设备间进行流转，例如可以在手机和 TV 设备进行流转，游戏的进度保持不变。

(3) 游戏过程中也可以把在下的棋盘发给其他设备玩家继续玩。

以下是手机屏幕的显示效果，如图 10-1 所示。

图 10-1　鸿蒙手机版五子棋图

接下来,看一下鸿蒙五子棋游戏界面的 TV 效果,如图 10-2 和图 10-3 所示。

图 10-2 TV 五子棋图

图 10-3 TV 五子棋提示图

界面采用媒体查询响应式设计,鸿蒙提供了基于 CSS 和原生接口的媒体查询,可以根据不同的设备选择不同的媒体,这样方便不同屏幕的适配能力。

10.2　五子棋游戏技术要点

上面介绍了鸿蒙五子棋游戏的核心功能，接下来看一看需要使用的技术要点有哪些。

(1) canvas 画布：通过鸿蒙 canvas 完成五子棋游戏的绘制。

(2) 鸿蒙分布式调度 API 的使用：鸿蒙分布式调度在 JavaScript 框架中使用非常简单，后面我们会介绍如何使用 FeatureAbility 实现鸿蒙应用的迁移。

10.3　五子棋游戏界面实现

为了在不同的分辨率下适配，首先需要修改 config.json 文件中 window 的配置。

当 autoDesignWidth 为 true 时，逻辑像素 px 将按照屏幕密度进行缩放，如 100px 在屏幕密度为 3 的设备上，实际渲染为 300 物理像素。当应用需要适配多种设备时，建议采用此方法，代码如下：

```
"js": [
  {
"pages": [],
"name": "default",
"window": {
"designWidth": 720,
"autoDesignWidth": true        //设置为 true
    }
  }
],
```

10.3.1　游戏界面布局

五子棋游戏的界面需要响应式设计才能够满足在不同的屏幕上显示不同的内容，所以这里采用媒体查询、FlexBox、Position 定位的方式实现。

通过 canvas 实现棋盘的绘制，游戏的角色都是在 canvas 上完成的，如代码示例 10-1 所示。

代码示例 10-1　游戏界面布局

```
<div class = "container">
<div class = "left">
<div class = "avatar">
<image src = "/common/imgs/me.png"></image>
</div>
<text>
            我
```

```
</text>
</div>
< canvas id = " chess " class = " chess " style = " width: {{ canvasWidth }} px; height:
{{canvasHeight}}px;"></canvas>
<div class = "right">
<div class = "avatar">
<image src = "/common/imgs/me.jpg"></image>
</div>
<text>
            AI
</text>
</div>
<div class = "send">
<image src = "/common/imgs/37bf.png" alt = "流转屏幕"></image>
</div>
<div class = "send2">
<image src = "/common/imgs/arrow0.png" alt = "查找联机"></image>
</div>
<div class = "send2_title">
<text>连接其他设备</text>
</div>
</div>
```

这里样式的处理也非常简单，通过媒体查询处理不同设备的适配，如代码示例 10-2 所示。

代码示例 10-2　游戏界面样式

```
.container {
    display: flex;
    justify-content: center;
    align-items: center;
    left: 0px;
    top: 0px;
    width: 100%;
    height: 100%;
    background-image: URL("/common/imgs/ee.png");
    flex-direction: row;
}
@media screen and (device-type: tv) {
    .left {
        display: flex;
        justify-content: center;
        align-items: center;
        flex-direction: column;
        height: 250px;
```

```
        padding: 10px;
        margin-right: 3px;
        background-color: gray;
        opacity: 1;
    }

    .right {
        display: flex;
        justify-content: center;
        align-items: center;
        flex-direction: column;
        height: 250px;
        padding: 10px;
        margin-left: 3px;
        background-color: gray;
        opacity: 1;
    }

    .avatar {
        width: 100px;
        height: 100px;
        border-radius: 50px;
    }
}

@media screen and (device-type: phone) {
    .left {
        display: none;
    }
    .right {
        display: none;
    }
}

.send {
    width: 50px;
    height: 50px;
    border-radius: 25px;
    position: absolute;
    top: 50px;
    left: 50px;
}

.send2 {
    width: 120px;
    height: 120px;
```

```
    position: absolute;
    top: 50px;
    right: 100px;
}

.send2_title{
    width: 120px;
    height: 120px;
    position: absolute;
    top: 30px;
    right: 70px;
}

.send2_title text {
    font-size: 12px;
    color: black;
}
```

10.3.2 画棋盘的网格

首先,需要在 canvas 上绘制出 15 行乘 12 列的长方形棋盘。在 data 中定义几个数据结构,如棋盘的列数和行数,方便后期统一修改,代码如下:

```
data: {
 cxt: {},
   rowLineNum:15,      //行数
   colLineNum:12,      //列数
   boxSize:30,         //宽和高一样
   canvasWidth:12 * 30,
   canvasHeight:15 * 30,
   lineNum : 12,       //行数和列数一样
   canvasSize:360,     //宽和高一样
   boxPadding: 15      //内边距
}
```

在 onShow 方法中获取 canvas 的 2d Context 对象,代码如下:

```
onShow() {
this.cxt = this.$element("chess").getContext("2d");
this.drawRectangleBoard();
 },
```

这里绘制的是 15×12 的方格线棋盘,如代码示例 10-3 所示。绘制棋盘的网格线,rowLineNum 为循环的行数,colLineNum 为循环的列数。boxPadding 表示第一条网格线

距离 cavans 的外边距的距离，这里设置为 15px。

代码示例 10-3 绘制 15×12 的方格线棋盘

```
drawRectangleBoard() {
        for (let index = 0; index < this.rowLineNum; index++) {
            //横线,X:轴不变,Y:轴变化
            this.cxt.beginPath();

this.cxt.moveTo(this.boxPadding,this.boxPadding + index * this.boxSize);
this.cxt.lineTo(this.canvasSize - this.boxPadding,
this.boxPadding + index * this.boxSize);
            this.cxt.closePath();
            this.cxt.stroke()
        }

        for (let index = 0; index < this.colLineNum; index++) {
            //竖线,X:轴变化,Y:轴不变
            this.cxt.beginPath();

this.cxt.moveTo(this.boxPadding + index * this.boxSize,this.boxPadding);
this.cxt.lineTo(this.boxPadding + index * this.boxSize,
this.canvasHeight - this.boxPadding);
            this.cxt.closePath();
            this.cxt.stroke()
        }
},
```

10.3.3 绘制棋盘背景

为了使棋盘的效果更加逼真，接下来为棋盘添加背景，如图 10-4 所示，通过在 canvas 上绘制图片的方法实现棋盘的背景，这里使用 drawImage 方法来绘制整个棋盘，如代码示例 10-4 所示。

代码示例 10-4 绘制棋盘背景

```
onShow() {
    this.cxt = this.$element("chess").getContext("2d");
    this.drawBoardBackground();
 },
 drawBoardBackground() {
      var img = new Image();
      img.src = "/common/imgs/6815604.jpg";
      img.onload = () =>{
this.cxt.drawImage(img,0,0,this.canvasWidth,this.canvasHeight + this.boxSize);
        this.drawRectangleBoard();
    }
}
```

图 10-4　TV 五子棋棋盘

10.4　五子棋逻辑实现(AI 篇)

本节介绍五子棋的 AI 逻辑,计算机 AI 落子实际上需要判断计算机最佳的落子坐标位置。AI 落子坐标位置需要通过对棋盘上所有未下棋子的坐标位置进行统计和评分,评分最高的位置就是计算机 AI 落子的位置。

10.4.1　在棋盘画棋子

通过在 canvas 上单击,获取单击事件的坐标位置,这里需要向 canvas 添加 ontouchstart 事件监听,代码如下:

```
<canvas ontouchstart = "drawChess"
    id = "chess"
    class = "chess"
    style = "width:{{canvasWidth}}px; height: {{canvasHeight}}px;">
</canvas>
```

通过 touchstart 事件 e 对象获取位置信息。在棋盘上单击落子,单击的位置可能偏离棋盘交叉点的位置,这个时候需要通过 floor()方法执行向下取整计算,它返回的值小于或等于函数参数,并且是与之最接近的整数。

棋子可通过 arc 方法绘制半径为 13px 的圆形,如代码示例 10-5 所示。

代码示例 10-5　画棋子

```
drawChess(e) {
        let x = Math.floor(e.touches[0].localX  / this.boxSize)
        let y = Math.floor(e.touches[0].localY/ this.boxSize)
        this.cxt.beginPath();
        this.cxt.arc( this.boxPadding  + x * this.boxSize,
this.boxPadding  + y * this.boxSize,13,0,2 * Math.PI)
        this.cxt.closePath();
        this.cxt.fill();
}
```

10.4.2　实现落子判断

落子分为自己的棋子和对方的棋子，这里有三点需要注意。第一，判断游戏是否已结束；第二，是否自己已经下过棋子，如果下过则不能再下；第三，判断棋盘哪些位置可以放棋子，已经下过棋子的位置不能再放棋子。在 data 中添加数据结构，代码如下：

```
me:true,            //判断是否是自己
over:false,         //判断游戏是否已结束
chessboard:[],      //棋盘数组
myWin:[],           //自己的赢法统计
computerWin:[]      //对方的赢法统计
wins: [],           //所有的赢法三维数组
count: 0,           //赢法统计
```

接下来在 onInit 方法中初始化上面的数据，onInit 方法在页面加载时调用一次。这里需要初始化的数据包括：棋盘数组，即用来保存整个棋盘状态的二维数组；自己的赢法数组，即用来记录自己在每个赢法位置上的落子数；AI 的赢法数组，即用来记录计算机在每个赢法位置上的落子数，如代码示例 10-6 所示。

代码示例 10-6　初始化每个赢法位置上的棋子数

```
onInit(){
    this.initChessData()
},
initChessData() {
    for (let i = 0; i < this.rowLineNum; i++) {
        this.chessboard[i] = []
        for (let j = 0; j < this.rowLineNum; j++) {
            this.chessboard[i][j] = 0
        }
    }
```

```
        //初始化每组的统计
        for (let index = 0; index < this.count; index++) {
            this.myWin[index] = 0;
            this.computerWin[index] = 0;
        }
    },
```

onPressChess 方法绑定在 canvas 的 ontouchstart 事件上，如代码示例 10-7 所示，需要处理的逻辑有以下几个：

（1）绘制棋子前判断游戏是否已结束，如果游戏已结束则返回，这里通过 over 属性控制游戏是否结束。

（2）判断我是否已经下过棋子，通过 me 属性来判断，这个属性的值在落子后会被设置为 false，这个是五子棋的游戏规则。

（3）通过棋盘数组判断当前的棋盘落子位置是否已经落过棋子，如果 this.chessboard[x][y]为 0，则表示当前位置没有落棋子。

（4）落完棋子后，需要把棋盘数组的 x 和 y 的位置设置为 1，表示已经落子，不能再落棋子了。

（5）在绘制完棋子后，把棋盘数组的 x 和 y 的位置设置为 1。

（6）如果游戏没有结束，则需要把 me 的值设置为 false，以便 AI 计算机继续下棋子。

代码示例 10-7　下棋

```
    onPressChess(e) {
        //游戏是否已结束
        if (this.over) {
            return
        }

        //如果不是我,则直接返回
        if (!this.me) {
            return
        }

        let x = Math.floor(e.touches[0].localX  / this.boxSize)
        let y = Math.floor(e.touches[0].localY/ this.boxSize)

        //判断当前位置是否有棋子,0 表示没有,1 表示有
        if (this.chessboard[x][y] == 0) {
//在 x 和 y 这个位置画棋子
            this.drawChess(x,y);

            //判断是否赢了
```

```
            //如果游戏没有结束,则当前是个开关,AI 计算机可继续下棋子
            if (!this.over) {
                this.me = !this.me;
            }
        }
    },

    drawChess(x,y) {
    //每次落子之前,把 x 和 y 的位置设置为 1,这个位置下次不能落子,这个很重要
    this.chessboard[x][y] = 1
        var px = this.boxPadding + x * this.boxSize;
        var py = this.boxPadding + y * this.boxSize
        this.cxt.beginPath();
        this.cxt.arc(px,py,13,0,2 * Math.PI)
        this.cxt.closePath();
        if (!this.me) {
            this.cxt.fillStyle = "white"
        } else {
            this.cxt.fillStyle = "black"
        }
        this.cxt.fill();
    }
```

10.4.3 赢法数组

判断输赢的关键就是判断谁的棋子落子后可以5个棋子连成一条线上了。怎么判断5个棋子已经连在一条线上了呢?

赢法数组是用来存放棋盘上所有可能连成一条线的5个棋子,5个棋子一组,在棋盘上有可以连成5个棋子的一组连续棋子组成一个赢法,那么怎样统计这些赢法呢?

这里要讲的是赢法数组,此数组是一个定义好的三维数组。

可以把一个棋盘分为横向、纵向、正斜线和反斜线。从这4个方向统计可能的赢法。

横向如何统计呢?需要一行一行地统计,其实就是循环从第一个位置向后4位,再从第二位向后4位,直到行的总列减4的位置为终点,这个位置后面就不能再移位了,因为再移位就不足5位数了,如图10-5所示,代码如下:

```
//记录 x 赢法数组
for (let i = 0; i < this.rowLineNum; i++) {
    for (let j = 0; j < this.rowLineNum - 4; j++) {
        for (let k = 0; k < 5; k++) {
            this.wins[j + k][i][this.count] = true
        }
        this.count++
    }
}
```

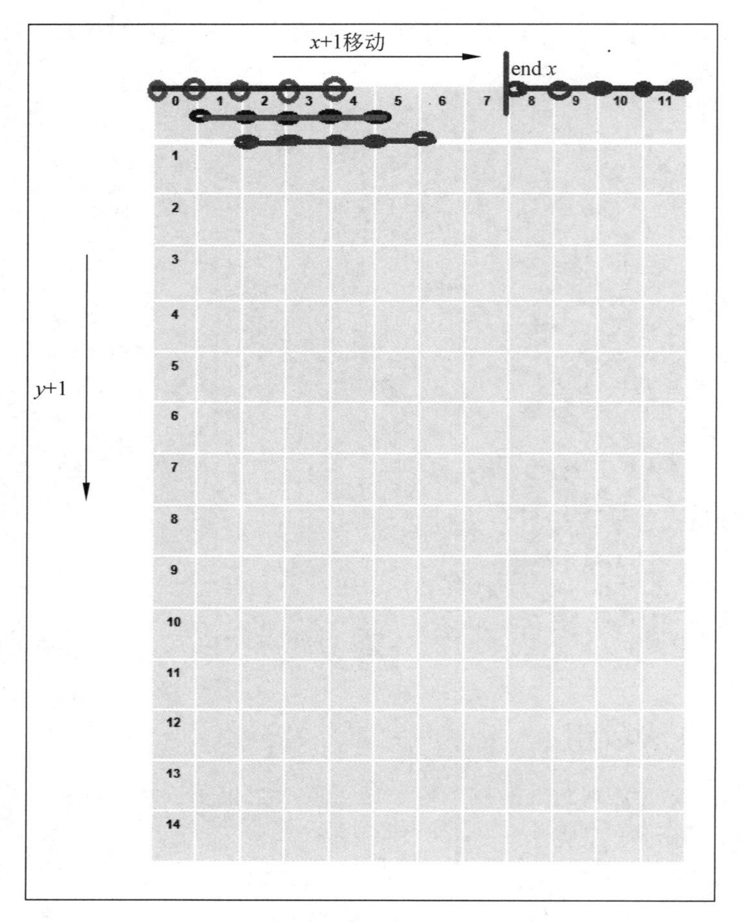

图 10-5 横向统计所有赢法

纵向需要一列一列地统计，从第一列的第一位开始，直到这列的总行数减 4 的位置为终点，如图 10-6 所示。

```
//记录 y 赢法数组
for (let i = 0; i < this.rowLineNum; i++) {
    for (let j = 0; j < this.rowLineNum - 4; j++) {
        for (let k = 0; k < 5; k++) {
            this.wins[i][j + k][this.count] = true
        }
        this.count++
    }
}
```

正斜线可以从棋盘的第一个交叉点开始统计，斜线的方式连成 5 个棋子。第一次循环

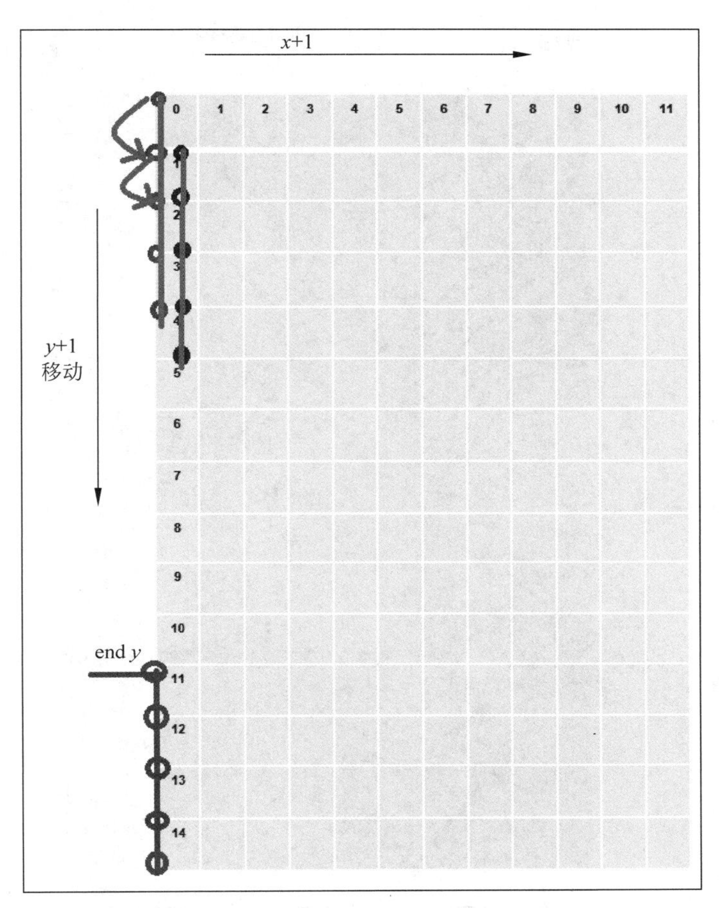

图 10-6 纵向统计所有赢法

x 从 0 坐标开始，y 轴加 1 循环，y 轴循环到 this.rowLineNum －4 位置。重新循环 x 轴直到 this.rowLineNum －4 坐标的位置，如图 10-7 所示。

```
for (var i = 0;i < this.rowLineNum - 4; i++) {
    for (var j = 0;j < this.rowLineNum - 4; j++) {
        //k 棋子
        for (var k = 0;k < 5; k++) {
            this.wins[i + k][j + k][this.count] = true
        }
        this.count++;
    }
}
```

反斜线可以从棋盘底部的第一个交叉点开始统计，以斜线的方式连成 5 个棋子，第一次循环 x 从 15 坐标开始，y 轴减 1 循环，y 轴循环到 3 的位置。重新循环 x 轴直到 this.

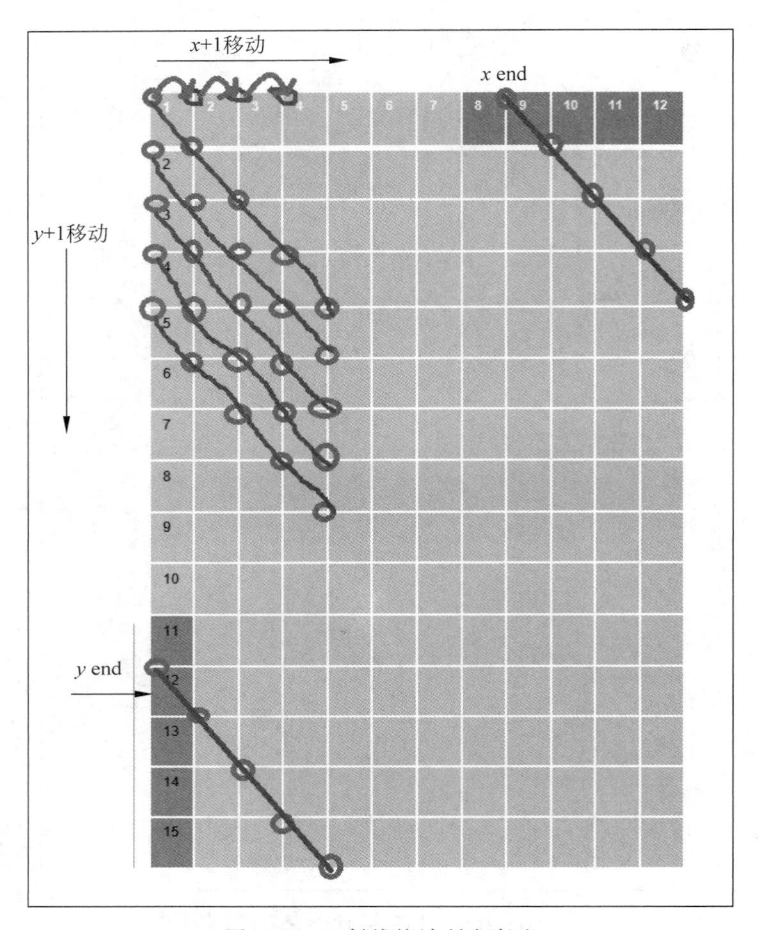

图 10-7 正斜线统计所有赢法

rowLineNum −4 坐标的位置,如图 10-8 所示。

```
//统计反斜线
for (var i = 0;i < this.rowLineNum - 4; i++) {
    for (var j = this.rowLineNum - 4;j > 3; j--) {
        //k 棋子
        for (var k = 0;k < 5; k++) {
            this.wins[i + k][j - k][this.count] = true
        }
        this.count++;
    }
}
```

通过 initWinGroup 方法初始化 x 和 y,统计斜线方向的赢法,放到赢法数组 wins 中,如代码示例 10-8 所示。

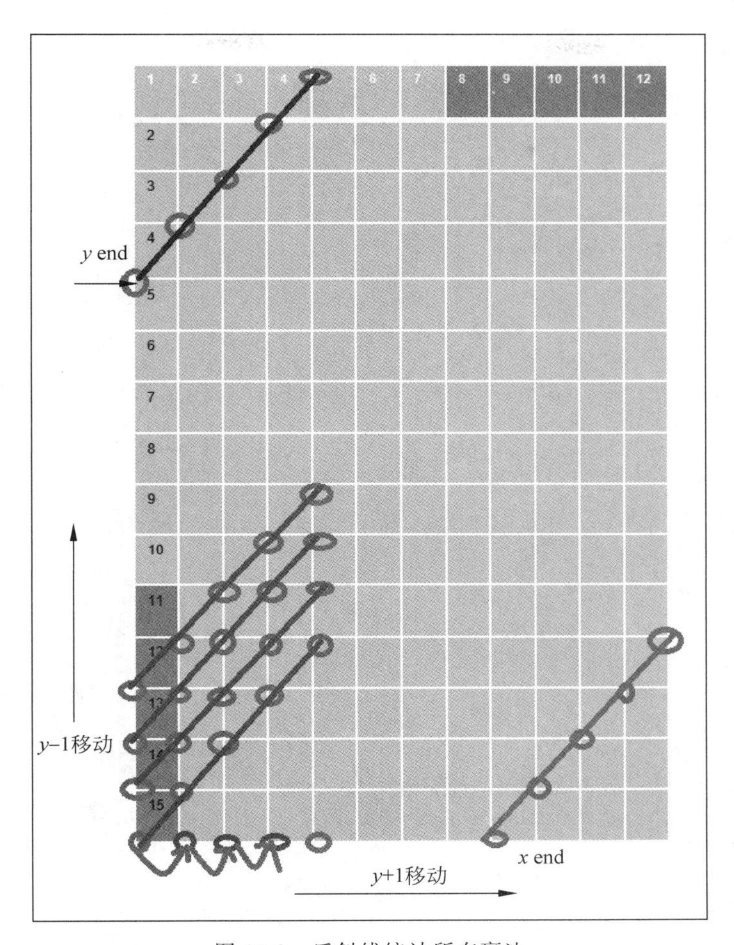

图 10-8 反斜线统计所有赢法

代码示例 10-8 初始化赢法数组

```
onInit() {
    this.initWinGroup();
    this.initChessData();
},
initWinGroup() {

    //初始化赢法数组 x、y、groupid(第几组赢法)
    for (let i = 0; i < this.rowLineNum; i++) {
        this.wins[i] = []
        for (let j = 0; j < this.rowLineNum; j++) {
            this.wins[i][j] = []
        }
```

```
        }

        //记录 x 赢法数组
        for (let i = 0; i < this.rowLineNum; i++) {
            for (let j = 0; j < this.rowLineNum - 4; j++) {
                for (let k = 0; k < 5; k++) {
                    this.wins[j + k][i][this.count] = true
                }
                this.count++
            }
        }

        //记录 y 赢法数组
        for (let i = 0; i < this.rowLineNum; i++) {
            for (let j = 0; j < this.rowLineNum - 4; j++) {
                for (let k = 0; k < 5; k++) {
                    this.wins[i][j + k][this.count] = true
                }
                this.count++
            }
        }

        //正斜线赢法数组
        //统计正斜线
        for (var i = 0;i < this.rowLineNum - 4; i++) {
            for (var j = 0;j < this.rowLineNum - 4; j++) {
                //k 棋子
                for (var k = 0;k < 5; k++) {
                    this.wins[i + k][j + k][this.count] = true
                }
                this.count++;
            }
        }

        //统计反斜线
        for (var i = 0;i < this.rowLineNum - 4; i++) {
            for (var j = this.rowLineNum - 4;j > 3; j--) {
                //k 棋子
                for (var k = 0;k < 5; k++) {
                    this.wins[i + k][j - k][this.count] = true
                }
                this.count++;
            }
        }
    },
    initChessData() {
```

```
    for (let i = 0; i < this.rowLineNum; i++) {
        this.chessboard[i] = []
        for (let j = 0; j < this.rowLineNum; j++) {
            this.chessboard[i][j] = 0
        }
    }

    //初始化每组的统计
    for (let index = 0; index < this.count; index++) {
        this.myWin[index] = 0;
        this.computerWin[index] = 0;
    }

},
```

10.4.4 判断是否赢棋

落子后,程序需要判断当前是否有连续5个棋子连成一条线的情况,那么怎样判断是否有5个棋子连成一条线呢?这里通过两个数组:一个是自己的赢法数组,另一个是AI的赢法数组。这两个数组都记录了各自的赢法位的棋子数,当一方的棋子在赢法位上等于5时,那么就是谁获胜,如代码示例10-9所示。

代码示例 10-9 判断是否赢棋

```
checkWin(x, y) {
    for (let index = 0; index < this.count; index++) {
        if (this.wins[x][y][index]) {
            //如果是自己,则在自己的统计中添加次数
            if (this.me) {
                this.myWin[index] += 1;
                if (this.myWin[index] == 5) {
                    this.over = true;
                    prompt.showToast({
                        message: "恭喜,你赢了!",
                        duration: 3000
                    })
                }
            } else {
                this.computerWin[index] += 1;
                if (this.computerWin[index] == 5) {
                    this.over = true;

                    prompt.showDialog({
                        title: 'ooss,你输了!',
```

```
                        message: 'ooss,你输了!',
                        buttons: [
                            {
                                text: '重新再来',
                                color: '#666666',
                            },
                        ],
                        success: function (data) {
                            router.push({
                                uri: "pages/chess/chess"
                            })
                        },
                        cancel: function () {
                            console.log('dialog cancel callback');
                        },
                    });
                }
            }
        }
    }
},
```

10.4.5 实现计算机 AI 落子

计算机 AI 落子,实际上需要判断计算机最佳的落子坐标位置,如代码示例 10-10 所示。

代码示例 10-10 计算机 AI 落子实现

```
computerAI() {

    //计算每个子所在的分值
    var myScore = []
    var computerScore = []
    for (var i = 0;i < this.rowLineNum; i++) {
        myScore[i] = []
        computerScore[i] = []
        for (var j = 0;j < this.rowLineNum; j++) {
            myScore[i][j] = 0;
            computerScore[i][j] = 0;
        }
    }

    //打分
    var maxScore = 0;
    var maxX = 0;
    var maxY = 0;
```

```
//计算 maxscore
for (var i = 0;i < this.rowLineNum; i++) {
    for (var j = 0;j < this.rowLineNum; j++) {
        //计算计算机最优的下子位置
        //棋盘上没有下子的位置
        if (this.chessboard[i][j] == 0) {

            for (var k = 0; k < this.count; k++) {
                if (this.wins[i][j][k]) {

                    //me
                    if (this.myWin[k] == 1) {
                        myScore[i][j] += 200
                    } else if (this.myWin[k] == 2) {
                        myScore[i][j] += 400
                    } else if (this.myWin[k] == 3) {
                        myScore[i][j] += 1000
                    } else if (this.myWin[k] == 4) {
                        myScore[i][j] += 10000
                    }

                    //computer
                    if (this.computerWin[k] == 1) {
                        computerScore[i][j] += 220
                    } else if (this.computerWin[k] == 2) {
                        computerScore[i][j] += 440
                    } else if (this.computerWin[k] == 3) {
                        computerScore[i][j] += 1600
                    } else if (this.computerWin[k] == 4) {
                        computerScore[i][j] += 30000
                    }
                }
            }

            //拦截
            if (myScore[i][j] > maxScore) {
                maxScore = myScore[i][j];
                maxX = i;
                maxY = j;
            } else if (myScore[i][j] == maxScore) {
                if (computerScore[i][j] > maxScore) {
                    maxX = i;
                    maxY = j;
                }
            }
```

```
                //落子
                if (computerScore[i][j] > maxScore) {
                    maxScore = computerScore[i][j];
                    maxX = i;
                    maxY = j;
                } else if (computerScore[i][j] == maxScore) {
                    if (myScore[i][j] > maxScore) {
                        maxX = i;
                        maxY = j;
                    }
                }
            }
        }
    }
    this.drawChess(maxX,maxY);
    this.checkWin(maxX,maxY)

    if (!this.over) {
        this.me = !this.me
    }
},
```

这个最佳的坐标位置,需要通过对棋盘上所有未下棋子的坐标位置进行统计和评分,评分最高的位置就是计算机 AI 落子的位置。

在每次落棋子的时候,需要有两个数组来记录棋子,一个是自己的棋子坐标,另一个是计算机的棋子坐标。

那么如何对棋盘上未下棋子的坐标位置评分呢?这里需要考虑两种情况,一种是拦截子,拦截子是拦截别人赢棋的棋子,另一种是计算机可能会赢棋的棋子,这个棋子是计算机最有可能连接 5 个子的坐标位置。

首先需要对所有赢法数组进行遍历,统计每组赢法上的棋子数,棋子数越多,那么这个赢法组的空白棋子位置的分值越高。

10.5 五子棋逻辑实现(鸿蒙篇)

通过 10.4 节所示的步骤,我们完成了一个五子棋的基本功能,接下来,我们需要给五子棋添加鸿蒙操作系统间多设备流转和操作能力。

10.5.1 多设备流转需要满足的条件

要使用鸿蒙的分布式能力,首先需要在 config.json 中配置相应的权限,代码如下:

```
"reqPermissions": [
  {
"name": "ohos.permission.DISTRIBUTED_DATASYNC"
  },
  {
"name": "ohos.permission.servicebus.ACCESS_SERVICE"
  },
  {
"name": "ohos.permission.servicebus.BIND_SERVICE"
  },
  {
"name": "ohos.permission.DISTRIBUTED_DEVICE_STATE_CHANGE"
  },
  {
"name": "ohos.permission.GET_DISTRIBUTED_DEVICE_INFO"
  },
  {
"name": "ohos.permission.GET_BUNDLE_INFO"
  }
]
```

同时需要在 MainAbility 中获取权限,如代码示例 10-11 所示。

代码示例 10-11 获取权限

```
public class MainAbility extends AceAbility {
    @Override
    public void onStart(Intent intent) {
        setInstanceName("control");
        super.onStart(intent);
        requestPermission();
    }

    //获取权限
    private void requestPermission() {
        String[] permission = {
"ohos.permission.READ_USER_STORAGE",
"ohos.permission.WRITE_USER_STORAGE",
"ohos.permission.DISTRIBUTED_DATASYNC",
"ohos.permission.MICROPHONE",
"ohos.permission.GET_DISTRIBUTED_DEVICE_INFO",
"ohos.permission.KEEP_BACKGROUND_RUNNING",
"ohos.permission.NFC_TAG"};
        List<String> applyPermissions = new ArrayList<>();
        for (String element : permission) {
            if (verifySelfPermission(element) != 0) {
```

```
                if (canRequestPermission(element)) {
                    applyPermissions.add(element);
                }
            }
        }
        requestPermissionsFromUser(applyPermissions.toArray(new String[0]), 0);
    }

    @Override
    public void onStop() {
        super.onStop();
    }
}
```

除了以上多权限配置外,还需要使用同一个华为账号登录多台需要流转应用的设备。

10.5.2 多设备间游戏流转实现

现在给五子棋游戏添加在多个设备间进行流转的能力,例如可以在手机和 TV 设备间进行流转。利用鸿蒙操作系统提供的分布式软总线能力,可以实现在电视与手机间的应用切换,流转效果如图 10-9 所示。

图 10-9 TV 五子棋与手机进行流转

流转只需调用 FeatureAbility.continueAbility(0,null),0 表示第一个附件设备的索引号。

在流转应用前,首先需要在 onSaveData 方法中临时缓存一下需要恢复的数据,这个我们把它叫作存盘,当程序成功流转到其他设备的时候,会调用 onRestoreData 回调方法,这个时候把缓存的数据取出来，绑定到 data 中就可以恢复数据了，如代码示例 10-12 所示。

代码示例 10-12　多设备间数据流转

```
transfer: async function transfer() {
    try {
        await FeatureAbility.continueAbility(0,null)
    } catch (e) {
        console.error("迁移出错: " + JSON.stringify(e))
    }
},
onStartContinuation: function onStartContinuation() {
    return true
},
onSaveData: function onSaveData(saveData) {
    //这个地方需要对象复制
    Object.assign(saveData, "这个地方需要对象复制")
    return true
},
//在 oninit 方法之前调用 onRestoreData 方法
onRestoreData: function onRestoreData(restoreData) {
    this.title = restoreData
    return true
},
```

提示：流转只需调用 FeatureAbility.continueAbility(0,null)。这种方法使用了 ES 高级语法，需要在代码中引入 pollyfill

```
const injectRef = Object.getPrototypeOf(global) || global
injectRef.regeneratorRunTime = require('@babel/RunTime/regenerator')
```

这行代码必须加上，否则会报错。

单击“连接其他设备”按钮时，完成手机设备的流转，如图 10-10 所示。下面介绍一下实现方式。

图 10-10　将应用流转到其他连接设备上

要实现流转屏幕，需要使用上面介绍的几个回调方法，onSaveData 方法用来在流转前保存需要流转后恢复的数据，流转成功后调用 onRestoreData 方法，从参数中取出之前保存的数据对象。

这里需要获取在 onSaveData 中保存的数据,此数据是棋盘数组,在这个数组中记录的是每个坐标位置上是否有棋子,如果有棋子则为 1,如果没有棋子则为 0。另外还有一个用来保存棋子是谁下的数组,有了这两个数组,就可以在 onRestoreData 方法调用后重新恢复棋盘上的棋子了。

说明:这里需要注意的是,重绘棋盘的时机,因为棋子重绘需要使用 canvas 对象,所以只能在 onShow 方法实现。

下面具体介绍如何为五子棋游戏添加游戏流转功能,具体实现步骤如下。

步骤 1:定义 restoreAll 数据结构,用来存放恢复的数据。

```
data: {
restoreAll:{} //流转后恢复的数据
}
```

步骤 2:在 onSaveData 中保存两个数组数据:第 1 个,chessboard 数组,即键盘上的棋子数组,0 为没有棋子,1 为有棋子;第 2 个,chessboardName 数组,即用来保存棋盘上的黑子和白子信息的数组,1 表示我,0 表示机器 AI,如代码示例 10-13 所示。

代码示例 10-13 在 onSaveData 中保存两个数组数据

```
onStartContinuation: function onStartContinuation() {
        return true
},
onSaveData: function onSaveData(saveData) {
        saveData.data = this.chessboard;
        saveData.names = this.chessboardName;
        return true
},
onRestoreData: function onRestoreData(restoreData) {
          //注意: 这种方法执行在 oninit 方法之前
      this.restoreAll = {
          data:restoreData.data,
          names: restoreData.names
}
      prompt.showToast({
          message: "切换到了其他屏幕!",
          duration: 3000
      })
      return true
},
```

步骤3：根据棋盘数据，重新绘制棋盘。在流转后恢复棋盘的时候，要注意canvas绘制的先后顺序，并确保棋子的恢复绘制是在棋盘绘制完成后。避免出现canvas后绘制的棋盘覆盖前面绘制的棋子。

流转后，onRestoreData方法的调用是在onShow方法之前，所以只需要在onShow方法中判断是否存在恢复的数据，如果有需要恢复的数据，则需重绘棋盘，如代码示例10-14所示。

代码示例10-14 存在恢复的数据，重绘棋盘

```
onShow() {
    this.cxt = this.$element("chess").getContext("2d");
    this.drawBoardBackground();
},
drawBoardBackground() {
    var img = new Image();
    img.src = "/common/imgs/6815604.jpg";
    img.onload = () =>{
        this.cxt.drawImage(img,0,0,this.canvasWidth,this.canvasHeight + this.boxSize)
        this.drawRectangleBoard()
        //恢复数据后画棋子,注意canvas后绘制的棋盘覆盖前面绘制的棋子
        if(this.restoreAll.data.length > 0){
            this.repaitChessBoard()
        }
    }
},
reDrawChessBoard(x,y,isMe) {
    var px = 15 + x * 30;
    var py = 15 + y * 30
    this.cxt.beginPath();
    this.cxt.arc(px,py,13,0,2 * Math.PI)
    this.cxt.closePath();
    if(isMe){
        this.cxt.fillStyle = "black"
    }else{
        this.cxt.fillStyle = "white"
    }

    this.cxt.fill();
},
repaitChessBoard() {
    let isMe = true;
    for (let i = 0; i < this.rowLineNum; i++) {
        for (let j = 0; j < this.rowLineNum; j++) {
            if (this.restoreAll.data[i][j] == 1) {
```

```
                    if(this.restoreAll.names[i][j] == 1){
                        isMe = true
                    }else {
                        isMe = false
                    }
                    this.reDrawChessBoard(i,j,isMe)
                }
            }
        }
    },
```

到此就可以实现鸿蒙操作系统间应用的流转了,鸿蒙应用流转效果如图 10-11 所示。

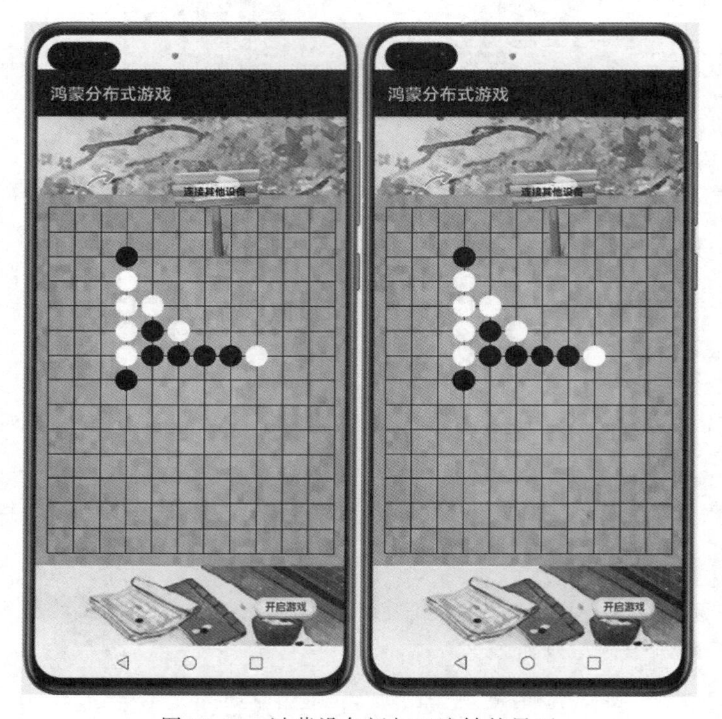

图 10-11　鸿蒙设备间相互流转的界面

10.6　本章小结

本章通过开发一款鸿蒙风格的五子棋游戏,让开发者可以学习如何开发多设备间流转的五子棋游戏,在这个游戏中涉及了鸿蒙操作系统的分布式流转 API 的用法,读者应特别注意鸿蒙操作系统的应用流转是需要先决条件的。

第 11 章

多设备应用开发案例（Java＋JavaScript 版）

本章介绍如何实现一个可供多设备共享涂鸦的画板：①实现在画板上随意绘制及涂鸦；②把涂鸦作品分享给其他人进行查看修改；③也可以把修改的成果返回给涂鸦者；④让涂鸦者在修改的基础上继续涂鸦。

11.1 鸿蒙涂鸦画板介绍

这里我们介绍一下基于鸿蒙操作系统的多设备共享涂鸦画板需要实现的一些功能：

- 涂鸦者可以选择不同的画笔及颜色进行自由涂鸦，也可以清除画板。
- 涂鸦者也可以选择图片进行涂鸦。
- 涂鸦者可以一键寻找附近的手机或者其他设备，并连接希望连接的设备。
- 涂鸦者选择好希望连接的设备后，可以直接把涂鸦成果流转给对应的设备。
- 其他设备接收流转的涂鸦后，可以在涂鸦的基础上添加涂鸦或者修改。
- 修改后的涂鸦可以继续流转给涂鸦者，或者流转到其他设备上，如电视上。
- 涂鸦者可以开启实时画板功能，可以与同 WiFi 下的其他设备进行实时涂鸦。

接下来，看一下共享涂鸦画板效果，如图 11-1 所示。

11.2 共享涂鸦画板技术要点

在涂鸦画板的游戏中我们将使用如下技术。

(1) canvas 的使用：这里使用 Canvas JavaScript API 实现画板的绘制。

(2) Service Ability 实现页面数据订阅与同步，实现远程设备的数据同步。

(3) 分布式数据库的使用：通过分布式数据库保存和共享数据。

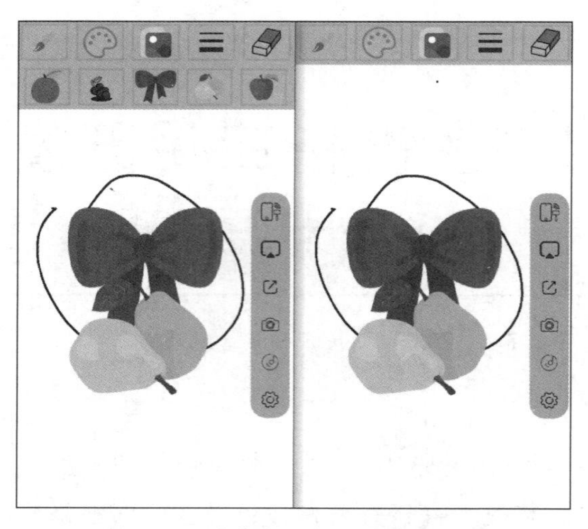

图 11-1 鸿蒙多设备共享涂鸦画板

11.3 涂鸦画板的界面实现

图 11-1 展示了涂鸦画板的界面效果，整个界面分为三部分：头部的菜单、右侧的导航菜单和底部的 canvas 画板。

11.3.1 涂鸦画板的界面布局

这里采用 Flex 加定位的布局方式，如代码示例 11-1 所示。

代码示例 11-1 涂鸦画板界面布局

```
<div class = "container">
<canvas class = "board" id = "board" ontouchmove = "paint" ontouchend = "paintEnd" ontouchstart
 = "painStart"></canvas>
<div class = "tools">
<div class = "item" ontouchstart = "showPanel(1)">
< image src = "/common/fonts/img/yanse.png"></image>
</div>
<div class = "item" ontouchstart = "showPanel(2)">
< image src = "/common/fonts/img/tx - changfangxing.png"></image>
</div>
<div class = "item">
< image src = "/common/fonts/img/xianduan.png"></image>
</div>
<div class = "item">
< image src = "/common/fonts/img/maobi.png"></image>
</div>
```

```
<div class = "item" ontouchstart = "clearCanvas">
<image src = "/common/fonts/img/xiangpica.png"></image>
</div>
</div>

<div class = "yanse" id = "yanse" if = "isYanse">
<div class = "cc" ontouchstart = "selectColor(index)" for = "{{(index,c) in colors}}" tid =
"index">
<div class = "bg" style = "background-color: {{c.bg}}; border:3px solid {{c.bc}}">
</div>
</div>
</div>

<div class = "xz" id = "xz" if = "isxz">
<div class = "cc">
<image src = "/common/fonts/img/xingzhuang-sanjiaoxing.png"></image>
</div>
<div class = "cc">
<image src = "/common/fonts/img/tx-changfangxing.png"></image>
</div>
</div>


<div class = "linksb"   ontouchstart = "linkDevice">
<image if = "isLink" src = "/common/fonts/img/ziyuan_active.png" class = "img"></image>
<image else src = "/common/fonts/img/ziyuan.png" class = "img"></image>
</div>
<div class = "changeX" ontouchstart = "changeLocal">
<image if = "localAct" src = "/common/fonts/img/shuangmoshiqiehuan_active.png" class = "img">
</image>
<image else src = "/common/fonts/img/shuangmoshiqiehuan.png" class = "img"></image>
</div>
</div>
```

11.3.2 涂鸦画板的界面样式

这里采用 div 进行布局，结合 Image 和 Canvas 组件进行界面绘制，如代码示例 11-2 所示。

代码示例 11-2 涂鸦画板样式

```
.container {
    display: flex;
    justify-content: center;
    align-items: center;
```

```
    left: 0px;
    top: 0px;
    width: 100 % ;
    height: 100 % ;
    color: #585858;
    font - family: Source Sans Pro,Helvetica,sans - serif;
}

.tools {
    position: absolute;
    top: 0px;
    left: 0px;
    height: 60px;
    width: 100 % ;
    background - color: #ccc;
    display: flex;
    flex - direction: row;
    justify - content: center;
    align - items: center;
}

.item {
    width: 100px;
    height: 100 % ;;
    border: 3px solid silver;
    margin: 5px;
    justify - content: center;
    align - items: center;
}

image {
    object - fit: contain;
}
.item:active {
    border: 3px solid gold;
}
.bg {
    width: 100 % ;
    height: 100 % ;
}

.board {
    width: 100 % ;
    height: 100 % ;
    background - color: #fff;
}
```

```
.yanse ,.xz {
    position: absolute;
    top: 60px;
    left: 0px;
    height: 60px;
    width: 100%;
    background-color: #ccc;
    display: flex;
    flex-direction: row;
    justify-content: center;
    align-items: center;
    flex-wrap: wrap;
}

.cc {
    width: 60px;
    height: 100%;;
    border: 3px solid silver;
    margin: 5px;
    justify-content: center;
    align-items: center;
}
.cc:active {
    border: 3px solid gold;
}

.linksb {
    position: absolute;
    bottom: 10px;
    right: 0px;
    height: 50px;
    width: 50px;
    display: flex;
    flex-direction: row;
    justify-content: center;
    align-items: center;
    border-radius: 20px;
}

.changeX {
    position: absolute;
    bottom: 10px;
    left: 0px;
```

```
    height: 50px;
    width: 50px;
    display: flex;
    flex-direction: row;
    justify-content: center;
    align-items: center;
    border-radius: 25px;
}
.changeX:active {
    border: 3px solid gold;
}
.img {
    width: 50%;
    height: 50%;
}
```

以上的样式并不复杂，读者可以参考实现，整体界面并不复杂，因此可以通过 FlexBox 加 position:absolute 实现，当然也可使用 Stack 组件。

11.4 涂鸦画板核心代码实现

本节介绍分布式设备间如何共享涂鸦画板的核心功能。

提示：本书中涉及的手机之间的应用流转，需要开发者注册华为开发者账号，同时在手机设备上进行登录，否则无法流转。

11.4.1 实现画板的自由绘制

画板的自由绘制，实际上是通过不断检测 canvas 上的坐标位置来不断连接移动坐标点实现绘制线条的，这里需要检测 canvas 中的 3 个事件 touchStart、touchMove 和 touchEnd，这 3 个事件中的事件对象封装了 touch 的坐标信息，利用这个坐标信息就可以实现线条的自由绘制了。

首先，需要通过 onShow 方法获取 canvas 对象，并获取 canvas 对象的二维上下文对象 Context，代码如下：

```
onShow() {
//这里需要把获取的二维 Context 对象放到 data 中,方便后面的方法使用
  this.cxt = this.$element("board").getContext("2d");
},
```

实现在 canvas 上自由绘制图形，需要在 cavans 上通过 touchstart、touchmove 和 touchend 绑定 3 个事件方法。这里的 3 个事件的事件 e 对象都封装了 touch 事件的坐标信息，这里使用 localX 和 localY 这两个对象的坐标信息。localX：距离被触摸组件左上角的横向距离，组件的左上角为原点；localY：距离被触摸组件左上角的纵向距离，组件的左上角为原点。代码如下：

```
touchStart(e) {
    this.cxt.beginPath();
    this.cxt.strokeStyle = this.selColor
    this.cxt.lineWidth = 10
    this.cxt.lineCap = "round"
    this.cxt.lineJoin = "round"
    this.cxt.moveTo(e.touches[0].localX,e.touches[0].localY)
},
touchMove(e) {
     this.cxt.lineTo(e.touches[0].localX,e.touches[0].localY);
     this.cxt.stroke();
},
paintEnd() {
     this.cxt.closePath();
},
```

使用涂鸦画板进行自由绘制，如代码示例 11-3 所示。

代码示例 11-3　涂鸦画板自由绘制

```
export default {
    data: {
        cxt: {},
        colors:[
            {bg:"red",bc:"white"},
            {bg:"blue",bc:"white"},
            {bg:"green",bc:"white"},
            {bg:"black",bc:"white"}
        ],
        isYanse:false,
        selColor:"black"
    },
    onInit() {
    },
    onShow() {
        this.cxt = this.$element("board").getContext("2d");
    },
    showPanel(){
```

```
            this.isYanse = !this.isYanse
        },
        painStart(e) {
            this.cxt.beginPath();
            this.cxt.strokeStyle = this.selColor
            this.cxt.lineWidth = 10
            this.cxt.lineCap = "round"
            this.cxt.lineJoin = "round"
            this.cxt.moveTo(e.touches[0].localX,e.touches[0].localY)
        },
        paint(e) {
            console.error(e.touches[0].localX);
            this.cxt.lineTo(e.touches[0].localX,e.touches[0].localY);
            this.cxt.stroke();
        },
        paintEnd() {
            this.cxt.closePath();
        },
        clearCanvas(){
            this.cxt.clearRect(0,0,1000,1000)
        },
        selectColor(index){
            this.selColor = this.colors[index].bg;
            this.colors.forEach(c =>{
                c.bc = "white"
            })
            this.colors[index].bc = "black";
        },
    }
```

涂鸦者可以选择不同的画笔、颜色进行自由涂鸦，也可以清除画板。这里选择不同的颜色是通过 this. cxt. strokeStyle 设置的，清除画板使用 this. cxt. clearRect(0，0，canvas. width，canvas. height)实现，非常简单。

11.4.2 选择图片进行涂鸦

通过上面的思路，选择图片进行涂鸦的方法，需要使用 canvas 的 drawImage 方法实现。首先选择需要涂鸦的图片，单击屏幕的位置，然后就可以绘制出所选的涂鸦的图片了。图片涂鸦的效果如图 11-2 所示。

选择图片进行涂鸦的方法如代码示例 7-4 所示。

图 11-2　图片涂鸦

代码示例 11-4　涂鸦画板绘制图形

```
drawGraph(index) {
    this.selDrawLine = false;
    switch (index) {
        case 1:
        this.selDrawPic = "/common/fonts/img/taozi.png"
        break;
        case 2:
        this.selDrawPic = "/common/fonts/img/putao.png"
        break;
        case 3:
        this.selDrawPic = "/common/fonts/img/hudiejie.png"
        break;
        case 4:
        this.selDrawPic = "/common/fonts/img/wodeshuiguoicon-.png"
        break;
        case 5:
        this.selDrawPic = "/common/fonts/img/wodeshuiguoicon_huabanfuben.png"
        break;
    }
},
```

单击屏幕绘制图片的方法可通过修改 paintCanvas 方法实现，如代码示例 11-5 所示。在 data 中添加 selDrawPic 属性，用来记录选择的图片，当单击画笔的按钮时，将

selDrawLine 设置为 true。

代码示例 11-5 涂鸦画板绘制图形方法

```
paintCanvas(point) {
    if (this.selDrawPic != "") {
        var img = new Image();
        img.src = this.selDrawPic;
        img.onload = () =>{
            if (point.flag == "start") {
                this.cxt.drawImage(img,point.x - 100,point.y - 100,200,200)
            }
        }
        this.selDrawPic = ""
    }
    if (this.selDrawLine != "") {
        if (point.flag == "start") {
            this.cxt.beginPath();
            this.cxt.strokeStyle = point.c
            this.cxt.lineWidth = this.lineHeight
            this.cxt.lineCap = "round"
            this.cxt.lineJoin = "round"
            this.cxt.moveTo(point.x,point.y);
        } else if (point.flag == "line") {
            this.cxt.lineTo(point.x,point.y);
            this.cxt.stroke()
        }
    }
},
```

11.4.3 查找附近的手机设备

涂鸦者可以一键寻找附近的手机或者其他设备，并连接希望连接的设备。如何发现和寻找附近的手机或者其他可以连接的设备呢？在鸿蒙操作系统的设备上，只需用同一个华为账号进行登录就可以通过一行代码实现查找、发现、连接的过程，如图 11-3 所示。

图 11-3 查找及发现附近的华为鸿蒙操作系统设备

要使用鸿蒙的分布式能力，首先需要在 config.json 中配置相应的权限，代码如下：

```
"reqPermissions": [
  {
"name": "ohos.permission.INTERNET"
  },
```

```
  {
"name": "ohos.permission.DISTRIBUTED_DATASYNC"
  },
  {
"name": "ohos.permission.servicebus.ACCESS_SERVICE"
  },
  {
"name": "ohos.permission.servicebus.BIND_SERVICE"
  },
  {
"name": "ohos.permission.DISTRIBUTED_DEVICE_STATE_CHANGE"
  },
  {
"name": "ohos.permission.GET_DISTRIBUTED_DEVICE_INFO"
  },
  {
"name": "ohos.permission.GET_BUNDLE_INFO"
  }
]
```

查找附近的已登录了同一华为账号的手机设备，需要通过 Service Ability 实现查找，我们来看一下如何在 Service Ability 中查找附近设备，这里使用的方法非常简单，如代码示例 11-6 所示。

代码示例 11-6 获取可连接的设备信息

```
//获取当前网络设备 Id、设备名称
public static List getDevice(){
    List deviceDataList = new ArrayList();
    List<DeviceInfo> deviceList = DeviceManager.getDeviceList(DeviceInfo.FLAG_GET_ONLINE_
DEVICE);
    deviceList.forEach(device->{
        HashMap<String, String> deviceInfo = new HashMap<String, String>();
        deviceInfo.put("deviceId",device.getDeviceId());
        deviceInfo.put("deviceName",device.getDeviceName());
        deviceDataList.add(deviceInfo);
    });
    return deviceDataList;
}
```

DeviceManager.getDeviceList(DeviceInfo.FLAG_GET_ONLINE_DEVICE) 返回的是附近发现的可以通信的华为鸿蒙设备信息列表。

有了这些附近的鸿蒙设备列表，还需要通过 Service Ability 返回 JavaScript 客户端，如代码示例 11-7 所示。

代码示例 11-7 获取鸿蒙设备列表

```
package com.cangjie.jsabilitydemo.services;

import com.cangjie.jsabilitydemo.utils.DeviceUtils;
import ohos.aafwk.ability.Ability;
import ohos.aafwk.content.Intent;
import ohos.data.distributed.common.*;
import ohos.data.distributed.user.SingleKvStore;
import ohos.rpc.*;
import ohos.hiviewdfx.HiLog;
import ohos.hiviewdfx.HiLogLabel;
import ohos.utils.zson.ZSONObject;
import java.util.*;


public class BoardServiceAbility extends Ability {

    @Override
    protected void onStart(Intent intent) {

    }

    private static final String TAG = "BoardServiceAbility";
    private MyRemote remote = new MyRemote();

    @Override
    protected IRemoteObject onConnect(Intent intent) {
        super.onConnect(intent);
        return remote.asObject();
    }

//定义一个内部类,代理远程调用
    class MyRemote extends RemoteObject implements IRemoteBroker {

        private static final int ERROR = -1;
        private static final int SUCCESS = 0;
        private static final int GHE_DEVICE = 1001;

        private Set<IRemoteObject> remoteObjectHandlers = new HashSet<IRemoteObject>();

        MyRemote() {
            super("MyService_MyRemote");
        }

        @Override
```

```
        public boolean onRemoteRequest(int code, MessageParcel data, MessageParcel reply,
MessageOption option) throws RemoteException {

            switch (code) {
                //获取设备
                case GHE_DEVICE:{
                    Map<String, Object> zsonResult = new HashMap<String, Object>();
                    List deviceList = DeviceUtils.getDevice();
                    zsonResult.put("data",deviceList);
                    zsonResult.put("code", SUCCESS);
                    reply.writeString(ZSONObject.toZSONString(zsonResult));
                    break;
                }
                default: {
                    reply.writeString("service not defined");
                    return false;
                }
            }
            return true;
        }

        @Override
        public IRemoteObject asObject() {
            return this;
        }
    }
}
```

这里封装了 util.js 工具模块，此外还封装了远程 Service Ability 的调用方法，如代码示例 11-8 所示。

代码示例 11-8　封装了远程 Service Ability 的调用方法

```
//异步方法引入所需文件
const globalRef = Object.getPrototypeOf(global) || global
//注入 regeneratorRunTime
globalRef.regeneratorRunTime = require('@babel/RunTime/regenerator')

//abilityType: 0 - Ability; 1 - Internal Ability
const ABILITY_TYPE_EXTERNAL = 0;
const ABILITY_TYPE_INTERNAL = 1;
//syncOption(Optional, default sync): 0 - Sync; 1 - Async
const ACTION_SYNC = 0;
const ACTION_ASYNC = 1;
const ACTION_MESSAGE_CODE_PLUS = 1001;
```

```
export const gameAbility = {
    callAbility: async function(data){
        let action = {};
        action.bundleName = 'com.cangjie.jsabilitydemo';
        action.abilityName = 'com.cangjie.jsabilitydemo.services.BoardServiceAbility';
        action.messageCode = data.code;
        action.data = data;
        action.abilityType = ABILITY_TYPE_EXTERNAL;
        action.syncOption = ACTION_SYNC;
        let result = await FeatureAbility.callAbility(action);
        return JSON.parse(result);
    },
    //根据 messeage_code 订阅不同事件
    subAbility: async function(messeage_code,callBack){
        let action = {};
        action.bundleName = 'com.cangjie.jsabilitydemo';
        action.abilityName = 'com.cangjie.jsabilitydemo.services.BoardServiceAbility';
        action.messageCode = messeage_code;
        action.abilityType = ABILITY_TYPE_EXTERNAL;
        action.syncOption = ACTION_SYNC;
        let result = await FeatureAbility.subscribeAbilityEvent(action, function(callbackData) {
            var callbackJson = JSON.parse(callbackData);
            this.eventData = JSON.stringify(callbackJson.data);
            callBack && callBack(callbackJson);
        });
        return JSON.parse(result);
    },
    //根据 message_code 取消订阅不同事件
   unSubAbility : async function(messeage_code){
       let action = {};
       action.bundleName = 'com.cangjie.jsabilitydemo';
       action.abilityName = 'com.cangjie.jsabilitydemo.services.BoardServiceAbility';
       action.messageCode = messeage_code;
       action.abilityType = ABILITY_TYPE_EXTERNAL;
       action.syncOption = ACTION_SYNC;
       let result = await FeatureAbility.unsubscribeAbilityEvent(action);
       return JSON.parse(result);
    }
}
```

最后，只需调用 util.js 中的 callAbility 方法，就可以返回所有在线的设备列表，再绑定到 devicelist 上就可以在页面进行绑定了。

```
linkDevice() {
    callAbility({
        code: 1001
    }).then(result =>{
        this.devicelist = result.data;
    })
    this.$element('simpledialog').show()
},
```

this.$element('simpledialog').show()方法用于打开页面中的dialog组件,以此弹出绑定的设备列表。

11.4.4　实现涂鸦作品发送至已连接手机

在涂鸦画板中有3个核心功能:

(1) 涂鸦者选择好希望连接的设备后,可以直接把涂鸦成果流转给对应的设备。

(2) 其他设备接收流转的涂鸦后,可以在涂鸦的基础上添加涂鸦或者修改。

(3) 修改后的涂鸦可以继续流转给涂鸦者,或者流转到其他设备上,如电视上。

这3个功能都是在连接好附近的设备后才能实现的。鸿蒙操作系统中对于室内网络的通信提供了对软总线的支持,软总线通过屏蔽设备的连接方式,采用一种最优的方式进行多设备的发现、连接和通信,如图11-4所示。

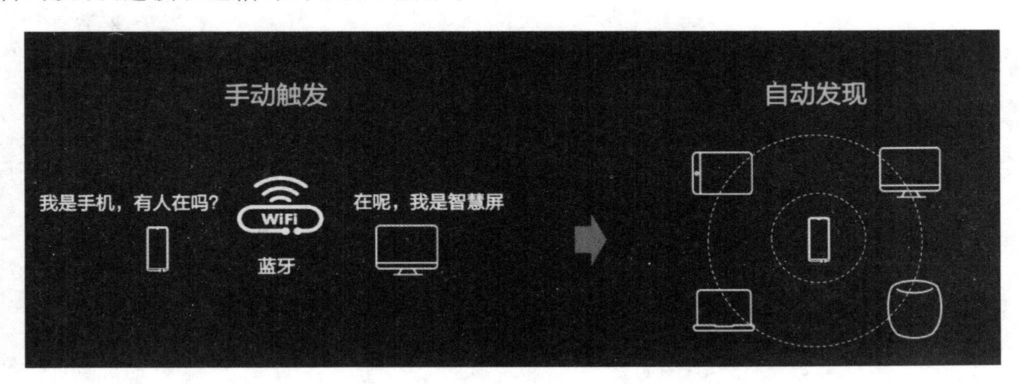

图11-4　华为鸿蒙分布式软总线

在JavaScript框架中提供了FeatureAbility.continueAbility方法,这种方法实现了设备间应用流转的所有功能。在需要流转的时候,只需保持好设备流转前的数据,在设备流转后,在其他设备上即可恢复流转前缓存的数据。

首先在需要流转的页面添加onStartContinuation和onSaveData这两种方法。流转后还需要用到的方法是onRestoreData和onCompleteContinuation,设备间流转触发的生命周期方法如图11-5所示,Ability流转方法如代码示例11-9所示。

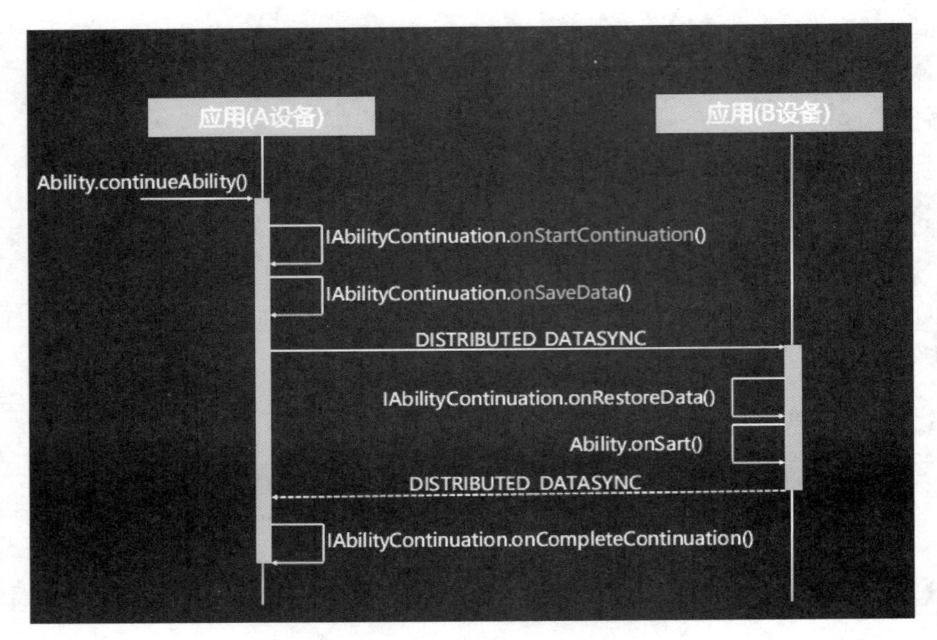

图 11-5 鸿蒙 Ability 流转图

代码示例 11-9 Ability 流转方法

```
transforAbility: async function transfer() {
    try {
        await FeatureAbility.continueAbility(0,null)
    } catch (e) {
        console.error("迁移出错: " + JSON.stringify(e))
    }
},
onStartContinuation: function onStartContinuation() {
    //判断当前的状态是不是适合迁移
    console.error("trigger onStartContinuation");
    return true;
},
onCompleteContinuation: function onCompleteContinuation(code) {
    //迁移操作完成,code 返回结果
    console.error("trigger onCompleteContinuation: code = " + code);
    return true
},
onSaveData: function onSaveData(saveData) {
    //将数据保存到 savedData 中进行迁移
    saveData.points = this.sharePoints;
    return true
},
```

```
onRestoreData: function onRestoreData(restoreData) {
    //收到迁移数据,进行数据恢复
    this.sharePoints = restoreData.points;
    return true
}
```

这里需要注意,如果需要实现应用流转,则 onStartContinuation、onCompleteContinuation、onSaveData 和 onRestoreData 的返回值都必须为 true,否则无法流转和恢复数据。

Ability 在流转后,数据需要恢复过来,所以需要在绘制坐标的时候,把所有的坐标保存到一个数组中,sharePoints 是用来保存 touchMove 中的所有坐标信息的数组,如代码示例 11-10 所示。

代码示例 11-10　实现流转

```
export default {
    data: {
        cxt: {},
        sharePoints: [],
lineHeight:3
    },
   onShow() {
        this.cxt = this.$element("board").getContext("2d");
//恢复数据后,重新绘制
        if (this.sharePoints.length > 0) {
            for (let index = 0; index < this.sharePoints.length; index++) {
                this.paintCanvas(this.sharePoints[index])
            }
        }
    },
//Touch start 事件
    painStart(e) {
        this.isShow = true;
        var p = {
            x: e.touches[0].localX,
            y: e.touches[0].localY,
            c: this.selColor,
            flag: "start"
        }
       this.sharePoints.push(p)
       //本地绘制
       this.paintCanvas(p)
 },
//Touch move 事件
 paint(e) {
```

```
            //坐标
            var p = {
                x: e.touches[0].localX,
                y: e.touches[0].localY,
                c: this.selColor,
                flag: "line"
            }
            this.sharePoints.push(p)
    this.paintCanvas(p)
    },
    //Touch事件结束
     paintEnd(e) {
        this.cxt.closePath();
    },
     //本地绘制自由线条
    paintCanvas(point) {
      if (point.flag == "start") {
         this.cxt.beginPath();
         this.cxt.strokeStyle = point.c
         this.cxt.lineWidth = this.lineHeight
         this.cxt.lineCap = "round"
         this.cxt.lineJoin = "round"
         this.cxt.moveTo(point.x,point.y);
       } else if (point.flag == "line") {
         this.cxt.lineTo(point.x,point.y);
         this.cxt.stroke()
     }
    },

    //启动流转
        setUpRemote: async function transfer() {
            try {
                await FeatureAbility.continueAbility(0,null)
            } catch (e) {
                console.error("迁移出错: " + JSON.stringify(e))
            }
        },
        onStartContinuation: function onStartContinuation() {
            //判断当前的状态是不是适合迁移
            return true;
        },
        onCompleteContinuation: function onCompleteContinuation(code) {
            //迁移操作完成,code返回结果
            return true
        },
        onSaveData: function onSaveData(saveData) {
```

```
        //将数据保存到 savedData 中进行迁移
        saveData.points = this.sharePoints;
        return true
    },
    onRestoreData: function onRestoreData(restoreData) {
        //收到迁移数据,进行数据恢复
        this.sharePoints = restoreData.points;
        return true
    }
}
```

11.4.5 实现画板实时共享功能

多设备间实现涂鸦画板数据同步,在鸿蒙操作系统中,需要通过分布式数据库来完成。同时需要 Service Ability 把分布式数据库中的数据推送给前端的 FA(JavaScript 页面)。

下面介绍如何为涂鸦画板添加实时共享功能,步骤如下。

步骤 1:通过分布式数据库实现数据共享。

定义一个 BoardServiceAbility 的 PA,如代码示例 11-11 所示,重写 onStart 方法,在 onStart 方法中引用和创建分布式数据服务,storeID 为数据库名称。

代码示例 11-11 创建分布式数据服务

```
public class BoardServiceAbility extends Ability {
private KvManagerConfig config;
private KvManager kvManager;
private String storeID = "board";
private SingleKvStore singleKvStore;

@Override
protected void onStart(Intent intent) {
   super.onStart(intent);
    //初始化 manager
    config = new KvManagerConfig(this);
    kvManager = KvManagerFactory.getInstance().createKvManager(config);

    //创建数据库
    Options options = new Options();
options.setCreateIfMissing(true).setEncrypt(false).setKvStoreType(KvStoreType.SINGLE_
VERSION);
    singleKvStore = kvManager.getKvStore(options, storeID);

}
}
```

步骤 2：Service Ability 实现页面数据订阅与同步。

把需要共享的数据保存到鸿蒙分布式数据库中，如果需要在页面中访问这些实时的数据，则需要通过创建 Service Ability 来推送分布式数据库中的数据。创建推送数据的逻辑如代码示例 11-12 所示。

代码示例 11-12　推送数据

```
package com.cangjie.jsabilitydemo.services;

import com.cangjie.jsabilitydemo.utils.DeviceUtils;
import ohos.aafwk.ability.Ability;
import ohos.aafwk.content.Intent;
import ohos.data.distributed.common.*;
import ohos.data.distributed.user.SingleKvStore;
import ohos.rpc.*;
import ohos.hiviewdfx.HiLog;
import ohos.hiviewdfx.HiLogLabel;
import ohos.utils.zson.ZSONObject;

import java.util.*;

public class BoardServiceAbility extends Ability {

    private KvManagerConfig config;
    private KvManager kvManager;
    private String storeID = "board";
    private SingleKvStore singleKvStore;

    @Override
    protected void onStart(Intent intent) {
        super.onStart(intent);
        //初始化 manager
        config = new KvManagerConfig(this);
        kvManager = KvManagerFactory.getInstance().createKvManager(config);

        //创建数据库
        Options options = new Options();
        options.setCreateIfMissing(true).setEncrypt(false).setKvStoreType(KvStoreType.
SINGLE_VERSION);
        singleKvStore = kvManager.getKvStore(options, storeID);

    }


    private static final String TAG = "BoardServiceAbility";
```

```
    private MyRemote remote = new MyRemote();

    @Override
    protected IRemoteObject onConnect(Intent intent) {
        super.onConnect(intent);
        return remote.asObject();
    }
}
```

步骤3：在多个分布式设备中进行数据同步，创建一个RPC代理类MyRemote，代理类用于远程通信连接。MyRemote类可以是BoardServiceAbility这个PA的内部类。

实现远程数据调用，这里最关键的是onRemoteRequest方法，这种方法的code参数是客户端发送过来的编号，通过这个编号便可以处理不同的客户端请求，如代码示例11-13所示。

代码示例11-13　远程代理

```
class MyRemote extends RemoteObject implements IRemoteBroker {


    private static final int ERROR = -1;
    private static final int SUCCESS = 0;
    private static final int SUBSCRIBE = 3000;
    private static final int UNSUBSCRIBE = 3001;
    private static final int SAVEPOINTS = 2000;  //保存绘制的坐标

    private Set<IRemoteObject> remoteObjectHandlers = new HashSet<IRemoteObject>();

    MyRemote() {
        super("MyService_MyRemote");
    }

    @Override
    public boolean onRemoteRequest (int code, MessageParcel data, MessageParcel reply,
MessageOption option) throws RemoteException {

        switch (code) {
            case SAVEPOINTS:{
                String zsonStr = data.readString();
                BoardRequestParam param = ZSONObject.stringToClass(zsonStr, BoardRequestParam.
class);
                Map<String, Object> zsonResult = new HashMap<String, Object>();
                zsonResult.put("data", param.getPoint());
                try {
                    singleKvStore.putString("point", param.getPoint());
```

```
                } catch (KvStoreException e) {
                    e.printStackTrace();
                }
                zsonResult.put("code", SUCCESS);
                reply.writeString(ZSONObject.toZSONString(zsonResult));
                break;
            }
            //开启订阅,保存对端的 remoteHandler,用于上报数据
            case SUBSCRIBE: {
                remoteObjectHandlers.add(data.readRemoteObject());
                startNotify();
                Map<String, Object> zsonResult = new HashMap<String, Object>();
                zsonResult.put("code", SUCCESS);
                reply.writeString(ZSONObject.toZSONString(zsonResult));
                break;
            }
            //取消订阅,置空对端的 remoteHandler
            case UNSUBSCRIBE: {
                remoteObjectHandlers.remove(data.readRemoteObject());
                Map<String, Object> zsonResult = new HashMap<String, Object>();
                zsonResult.put("code", SUCCESS);
                reply.writeString(ZSONObject.toZSONString(zsonResult));
                break;
            }
            default: {
                reply.writeString("service not defined");
                return false;
            }
        }
        return true;
    }

    public void startNotify() {
        new Thread(() -> {
            while (true) {
                try {
                    Thread.sleep(5);   //每 5ms 发送一次
BoardReportEvent();

                } catch (RemoteException | InterruptedException e) {
                    break;
                }
            }
        }).start();
    }
```

```
    private void BoardReportEvent() throws RemoteException {
        String points = "";
        try {
            points = singleKvStore.getString("point");
            singleKvStore.delete("point");
        } catch (KvStoreException e) {
            e.printStackTrace();
        }
        MessageParcel data = MessageParcel.obtain();
        MessageParcel reply = MessageParcel.obtain();
        MessageOption option = new MessageOption();
        Map<String, Object> zsonEvent = new HashMap<String, Object>();
        zsonEvent.put("point", points);
        data.writeString(ZSONObject.toZSONString(zsonEvent));
        for (IRemoteObject item : remoteObjectHandlers) {
            item.sendRequest(100, data, reply, option);
        }
        reply.reclaim();
        data.reclaim();
    }

    @Override
    public IRemoteObject asObject() {
        return this;
    }
}
```

这里需要向页面订阅的方法提供推送的数据,需要定义 startNotify 方法启动一个线程来推送数据,推送的数据是在 BoardReportEvent 中定义的,推送的数据是从分布式数据库中获取的,通过 singleKvStore.getString("point")方法获取共享的数据。

这里还需要创建 BoardRequestParam 类,用于序列化页面传过来的字符串,并将字符串转换成对象。Point 是坐标信息,如{x:1,y:2,flag:"start"},是从前端的 FA 中传递过来的数据,如代码示例 11-14 所示。

代码示例 11-14 序列化页面传过来的字符串 BoardRequestParam.java

```
package com.cangjie.jsabilitydemo.services;

public class BoardRequestParam {
    private int code;          //code
    private String point;    //坐标

    public int getCode() {
        return code;
    }
```

```
    public void setCode(int code) {
        this.code = code;
    }

    public String getPoint() {
        return point;
    }

    public void setPoint(String point) {
        this.point = point;
    }
}
```

这样就完成了画板的 Service Ability 的创建，接下来需要改造前面的 JavaScript 页面代码。

(1) 首先需要修改页面 canvas 的 touchStart、touchMove 和 touchEnd 事件中的绘制方法。要实现实时同步共享画板，就不能直接在 touch 事件中绘制线了。

需要在 touchStart 中将起点的坐标发送到上面定义的 Service Ability 中，PA 接收到请求后，把这个坐标点放到分布式数据服务中。同时页面可以订阅 PA，以便页面获取推送的坐标信息。

touchMove 将终点的坐标发送到上面定义的 Service Ability 中，PA 接收到请求后，把这个坐标点放到分布式数据服务中。同时页面可以订阅 PA，以便页面获取推送的坐标信息，如代码示例 11-15 所示。

代码示例 11-15　向 PA 发送需要同步的 Point

```
//起点
var p = {
    x: e.touches[0].localX,
    y: e.touches[0].localY,
    c: this.selColor,
    flag: "start"
}
//终点
var p = {
x: e.touches[0].localX,
    y: e.touches[0].localY,
    c: this.selColor,
    flag: "line"
  }
//向 PA 发送需要同步的 Point
```

```
sendPoint(p) {
gameAbility.callAbility({
        code: 2000,
        point: p
}).then(result =>{
})
}
```

(2) 实现多设备共享绘制效果,如代码示例 11-16 所示,需要通过鸿蒙操作系统的软总线实现,在页面中,需要在 onInit 中订阅 Service Ability 中的回调事件以便获取共享的坐标信息。

代码示例 11-16

```
onInit() {
     //订阅 PA 事件
     this.subscribeRemoteService();
 },
//订阅 PA
 subscribeRemoteService() {
     gameAbility.subAbility(3000,(data) =>{
         console.error(JSON.stringify(data));
         if (data.data.point != "") {
             var p = JSON.parse(data.data.point)
             this.paintCanvas(p)
             prompt.showToast({
                 message: "POINTS: " + JSON.stringify(p)
             })
         }
     }).then(result =>{
         //订阅状态返回
         if (result.code != 0)
         return console.warn("订阅失败");
         console.log("订阅成功");
     });
 },
```

这个订阅方法,只需在页面启动后调用一次就可以了,所以放在 onInit 方法中进行调用。

这里通过 subscribeRemoteService 方法来订阅从 Service Ability 中推送过来的数据,gameAbility.subAbility 方法就是订阅 Service Ability 推送的方法,3000 是调用 Service Ability 接收的编号。鸿蒙多设备共享涂鸦画板的效果如图 11-6 所示。

图 11-6 鸿蒙多设备共享涂鸦画板同步

11.5 本章小结

本章使用 JavaScript UI 结合 Java Service Ability 实现了一个可以在多设备间同步共享数据的鸿蒙涂鸦画板的应用，提供了 JavaScript 调用 Java Service Ability 的多种方式，通过 Service Ability 实现一些 JavaScript 无法实现的功能。

第 12 章

鸿蒙应用签名与发布

通过前面几个章节的学习，已经可以开发一款鸿蒙风格的游戏了，完成开发后我们希望将此应用发布到鸿蒙应用市场。本章详细介绍应用发布前的签名准备和华为应用市场的发布流程。发布应用的流程如图 12-1 所示。

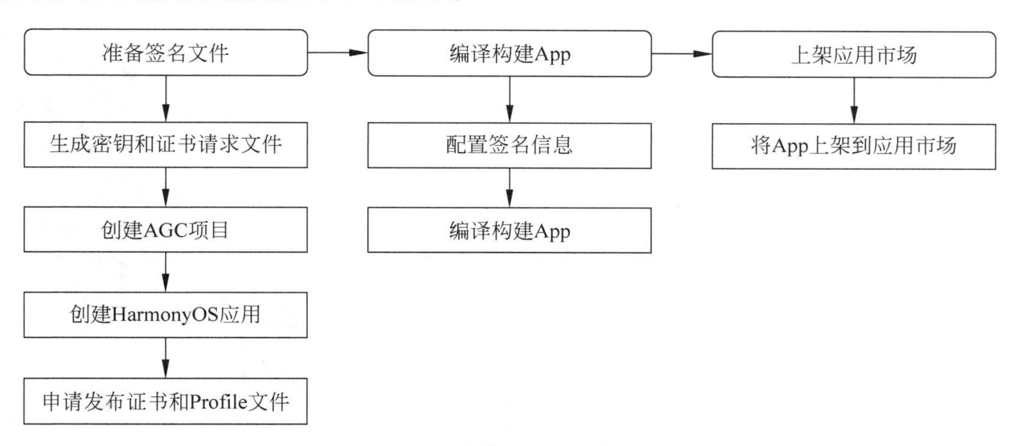

图 12-1　鸿蒙 HAP 发布流程

下面的小节将详细介绍鸿蒙签名证书申请、编译构建 HAP 包及将 HAP 上架应用市场的整个流程。

12.1　准备应用发布的签名文件

发布到华为应用市场的 HarmonyOS 应用，需要在 AppGallery Connect 中申请发布证书和 Profile 文件，并对应用进行签名。

鸿蒙应用程序发布签名证书与测试证书申请流程类似，下面介绍鸿蒙应用程序发布证书的申请流程。

12.1.1 生成密钥和证书请求文件

HarmonyOS 应用通过数字证书和授权文件来保证应用的完整性，在申请数字证书和 Profile 文件前，需要通过 DevEco Studio 来生成私钥（存放在.p12 文件中）和证书请求文件（.csr 文件）。同时，也可以使用命令行工具的方式来生成密钥和证书请求文件，用于构筑工程流水线。

（1）生成 p12 私钥文件。在主菜单栏选择 Build→Generate Key 命令，如图 12-2 所示。

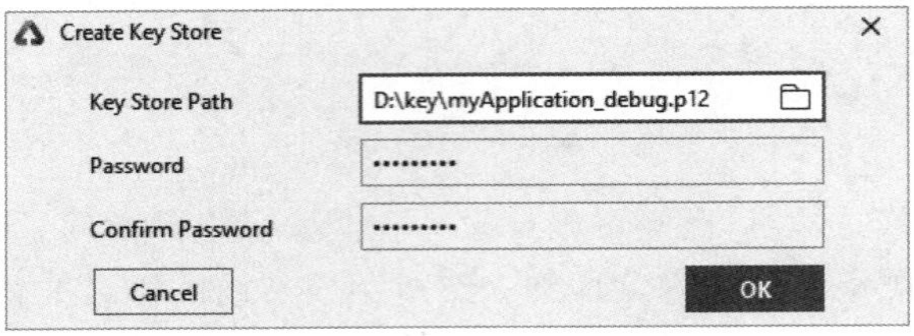

图 12-2 创建 p12 文件

（2）创建 CSR 证书请求文件。

在 Generate Key 界面中，填写完密钥信息后，单击 Generate Key and CSR 按钮，如图 12-3 所示。

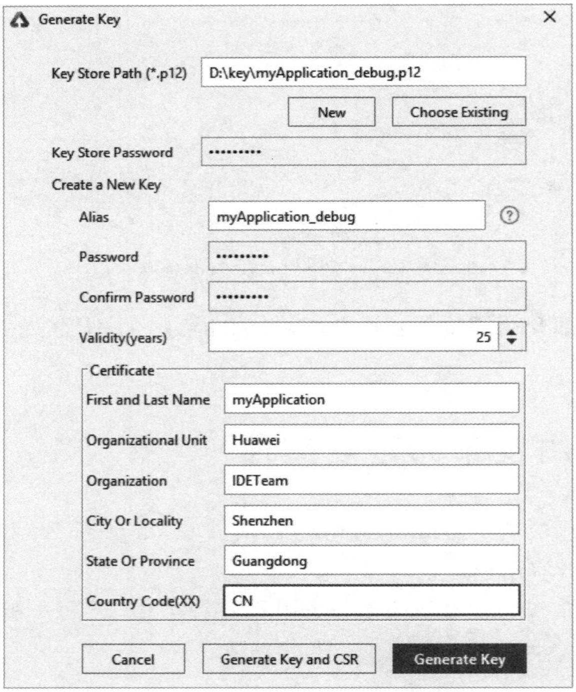

图 12-3 创建 CSR 证书文件

Alias：密钥的别名信息，用于标识密钥名称。需要记住该别名，后续进行签名配置时需要使用。

Password：输入密钥对应的密码，密钥密码需要与密钥库密码保持一致。需要记住该密码，后续进行签名配置时需要使用。

Confirm Password：再次输入密钥密码。

Validity：证书有效期，建议设置为25年及以上，以便覆盖应用的完整生命周期。

Certificate：输入证书基本信息，如组织、城市或地区、国家码等。

在弹出的窗口中，单击CSR File Path对应的图标，选择CSR文件存储路径，如图12-4所示。

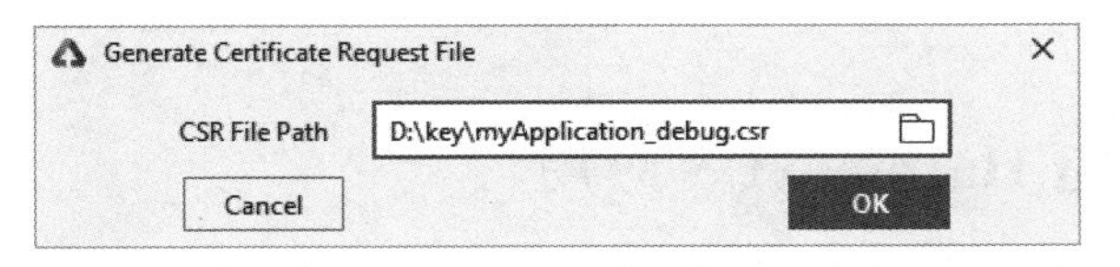

图12-4 保存CSR证书文件

单击OK按钮，便会成功创建CSR文件，工具会同时生成密钥文件(.p12)和证书请求文件(.csr)，如图12-5所示。

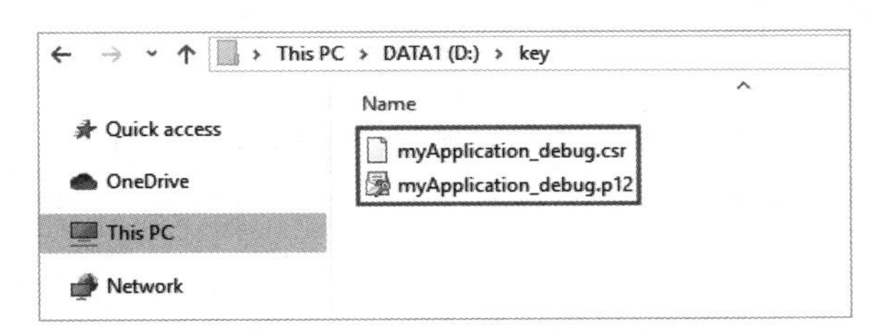

图12-5 p12和csr文件清单

12.1.2 创建AGC项目

登录AppGallery Connect网站，选择“我的项目”。打开并登录AppGallery Connect网站(如果可没有账号，则需要提前申请)，网址为https://developer.huawei.com/consumer/cn/service/josp/agc/index.html，在弹出的页面中选择“我的项目”。

在我的项目页面单击“添加项目”，如图12-6所示。

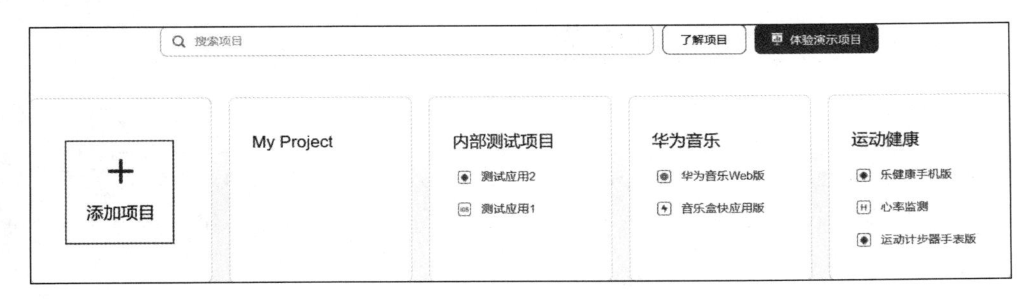

图12-6 添加项目

输入预先规划的项目名称，单击“确定”按钮，如图 12-7 所示。项目创建成功后，会自动进入“项目设置”页面。

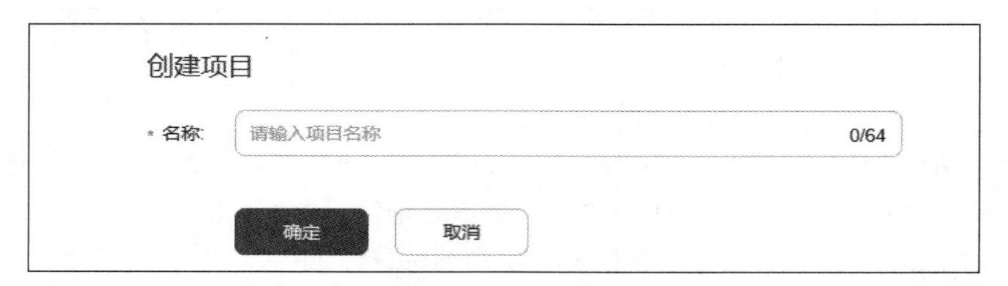

图 12-7 创建 AGC 项目

12.1.3 创建 HarmonyOS 应用

在应用列表首页，单击右侧的“新建”按钮，如图 12-8 所示。

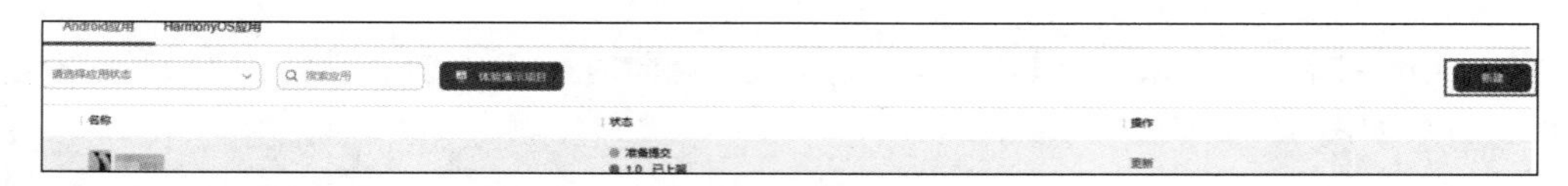

图 12-8 创建鸿蒙应用

填写应用信息，完成后单击“确定”按钮，如图 12-9 所示。

创建应用

* 软件包类型: APK(Android应用) RPK(快应用) APP(HarmonyOS应用)

* 支持设备: 手表 大屏 路由器

* 应用名称: 请填写新应用名称（限64字符）

* 应用分类: 请选择应用分类

* 默认语言: 请选择语言

添加到项目

确定 取消

图 12-9 填写应用信息

注意：当前只有受邀开发者才能创建 HarmonyOS 应用。如果 AGC 页面未展示“APP(HarmonyOS 应用)”选项，则需将开发者名称、申请背景及 Developer ID 发送至 agconnect@huawei.com，华为运营人员将在 1～3 个工作日为你安排对接人员。Developer ID 查询方法如下：登录 AGC 网站，单击“我的项目”，选择你的任意项目，在“项目设置→常规→开发者”下即可找到 Developer ID。

返回应用列表，在“HarmonyOS 应用”页签查看已创建的应用，如图 12-10 所示。

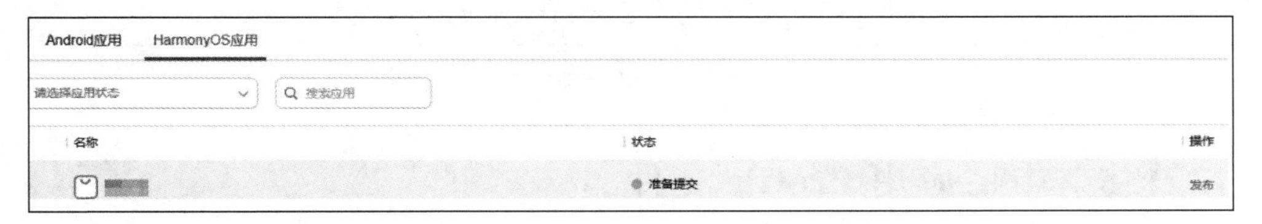

图 12-10　查看已创建的应用列表

12.1.4　申请应用发布证书

登录 AppGallery Connect 网站，选择“用户与访问”。打开并登录 AppGallery Connect 网站（如果可没有账号，则需要提前申请），网址为 https://developer.huawei.com/consumer/cn/service/josp/agc/index.html，在弹出的页面中选择“用户与访问”。

在左侧导航栏选择“证书管理”，进入证书管理页面，单击“新增证书”按钮，如图 12-11 所示。

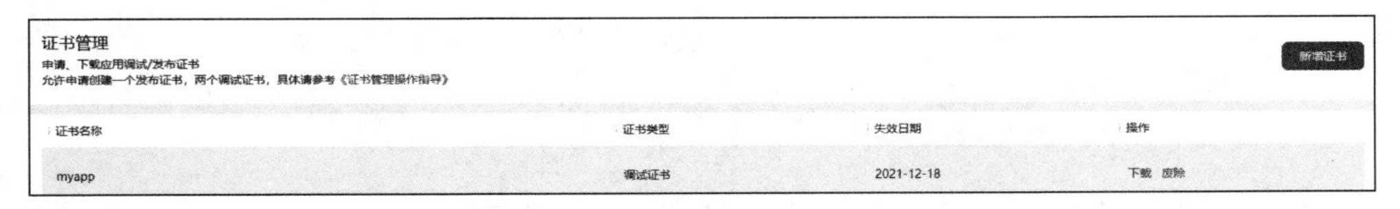

图 12-11　添加证书

在弹出的“新增证书”窗口，填写要申请的证书信息，单击“提交”按钮。证书类型需选择“发布证书”。每次最多可申请发布一个证书，如图 12-12 所示。

新增证书
* 证书名称：　0/100
* 上传证书请求文件(CSR)：　上传
* 证书类型：
提交　取消

图 12-12　填写证书信息

下载并保存证书，如图 12-13 所示。

证书管理
申请、下载应用调试/发布证书
允许申请创建一个发布证书，两个调试证书，具体请参考《证书管理操作指导》

新增证书

证书名称	证书类型	失效日期	操作
myapp	调试证书	2021-12-18	下载 废除

图 12-13　下载并保存证书

12.1.5　申请应用 Profile 文件

找到你的项目，单击所创建的 HarmonyOS 应用，如图 12-14 所示。

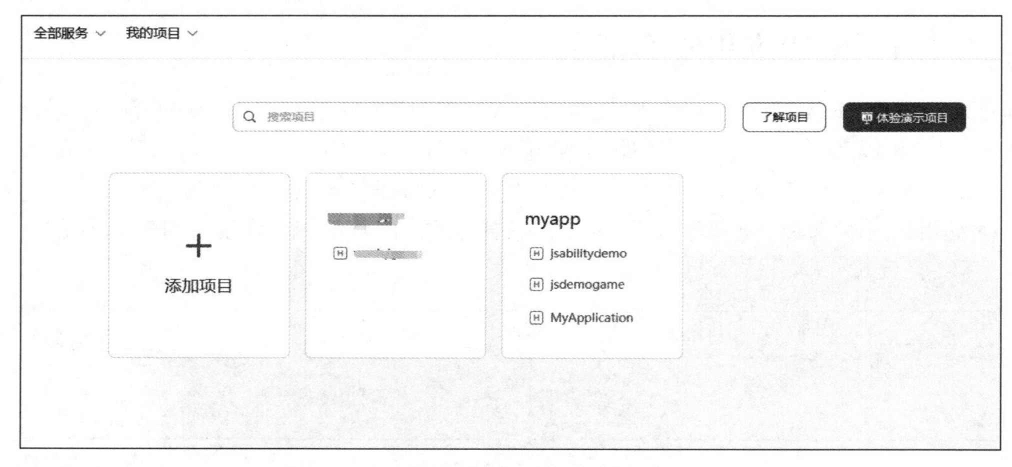

图 12-14　项目列表

单击"HarmonyOS 应用→HAP Provision Profile 管理"，进入"管理 HAP Provision Profile"页面，单击右上角的"添加"按钮，如图 12-15 所示。

管理HAP Provision Profile
添加、删除、下载Profile
允许申请创建100个Profile文件，具体请参考《Provision管理操作指导》

添加

名称	类型	证书	失效日期	操作
Provision _1	调试	CSR_1	2024-07-03	删除 下载 查看设备

图 12-15　添加 Profile 证书

在弹出的"HarmonyAppProvision 信息"窗口添加调试 Profile，如图 12-16 所示。

填写完 Profile 证书信息后，单击"提交"按钮，完成添加操作。在"操作"类别下单击"下载"按钮，可下载 Profile 文件，最多可申请 100 个 Profile 文件。

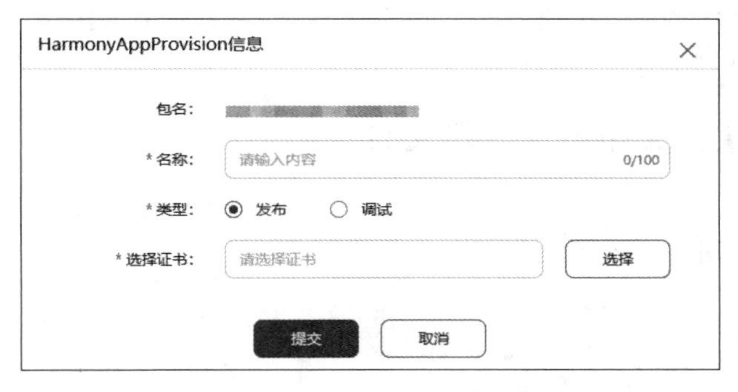

图 12-16 填写 Profile 证书信息

12.2 构建类型为 Release 的 HAP

12.1 节介绍了如何申请发布应用的证书文件，本节将介绍如何构建并编译出用于发布证书签名的 HAP 包。

12.2.1 配置签名信息

在 Project Structure 中的 Modules 中填写签名信息，如图 12-17 所示。如果一个工程

图 12-17 配置签名信息

目录下存在多个模块，当对单个模块进行构建时，则只需对指定的模块进行签名。如果对整个工程进行构建，则需要对所有的模块进行签名。

12.2.2 构建发布的 HAP 文件

在主菜单栏，单击 Build→Build APP(s)/Hap(s) →Build Release Hap(s)，生成已签名的 Release HAP，如图 12-18 所示。

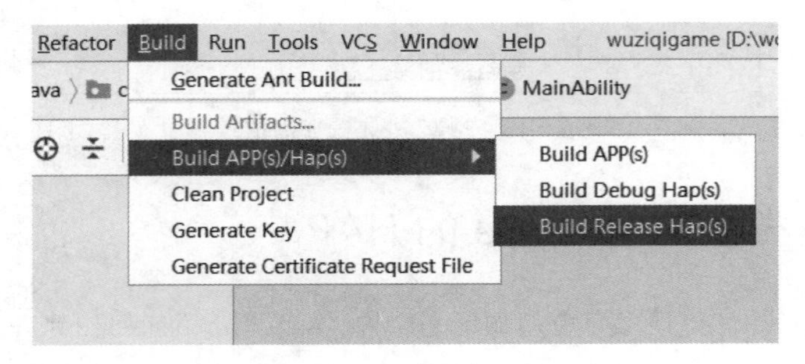

图 12-18 生成签名 HAP

生成的用于发布 HAP 包的目录，如图 12-19 所示。

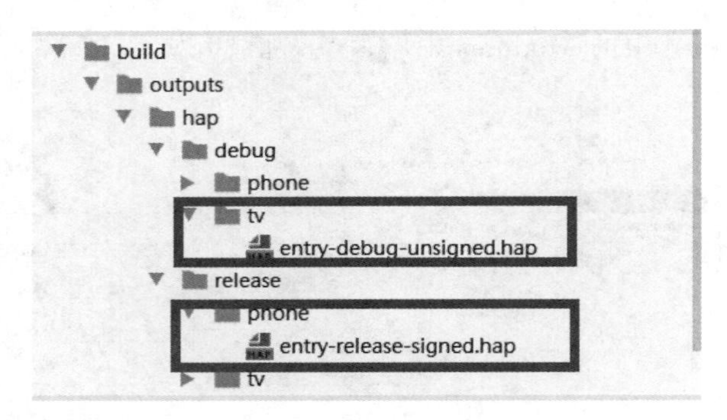

图 12-19 生成签过名的 HAP 包文件

12.3 将应用发布到华为应用市场

在 12.2 节，通过发布证书编译并构建了可以发布的 HAP 应用包，本节介绍如何把编译好的应用 HAP 包发布到华为应用市场。

12.3.1 登录 AppGallery Connect 网站

打开并登录 AppGallery Connect 网站（如果没有账号，则需要提前申请），网址为 https://developer.huawei.com/consumer/cn/service/josp/agc/index.html，选择“我的应用”，如图 12-20 所示，进入应用列表首页。

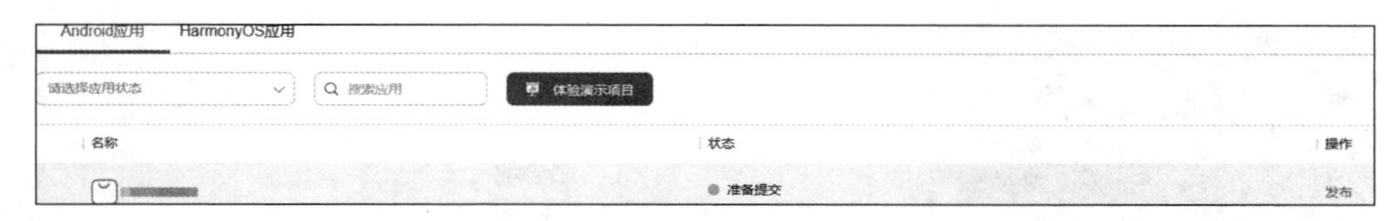

图 12-20　我的应用列表

12.3.2　完善应用发布信息

在“HarmonyOS 应用”列表中单击待发布的 HarmonyOS 应用名称，如图 12-21 所示，系统进入“应用信息”页面。

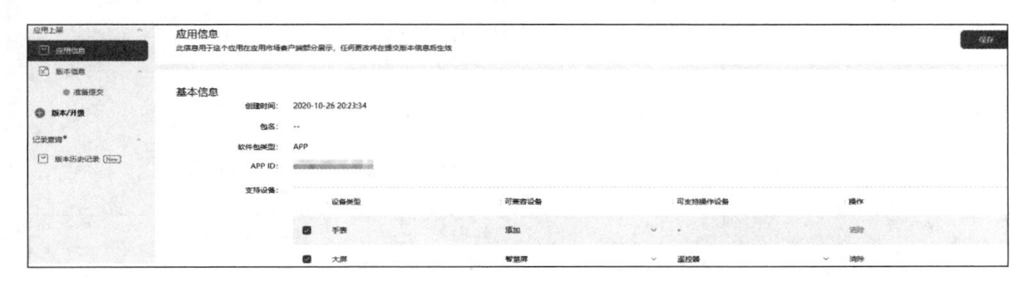

图 12-21　完善应用信息

完善应用信息，需要设置应用的图标(icon)图片，这一步需要把预先设计好的图标准备好，上传即可，如图 12-22 所示。

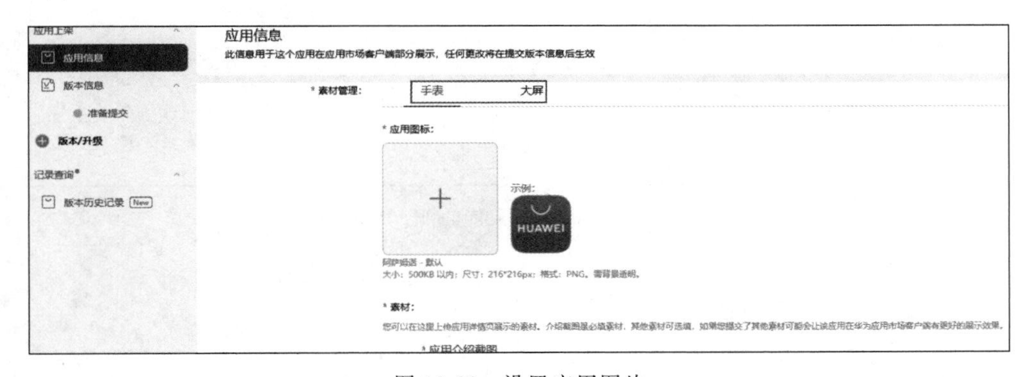

图 12-22　设置应用图片

12.3.3　设置版本信息

配置完成后单击“保存”按钮，在弹窗中单击“确定”按钮，进入“准备提交”页面，开始设置版本信息。

在“软件版本”下单击“软件包管理”按钮，如图 12-23 所示，然后在弹窗中单击“上传”按钮。

图 12-23　上传 HAP 包

12.3.4　添加上传 HAP 包

在"上传包"界面中单击"+"按钮上传应用的软件包,如图 12-24 所示。

图 12-24　添加需上传的 HAP 包

需确保 HarmonyOS 应用软件包为.app 格式,且大小限制如下。

- 运动手表:10MB 以内;
- 路由器:15MB 以内;
- 智能手表、大屏:150MB 以内。

如果所开发的应用支持多设备,则软件包大小上限为各设备所对应的软件包大小上限中的最大值。如果所开发的应用支持运动手表和智能手表,则软件包大小必须在 150MB 以内。

需确保上传的软件包签名与应用发布证书匹配,否则系统将提示错误,提示错误时需要重新上传。

12.3.5 填写应用隐私说明

针对每个“敏感权限项”填写“权限说明”，每条权限说明最大支持500个字符，如图12-25所示。

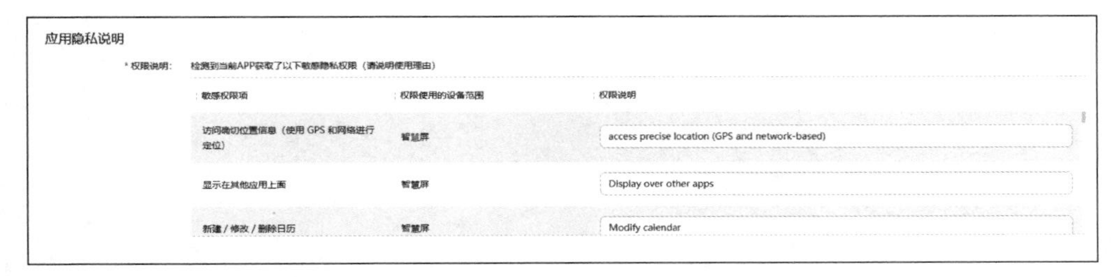

图 12-25 填写应用隐私说明

12.3.6 设置是否必须联网才可以使用

如果选择分发到路由器设备，则可设置是否必须联网才能使用，如图12-26所示。

图 12-26 设置是否必须联网才可以使用

通过以上步骤，单击“提交审核”按钮，确认版本号无误后单击“确认”按钮，到此就完成了将应用发布到华为应用市场的流程。

注意：目前的华为应用发布政策也会随着鸿蒙操作系统的迭代而升级，发布应用的政策可能也会有所调整，读者可参考鸿蒙开发者网站查阅最新的发布政策。

12.4 本章小结

本章讲解了鸿蒙应用上线的签名文件申请流程和华为应用市场的发布流程。通过本章的学习，读者可以把自己做好的应用发布到华为应用商店了。

第五篇　硬件开发篇

学习目标

通过本篇的学习，读者可以按照本篇的指导步骤，搭建好鸿蒙操作系统编译和开发环境。本篇将对 HiSpark 系列开发板的鸿蒙操作系统烧写，以及鸿蒙应用安装进行详细讲解，读者可以按照本书的步骤完成开发板的系统烧写。

主要内容如下：

- 鸿蒙（OpenHarmony）开发环境搭建；
- HiSpark 系列开发板套件开发入门。

如何学习本篇：

- 本篇的内容偏向实践，读者可以按本书内容作为参考反复安装实践；
- 本书中的内容有时效性，读者可以本篇作为实践参考，结合实践进行学习。

第 13 章 搭建 OpenHarmony 开发环境

OpenHarmony 是开放原子开源基金会（OpenAtom Foundation）旗下开源项目，其定位是一款面向全场景的开源分布式操作系统。

OpenHarmony 在传统的单设备系统能力的基础上，创造性地提出了基于同一套系统能力、适配多种终端形态的理念，支持在多种终端设备上运行，第一个版本支持在 128KB～128MB 设备上运行，如图 13-1 所示。

图 13-1　OpenHarmony 开源路标

针对设备开发者，OpenHarmony 采用了组件化的设计方案，可以根据设备的资源能力和业务特征进行灵活裁剪，满足不同形态的终端设备对于操作系统的要求。可运行在百千级别的资源受限设备和穿戴类设备，也可运行在百兆级别的智能家用摄像头/行车记录仪等相对资源丰富的设备。

13.1　OpenHarmony 编译环境准备

开源鸿蒙操作系统源代码编译目前需要在 Linux 下编译完成后才能在 Windows 中通过 DevEco Device Tool 工具进行一键烧录。本节将详细介绍 Linux 下鸿蒙源代码的配置、工具搭建、编译流程，如表 13-1 所示。

表 13-1 不同的操作系统编译环境

硬 件	说 明
Linux 主机	推荐实际物理机器，Ubuntu16.04 及以上 64 位系统，Shell 使用 bash。使用虚拟机也可以，如何安装这里不进行详细介绍
Windows 主机	Windows XP/Windows 7/Windows 10 系统
远程终端	推荐 MobaXterm（PuTTY、SecureCRT 等其他远程终端也可以），用于在 Windows 主机上登录 Linux 主机，进行源代码下载、编译等工作
USB 转串口芯片驱动	下载链接：http://www.hihope.org/download，USB-to-Serial Comm Port.exe 文件（AI Camera 和 DIY IPC 套件附赠的 USB 串口线中集成了 PL2302 芯片，需要安装此驱动才能识别）。WiFi IoT 主控芯片上集成的是 CH340G 芯片，可以选择联网自动安装驱动程序

13.1.1 虚拟机安装 Ubuntu 系统

下面我们介绍如何准备 OpenHarmony 编译环境。本节通过在虚拟机上安装 Linux 编译环境，通过 Samba 搭建 Linux 和 Windows 10 的代码共享环境。

步骤 1：下载的 Ubuntu 安装文件。

注意：这里尽量安装比较新的 Ubuntu 版本，如图 13-2 所示。

本书使用的是 Ubuntu 20.04.1 Destktop 版本。Ubuntu 安装镜像下载网址为 https://ubuntu.com/download/alternative-downloads，选择 18.04 或 20.04 版都可以。

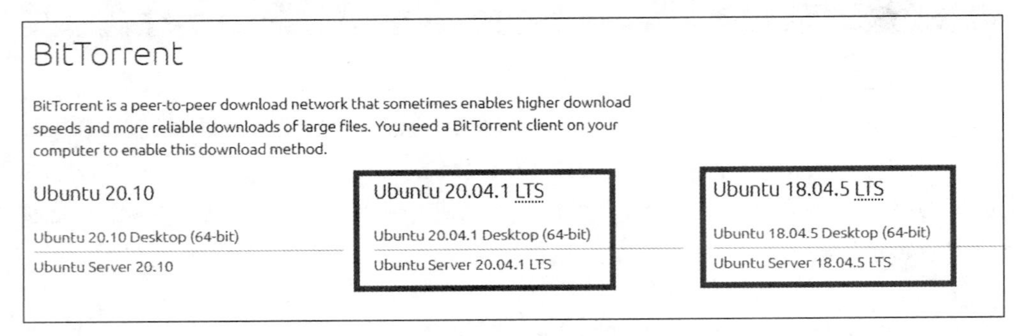

图 13-2 Ubuntu 下载

这里使用(VMware)来安装 Ubuntu 镜像文件，也可以使用 Oracle VM VirtualBox 进行安装。

步骤 2：安装 VMware，这里使用的是 Windows 10 操作系统，首先安装 VMware Workstation，如图 13-3 所示。

步骤 3：新建虚拟机，如图 13-4 所示，在虚拟机中安装 Ubuntu 镜像文件。

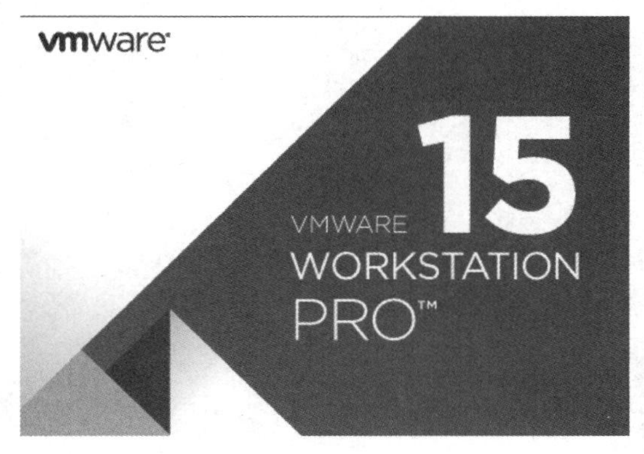

图 13-3 安装 VMware Workstation

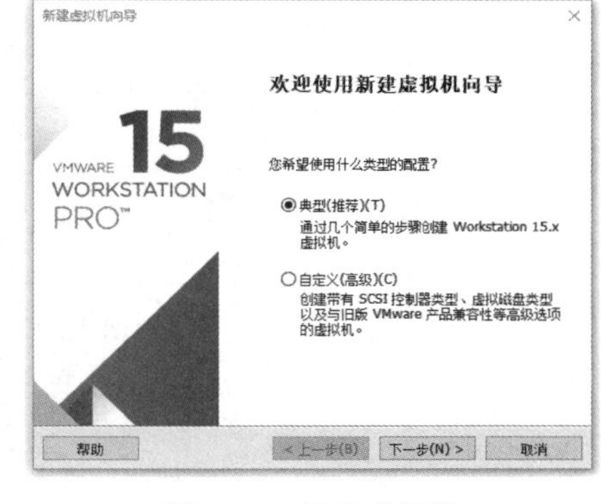

图 13-4 新建虚拟机

选择典型方式，安装光盘镜像文件，浏览到刚才所下载的 iso 镜像文件，如图 13-5 所示。

这里将默认用户名设置为 harmony，将密码设置为 harmonyos。可以用这个账号登录系统，如图 13-6 所示。

图 13-5 选择 iso 镜像文件

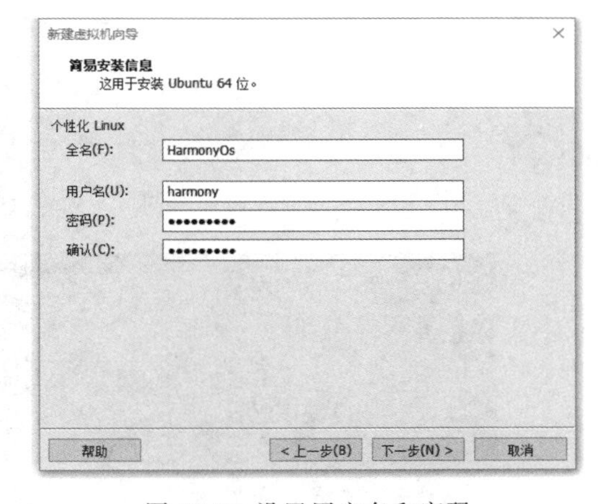

图 13-6 设置用户名和密码

设置虚拟机保存位置，选择默认即可，如图 13-7 所示。

初始分配磁盘空间为 20GB，后续不够可以再增加，如图 13-8 所示。

确认配置信息无误后，单击“完成”按钮，如图 13-9 所示。

单击完成后，一般需要 15min 左右才能完成安装，如图 13-10 所示。

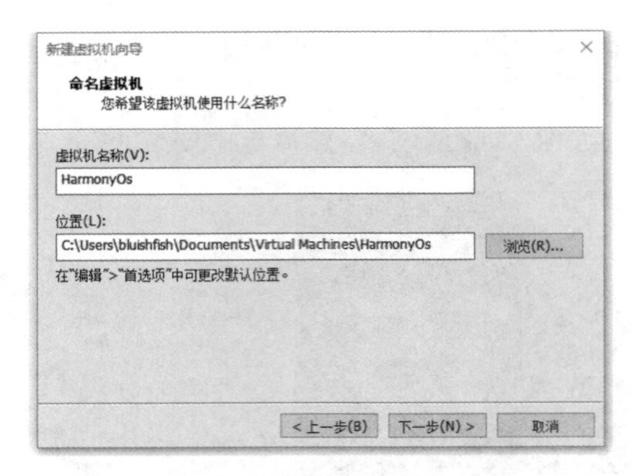

图 13-7　设置虚拟机保存位置

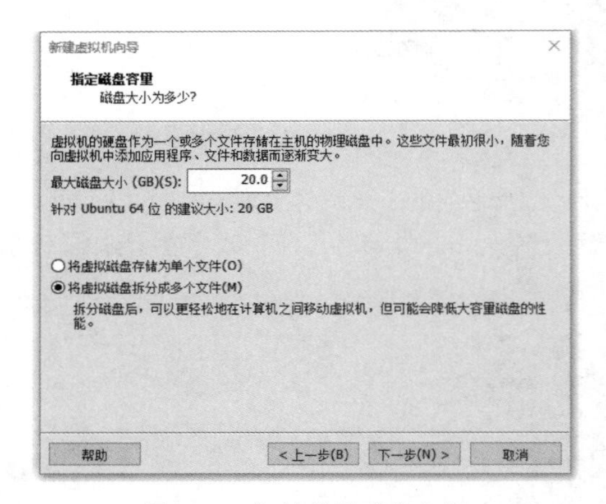

图 13-8　初始分配磁盘空间

新建虚拟机向导

已准备好创建虚拟机

单击"完成"创建虚拟机，并开始安装 Ubuntu 64 位 和 VMware Tools。

将使用下列设置创建虚拟机:

名称:	HarmonyOs
位置:	C:\Users\bluishfish\Documents\Virtual Machines\HarmonyOs
版本:	Workstation 15.x
操作系统:	Ubuntu 64 位
硬盘:	20 GB, 拆分
内存:	2048 MB
网络适配器:	NAT
其他设备:	CD/DVD, USB 控制器, 打印机, 声卡

自定义硬件(C)...

☑ 创建后开启此虚拟机(P)

< 上一步(B)　完成　取消

图 13-9　确定配置

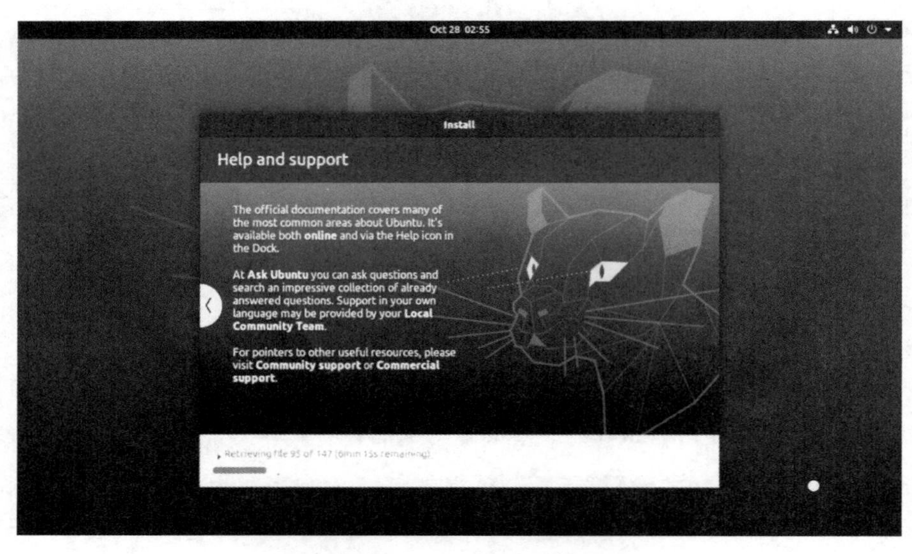

图 13-10　Ubuntu 安装过程

安装完毕并登录系统后，可更改为中文服务器，采用加速下载的方式更新系统，如图 13-11 所示。

图 13-11 更新系统

在命令行中输入更新命令，命令如下：

```
sudo apt - get update
sudo apt - get upgrade
```

虚拟机默认的内存为 2GB，处理器为单核，为了提高后续编译鸿蒙代码的效率，可以在虚拟机菜单→设置进行参数调整，如图 13-12 所示。

这里将内存调整为 8GB，将处理器调整为 4 核，可以根据自己主计算机的配置，相应调整虚拟机的可用资源。

步骤 4：安装 VMware Tools(可选)。

在虚拟机菜单→安装 VMware Tools 目录下，能够看见一个虚拟光盘文件，将 gz 文件解压到用户目录下：

```
cd /media/harmony/'VMware Tools'/
tar -zxvf VMwareTools - 10.3.2 - 9925305.tar.gz -C ~/
```

进入解压目录后，安装工具，命令如下：

```
cd vmware - tools - distrib/
sudo ./vmware - install.pl
```

这样一些增强功能，例如不同操作系统之间的文件拖曳复制，就能使用了，临时传些小

图 13-12 增加虚拟机资源

文件会很方便。遇到 Windows 10 和 Ubuntu 共用的文件，例如源代码文件，直接共享比复制会更方便些。

在虚拟机菜单→设置→选项→共享文件夹进行设置，如图 13-13 所示。

选择“添加”，把 Windows 10 系统内的目录加入文件夹列表，启用读写权限。我们往里面添加几个文件和目录，如图 13-14 所示。

这时，选择的目录会映射到 Ubuntu 系统中的/mnt/hgfs 目录，我们可以建立一个软链接，方便访问，命令如下：

```
ln -s /mnt/hgfs/HarmonyOS_Code ~/harmony/HarmonyOS_Code
```

若不需要，则可以删除软链接，命令如下：

```
rm -rf  ~/harmony/HarmonyOS_Code
```

进入目录就能看到 Windows 系统下的文件了，如图 13-15 所示。

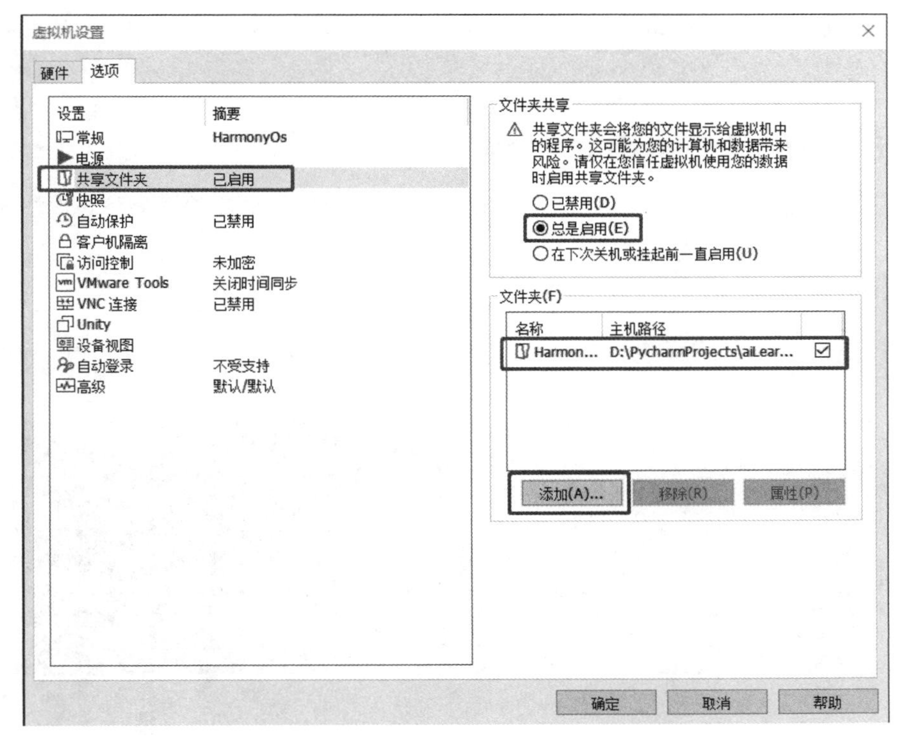

图 13-13 设置共享文件夹

图 13-14 把 Windows 10 系统内的目录加入文件夹列表

```
harmony@ubuntu:~$ cd harmony/
harmony@ubuntu:~/harmony$ ll
总用量 16
drwxrwxrwx  4 harmony harmony 4096 10月 28 14:50 ./
drwxr-xr-x 26 harmony harmony 4096 10月 30 13:03 ../
lrwxrwxrwx  1 harmony harmony   24 10月 28 14:22 HarmonyOS_Code -> /mnt/hgfs/HarmonyOS_Code/
drwxrwxrwx 19 harmony harmony 4096 10月 30 01:12 openharmony/
drwxrwxr-x  6 harmony harmony 4096 10月 30 01:06 tools/
harmony@ubuntu:~/harmony$ cd HarmonyOS_Code
harmony@ubuntu:~/harmony/HarmonyOS_Code$ ll
总用量 230882
drwxrwxrwx 1 root root         0 10月 28 14:50 ./
dr-xr-xr-x 1 root root      4192 10月 30  2020 ../
drwxrwxrwx 1 root root      4096 10月 28 14:40 code-1.0/
-rwxrwxrwx 1 root root 236414429 10月 27 13:42 code-1.0.tar.gz*
harmony@ubuntu:~/harmony/HarmonyOS_Code$
```

图 13-15 进入目录就能看到 Windows 系统下的文件

```
cd ~/harmony/HarmonyOS_Code
ll
```

13.1.2 配置 OpenHarmony 编译环境

配置 OpenHarmony 的编译环境比较复杂，下面介绍如何搭建编译环境。

步骤 1：将 Linux shell 改为 bash。

查看 shell 是否为 bash，如图 13-16 所示，在终端运行命令如下：

```
ls -l /bin/sh
```

```
harmony@ubuntu:~$ ls -l /bin/sh
lrwxrwxrwx 1 root root 4 10月 28 02:48 /bin/sh -> dash
harmony@ubuntu:~$ sudo dpkg-reconfigure dash
正在删除 通过 dash 从 /bin/sh 到 /bin/sh.distrib 的转移
正在添加 通过 bash 从 /bin/sh 到 /bin/sh.distrib 的转移
正在删除 通过 dash 从 /usr/share/man/man1/sh.1.gz 到 /usr/share/man/man1/sh.dist
rib.1.gz 的转移
正在添加 通过 bash 从 /usr/share/man/man1/sh.1.gz 到 /usr/share/man/man1/sh.dist
rib.1.gz 的转移
harmony@ubuntu:~$ ls -l /bin/sh
lrwxrwxrwx 1 root root 4 10月 28 12:43 /bin/sh -> bash
harmony@ubuntu:~$
```

图 13-16 查看 shell 是否为 bash

如果显示为/bin/sh→bash 则为正常，否则应按以下方式进行修改。在终端运行如下命令，然后选择 no：

```
sudo dpkg-reconfigure dash
```

步骤 2：安装 ssh server。

如果要通过 ssh 对外提供链接，则需要安装 ssh 服务，命令如下：

```
sudo apt-get install openssh-server
```

其余常用相关命令如下：

```
#查看 ssh 服务
sudo ps -e | grep ssh
#开启服务
sudo /etc/init.d/ssh start
#查看服务状态
sudo service ssh status
#关闭服务
sudo service ssh stop
#重启服务
sudo service ssh restart
```

步骤 3：配置 Python 环境。

Ubuntu 20.04.1 默认已经安装了 Python 3.8，如图 13-17 所示。

```
which python
/usr/bin/python3.8
```

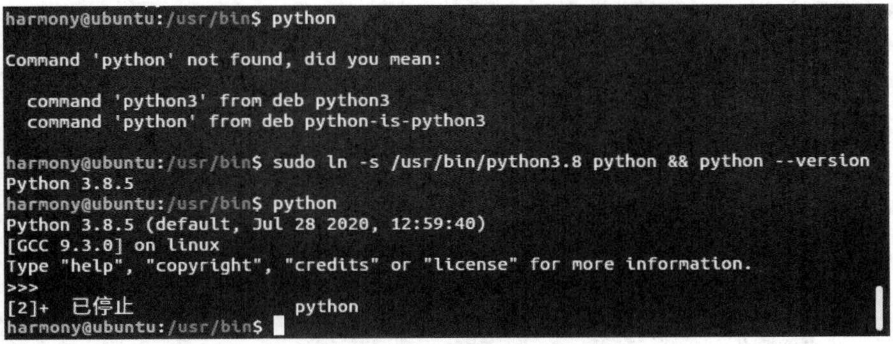

```
harmony@ubuntu:/usr/bin$ python

Command 'python' not found, did you mean:

  command 'python3' from deb python3
  command 'python' from deb python-is-python3

harmony@ubuntu:/usr/bin$ sudo ln -s /usr/bin/python3.8 python && python --version
Python 3.8.5
harmony@ubuntu:/usr/bin$ python
Python 3.8.5 (default, Jul 28 2020, 12:59:40)
[GCC 9.3.0] on linux
Type "help", "copyright", "credits" or "license" for more information.
>>>
[2]+  已停止                  python
harmony@ubuntu:/usr/bin$
```

图 13-17 配置 Python 环境

建立软链接，指向 Python，命令如下：

```
cd /usr/bin
sudo ln - s /usr/bin/python3.8 python && python -- version
```

安装并升级 Python 包管理工具(pip3)，命令如下：

```
sudo apt - get install python3 - setuptools python3 - pip - y
sudo pip3 install -- upgrade pip
```

通过换源配置 pip，新建 ～/.pip/pip.conf 文件，写入清华源，命令如下：

```
[global]
index - URL = https://pypi.tuna.tsinghua.edu.cn/simple
```

安装 Python 模块 setuptools，命令如下：

```
pip3 install setuptools
```

安装 GUI menuconfig 工具(Kconfiglib)，建议安装 Kconfiglib 13.2.0＋版本，命令如下：

```
sudo pip3 install kconfiglib
```

安装文件打包工具，命令如下：

```
sudo apt-get install dosfstools mtools mtd-utils
```

下载、配置编译工具链，命令如下：

```
mkdir -p ~/harmony/tools && cd ~/harmony/tools

#下载 gn/ninja/LLVM/hc-gen 包
URL_PREFIX=https://repo.huaweicloud.com/harmonyos/compiler
wget $URL_PREFIX/gn/1523/Linux/gn.1523.tar
wget $URL_PREFIX/ninja/1.9.0/Linux/ninja.1.9.0.tar
wget $URL_PREFIX/clang/9.0.0-34042/Linux/llvm-Linux-9.0.0-34042.tar
wget $URL_PREFIX/hc-gen/0.65/Linux/hc-gen-0.65-Linux.tar

#解压 gn/ninja/LLVM/hc-gen 包
tar -C ~/harmony/tools/ -xvf gn.1523.tar
tar -C ~/harmony/tools/ -xvf ninja.1.9.0.tar
tar -C ~/harmony/tools/ -xvf llvm-Linux-9.0.0-34042.tar
tar -C ~/harmony/tools/ -xvf hc-gen-0.65-Linux.tar

#向 ~/.bashrc 中追加 gn/ninja/LLVM/hc-gen 路径配置
cat << EOF >> ~/.bashrc
export PATH=~/harmony/tools/gn:\$PATH
export PATH=~/harmony/tools/ninja:\$PATH
export PATH=~/harmony/tools/llvm/bin:\$PATH
export PATH=~/harmony/tools/hc-gen:\$PATH
EOF

#使环境变量生效
source ~/.bashrc
```

安装虚拟环境，命令如下：

```
sudo pip3 install -U virtualenv
virtualenv -p python3 ~/my_envs/harmonyos
source ~/my_envs/harmonyos/bin/activate
pip install requests  #后续配置 repo 时需要
```

配置 repo 工具，命令如下：

```
sudo apt install cURL
cURL https://gitee.com/oschina/repo/raw/fork_flow/repo-py3 > ~/harmony/tools/repo
chmod +x ~/harmony/tools/repo
echo 'export PATH=~/harmony/tools:$PATH'  >> ~/.bashrc
source ~/.bashrc
```

13.1.3　使用 MobaXterm 远程登录 Ubuntu

MobaXterm 为 Windows 桌面提供了所有重要的远端网络工具，如 SSH、X11、RDP、VNC、FTP、MOSH 和 UNIX 命令(bash、ls、cat、sed、grep、awk、rsync……)。

步骤 1：安装 openssh-server，命令如下：

```
sudo apt - get install openssh - server
sudo /etc/init.d/ssh start
```

步骤 2：配置 MobaXterm，如图 13-18 所示。

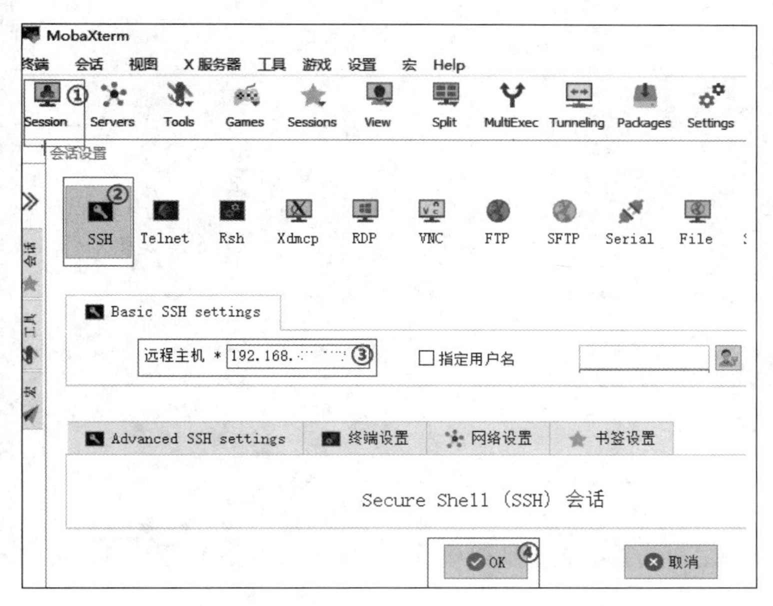

图 13-18　配置 MobaXterm

13.1.4　下载 OpenHarmony 源代码

下载 OpenHarmony 源代码有很多种途径，这里我们通过 git 下载 OpenHarmony 源代码，如图 13-19 所示，代码如下：

```
mkdir -p ~/harmony/openharmony && cd ~/harmony/openharmony
sudo apt install git
#配置用户信息
git config --global user.name "yourname"
git config --global user.email "your - email - address"
git config --global credential.helper store
repo init -u https://gitee.com/openharmony/manifest.git -b master --no - repo - verify
repo sync -c  #同步远程仓
```

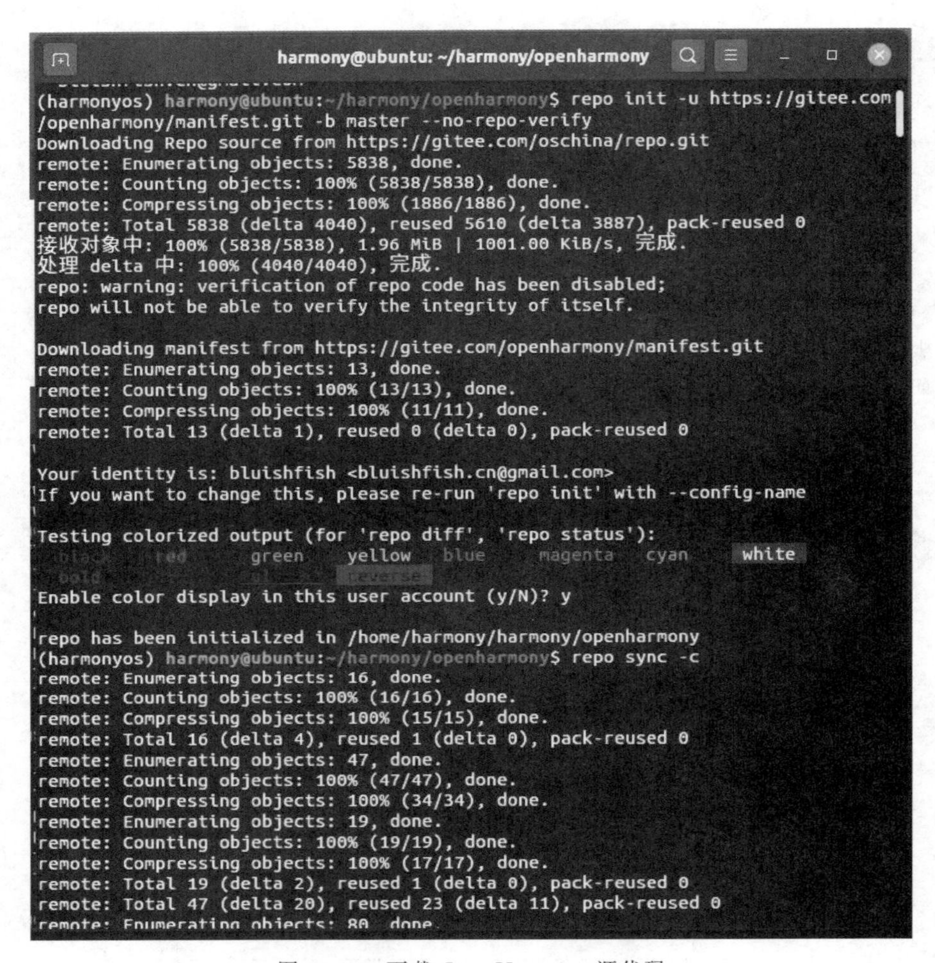

图 13-19 下载 OpenHarmony 源代码

13.1.5 编译 OpenHarmony 源代码

下面来一次性配置并编译 3 个目标平台（Hi3516、Hi3518 和 Hi3861）的编译环境，以及如何将源代码编译为 3 个目标平台的二进制文件。

在源代码根目录下（openharmony 目录）执行的代码如下：

```
 $ python build.py -h
usage:
  python build.py ipcamera_hi3516dv300
  python build.py ipcamera_hi3518ev300
  python build.py wifiiot Quickstart: https://device.harmonyos.com/cn/docs/start/introduce/
oem_start_guide-0000001054913231
```

```
positional arguments:
  product Name of the product

optional arguments:
  -h, --help show this help message and exit
  -b BUILD_TYPE, --build_type BUILD_TYPE release or debug version.
  -t [TEST [TEST ...]], --test [TEST [TEST ...]] Compile test suit
  -n, --ndk Compile ndk
```

可以查看支持哪些目标平台，目前可选的平台有3个。

编译Hi3516目标平台的命令如下：

```
python build.py ipcamera_hi3516dv300 -b debug
```

编译生成的Kernel、rootfs、userfs映像文件会被保存到out/ipcamera_hi3516dv300目录下，u-boot二进制文件会被保存在vendor目录(可以通过find vendor -name u-boot*.bin命令进行查找)。

编译Hi3518目标平台的命令如下：

```
python build.py ipcamera_hi_3518ev300 -b debug
```

编译生成的Kernel、rootfs、userfs映像文件会被保存到out/ipcamera_hi3518ev300目录下，u-boot二进制文件会被保存在vendor目录(可以通过find vendor -name u-boot*.bin命令进行查找)。

编译Hi3861目标平台的命令如下：

```
python build.py wifiiot
```

编译生成的二进制文件位于out/wifiiot/子目录下，刷机需要使用Hi3861_wifiiot_app_allinone.bin文件，如图13-20所示。

稍微等待一会儿，编译时间取决于虚拟机的资源，如果太慢，则可以多分配一些虚拟机的资源。

13.1.6 通过Samba共享Linux源代码

Samba最大的功能就是可以用于文件共享和打印共享，Samba既可以用于Windows与Linux之间的文件共享，也可以用于Linux与Linux之间的资源共享，由于NFS(网络文件系统)可以很好地完成Linux与Linux之间的数据共享，因而Samba较多地用在了Linux与Windows之间的数据共享上。

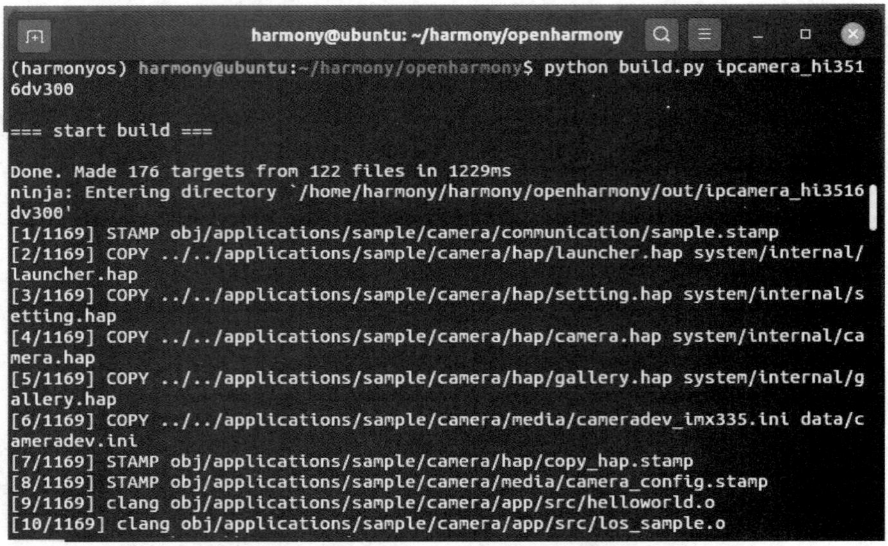

```
harmony@ubuntu: ~/harmony/openharmony
(harmonyos) harmony@ubuntu:~/harmony/openharmony$ python build.py ipcamera_hi351
6dv300

=== start build ===

Done. Made 176 targets from 122 files in 1229ms
ninja: Entering directory `/home/harmony/harmony/openharmony/out/ipcamera_hi3516
dv300'
[1/1169] STAMP obj/applications/sample/camera/communication/sample.stamp
[2/1169] COPY ../../applications/sample/camera/hap/launcher.hap system/internal/
launcher.hap
[3/1169] COPY ../../applications/sample/camera/hap/setting.hap system/internal/s
etting.hap
[4/1169] COPY ../../applications/sample/camera/hap/camera.hap system/internal/ca
mera.hap
[5/1169] COPY ../../applications/sample/camera/hap/gallery.hap system/internal/g
allery.hap
[6/1169] COPY ../../applications/sample/camera/media/cameradev_imx335.ini data/c
ameradev.ini
[7/1169] STAMP obj/applications/sample/camera/hap/copy_hap.stamp
[8/1169] STAMP obj/applications/sample/camera/media/camera_config.stamp
[9/1169] clang obj/applications/sample/camera/app/src/helloworld.o
[10/1169] clang obj/applications/sample/camera/app/src/los_sample.o
```

图 13-20　编译 OpenHarmony 源代码

步骤 1：安装命令，命令如下：

```
sudo apt - get install samba
sudo apt - get install samba - common
```

步骤 2：修改 Samba 配置文件，命令如下：

```
sudo vim /etc/samba/smb.conf
```

步骤 3：在最后加入以下内容：

```
[work]
comment = samba home directory
path = /home/harmony
public = yes
browserable = yes
read only = no
valid users = harmony
create mask = 0777
directory mask = 0777
available = yes
```

步骤 4：保存并退出后，输入如下命令，设置 Samba 密码，命令如下：

```
sudo smbpasswd -a harmony
```

步骤 5：重启 Samba 服务，命令如下：

```
sudo service smbd restart
```

步骤 6：在 Windows 上，右击“此计算机”，选择“映射网络驱动器”，如图 13-21 所示。

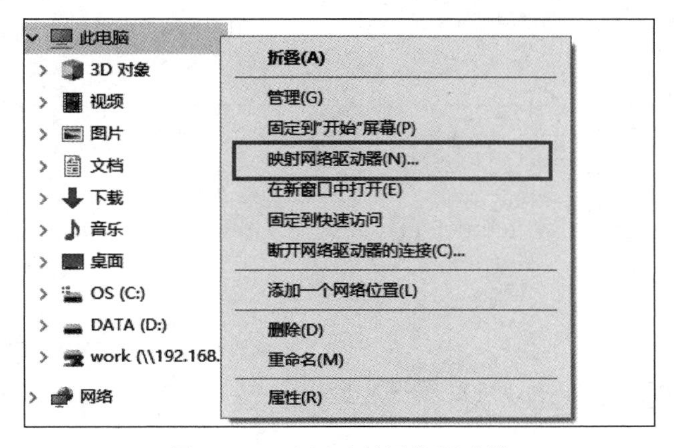

图 13-21 配置映射网络驱动器

步骤 7：选择一个未使用的驱动器，并设置 HarmonyOS 源代码所在的路径，格式为\\Linux IP 地址\共享文件夹名称，如：\\192.168.1.16\HarmonyOS_Code，单击“完成”按钮，如图 13-22 所示。

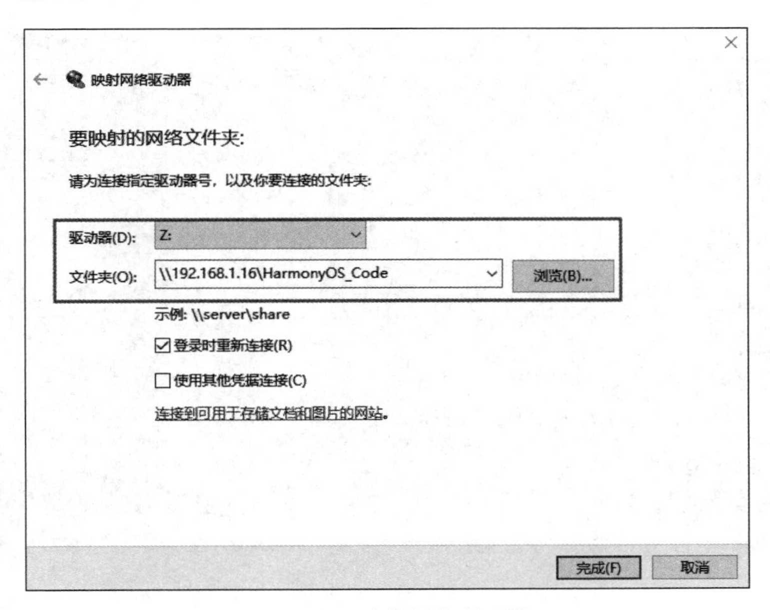

图 13-22 配置映射网络驱动器(N)

步骤 8：在 Windows 此计算机中就可以看到 Samba 共享的目录了，如图 13-23 所示。

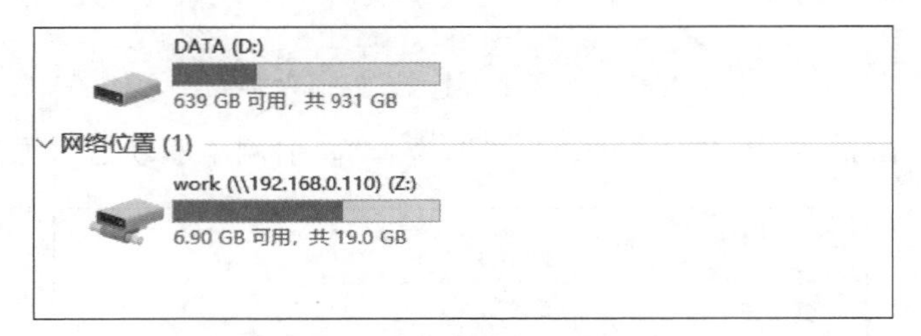

图 13-23　此计算机中可以看到 Samba 共享的目录

注意：第一次访问 \\192.168.1.16\HarmonyOS_Code 地址时，需要输入步骤 4 所设置的密码，然后才可以访问。

13.2　OpenHarmony 烧录环境准备

对于硬件开发者，鸿蒙提供了 HUAWEI DevEco Device Tool（以下简称 DevEco Device Tool），如图 13-24 所示，这是 HarmonyOS 面向智能设备开发者提供的一站式集成开发环境，支持 HarmonyOS 的组件按需定制，支持代码编辑、烧录和调试等功能，支持 C/C++ 语言，并采用插件的形式部署在 Visual Studio Code 上。

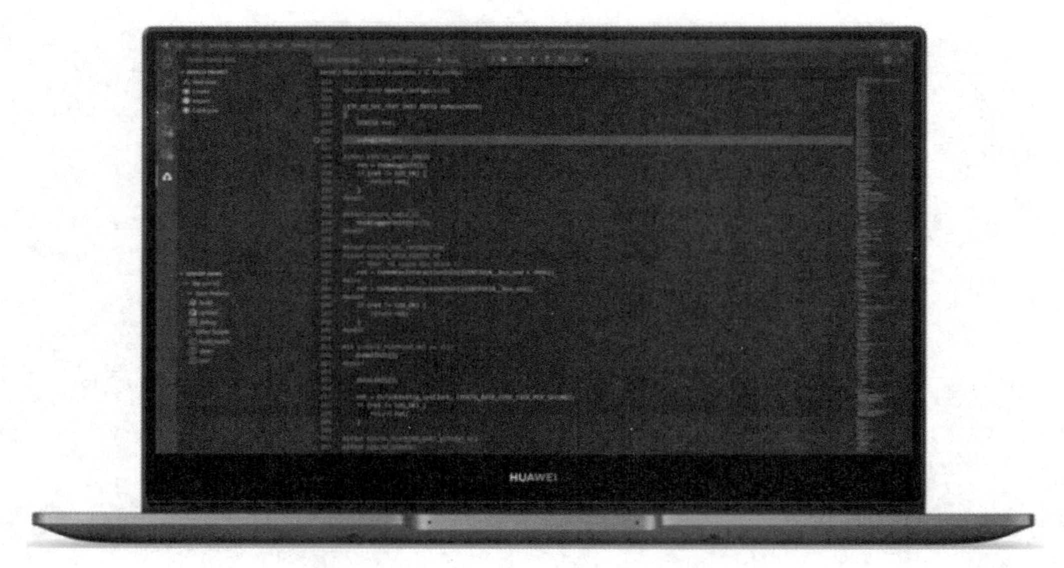

图 13-24　DevEco Device Tool 插件

DevEco Device Tool 工具具有以下特点：

- DevEco Device Tool 以 Visual Studio Code 插件的形式提供，体积小巧；
- 支持代码查找、代码高亮、代码自动补齐、代码输入提示、代码检查等，开发者可以轻松、高效地编码；
- 支持 ARM 架构的 Hi3516/Hi3518 系列和 RISC-V 架构的 Hi3861 系列开发板，提供可以一键式烧录和调试的 GUI 界面；
- 支持单步调试能力和查看内存、变量、调用栈、寄存器、汇编等调试信息。

本节详细介绍如何在 Windows 平台把编译好的代码烧写到芯片上。

13.2.1 安装 Visual Studio Code

登录 Visual Studio Code 官方网站，下载 Visual Studio Code 软件包，要求为 1.45.1 及以上版本。下载完成后，单击软件包进行安装。安装过程中，需勾选"添加到 PATH（重启后生效）"选项，如图 13-25 所示。

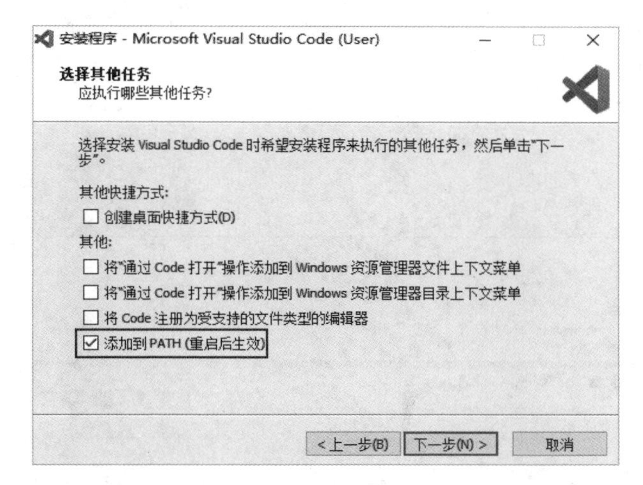

图 13-25　安装 Visual Studio Code

安装完成后，重启计算机，使 Visual Studio Code 的环境变量生效。

13.2.2 安装 Node.js

登录 Node.js 官方网站，下载 Node.js 软件包。需选择 LTS 版本 12.0.0 及以上，Windows 64 位对应的软件包如图 13-26 所示。

下载完成后，单击软件包进行安装，勾选图 13-27 的选项框，自动安装必要工具（如 Python、Visual Studio 构建工具链），如图 13-27 所示。

安装好 Node.js 后，需要在系统环境变量中添加 NODE_PATH，如图 13-28 所示。

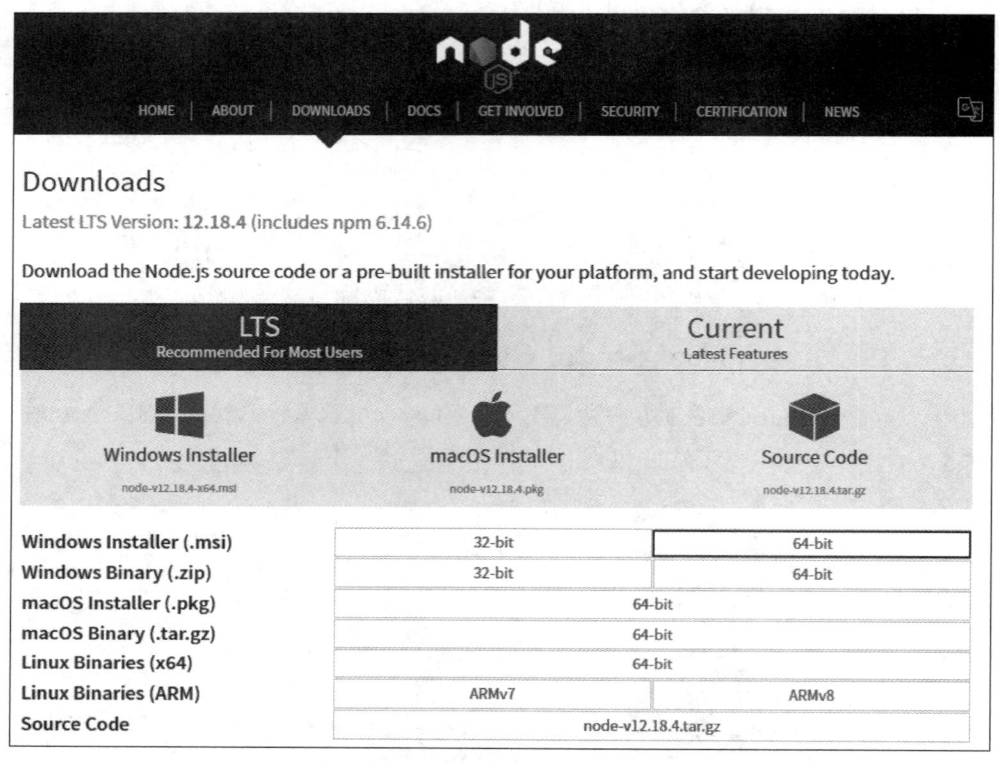

图 13-26　下载 Node.js

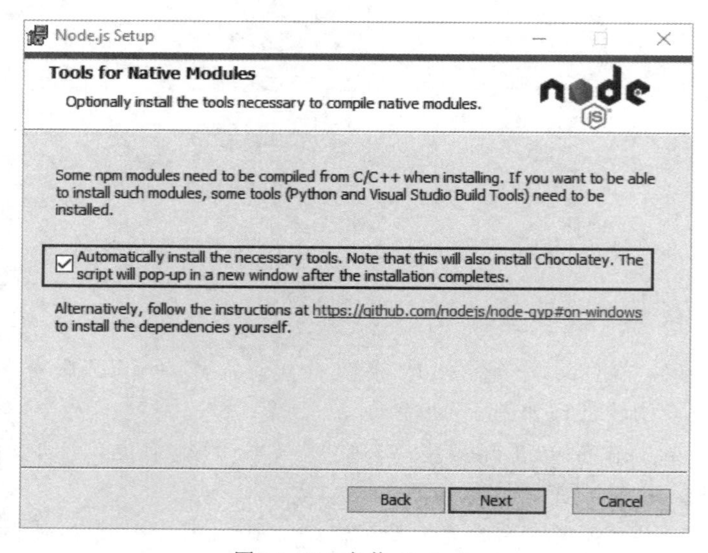

图 13-27　安装 Node.js

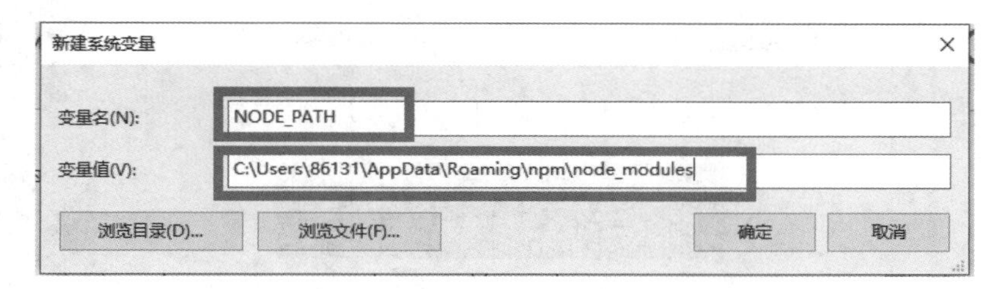

图 13-28 在系统环境变量中添加 NODE_PATH

13.2.3 安装 JDK

下载并安装 JDK，版本要求为 1.8，如图 13-29 所示。下载网址为 https://www.oracle.com/java/technologies/javase/javase-JDK8-downloads.html。

Linux x64 Compressed Archive	136.51 MB	jdk-8u271-linux-x64.tar.gz
macOS x64	205.46 MB	jdk-8u271-macosx-x64.dmg
Solaris SPARC 64-bit (SVR4 package)	125.94 MB	jdk-8u271-solaris-sparcv9.tar.Z
Solaris SPARC 64-bit	88.75 MB	jdk-8u271-solaris-sparcv9.tar.gz
Solaris x64 (SVR4 package)		x64.tar.Z
Solaris x64		x64.tar.gz
Windows x86		s-i586.exe
Windows x64		s-x64.exe

You must accept the Oracle Technology Network License Agreement for Oracle Java SE to download this software.

I reviewed and accept the Oracle Technology Network License Agreement for Oracle Java SE

You will be redirected to the login screen in order to download the file.

Download jdk-8u271-windows-x64.exe

图 13-29 下载完成后，安装 JDK-8u271-Windows-x64.exe

安装完成后可在命令行中查看版本，如图 13-30 所示。

```
C:\Users\86131>java -version
java version "1.8.0_241"
Java(TM) SE Runtime Environment (build 1.8.0_241-b07)
Java HotSpot(TM) 64-Bit Server VM (build 25.241-b07, mixed mode)
```

图 13-30 查看 JDK 版本

13.2.4 安装 HPM

HPM 全称为 HarmonyOS Package Manager，是 HarmonyOS 组件包的管理和分发工具。HPM 主要面向设备开发者，用于获取/定制 HarmonyOS 源代码，执行安装、编译、打包、升级等操作。

打开命令行工具，执行如下命令进行安装 HPM：

```
npm install -g @ohos/hpm-cli
```

安装完成后，执行如下命令查看 HPM 是否安装成功，如图 13-31 所示。

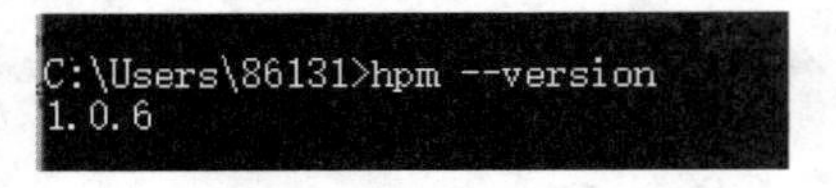

图 13-31　查看 HPM 版本

13.2.5　安装 DevEco Device Tool 插件

安装 DevEco Device Tool 步骤如下。

步骤 1：登录 HarmonysOS 设备开发门户，单击右上角的注册按钮，注册开发者账号，注册指导可参考注册华为账号。如果已有华为开发者账号，可直接单击“登录”按钮，如图 13-32 所示。

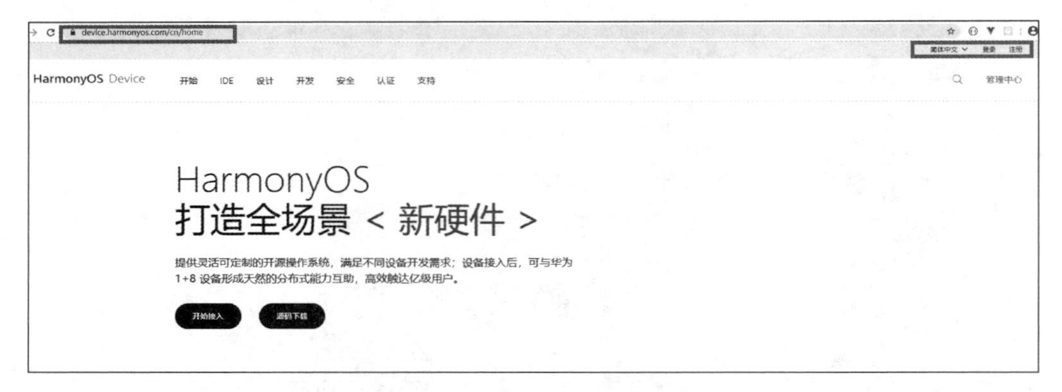

图 13-32　注册并登录开发者账号 https://device.harmonyos.com/cn/home

步骤 2：进入 HUAWEI DevEco Device Tool 产品页，下载 DevEco Device Tool 安装包。在产品页下载工具插件，如图 13-33 所示。

```
https://device.harmonyos.com/cn/ide
```

HUAWEI DevEco Device Tool 1.0 Beta1

Platform	DevEco Device Tool Package	Size	SHA-256 checksum	Download
Windows (64-bit)	devicetool-device-1.0.0.0.zip	20.3MB	893a1076e7f63759287ea23e3a57d50acd99102c450bc9fe4bf09a3f484af00c	⏬
Windows (64-bit)	devicetool-device-1.0.0.1.zip	33.8MB	eb432c87c3e19b61846f8dee8769c262bfc6d3d9217210d6df4e381aa6f9a186	⏬

图 13-33　下载 HUAWEI DevEco Device Tool

步骤 3：打开 Visual Studio Code 软件，如图 13-34 所示。

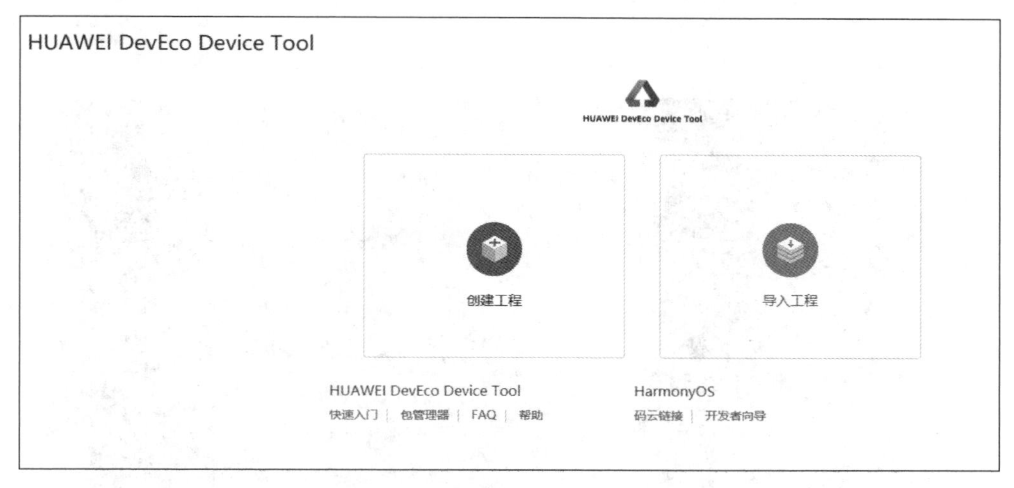

图 13-34　Visual Studio Code 软件

步骤 4：采用从本地磁盘进行安装的方式，安装 DevEco Device Tool，如图 13-35 所示。

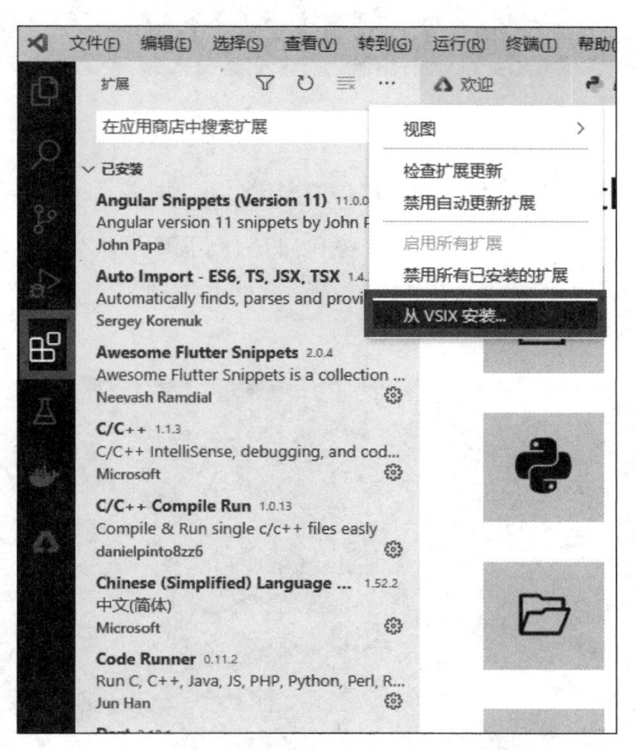

图 13-35　从本地磁盘进行安装 DevEco Device Tool

13.2.6 安装 C/C++插件

单击 Visual Studio 的插件扩展，搜索 C/C++，安装如图 13-36 所示的插件。

图 13-36 安装 C/C++插件

13.2.7 导入和配置 OpenHarmony 工程

HarmonyOS 暂不支持 Windows 系统源代码，因此需要从映射的 Linux 服务器的共享路径导入源代码。在 Visual Studio Code 中，单击 DevEco Device Tool 插件按钮图标，然后单击 Import 按钮，如图 13-37 所示。

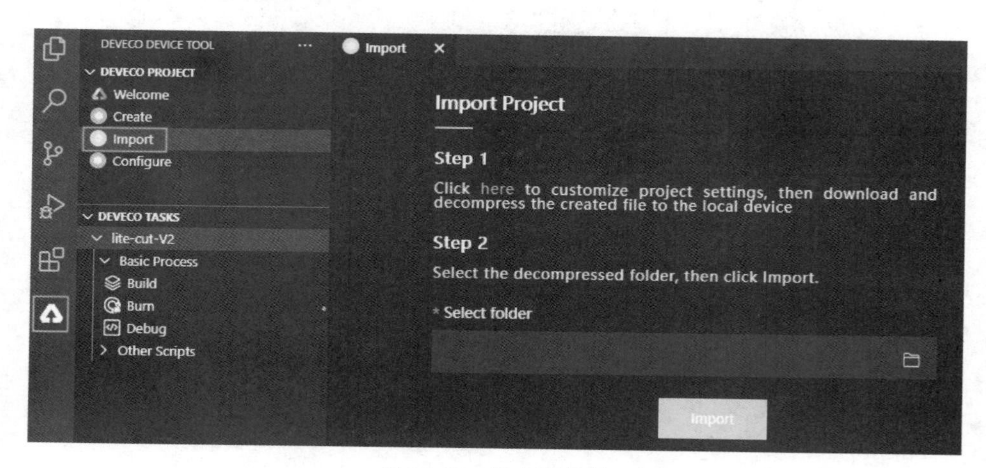

图 13-37 导入源代码

选择本地映射的 HarmonyOS 源代码文件夹，然后单击 Import 按钮导入工程，如图 13-38 所示。

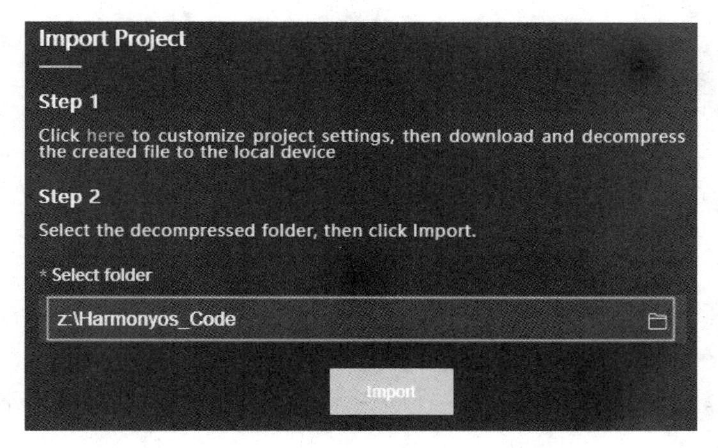

图 13-38　设置 Samba 源代码共享目录

导入工程源代码文件后，需要对工程进行配置，如工程烧录和调试依赖的工具等信息。

单击底部工具栏中的 Board 按钮，选择对应开发板的配置模板，例如 HI3516DV300，如图 13-39 所示。

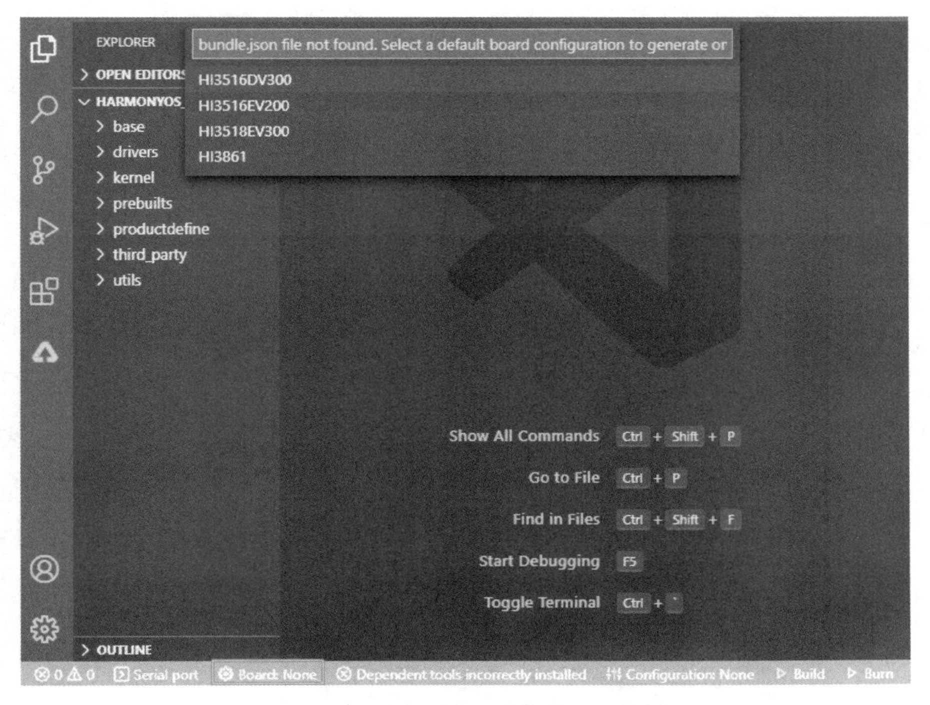

图 13-39　选择对应开发板的配置模板

在命令行工具中，分别执行如下命令，下载烧录依赖工具，命令如下：

```
npm install -g tftp
npm install -g serialport
npm install -g @serialport/parser-readline
npm install -g usb
npm install -g crc
```

npm install -g usb 命令仅用于 Hi3516/Hi3518 系列开发板在使用 USB 烧录时需要执行。

13.3 本章小结

本章详细介绍了 OpenHarmony 编译和烧录环境的搭建，读者可根据本章的步骤完成 OpenHarmony 开发编译环境搭建。

第 14 章 HiSpark 开发板开发入门

HarmonyOS 打造全场景新硬件，提供灵活可定制的开源操作系统，满足不同设备开发需求，设备接入后，可与华为 1+8 设备形成天然的分布式能力互助，高效触达亿级用户。

14.1 HiSpark 系列开发套件介绍

鸿蒙 HarmonyOS 设备开发提供了多种典型的开发场景，如图 14-1 所示。

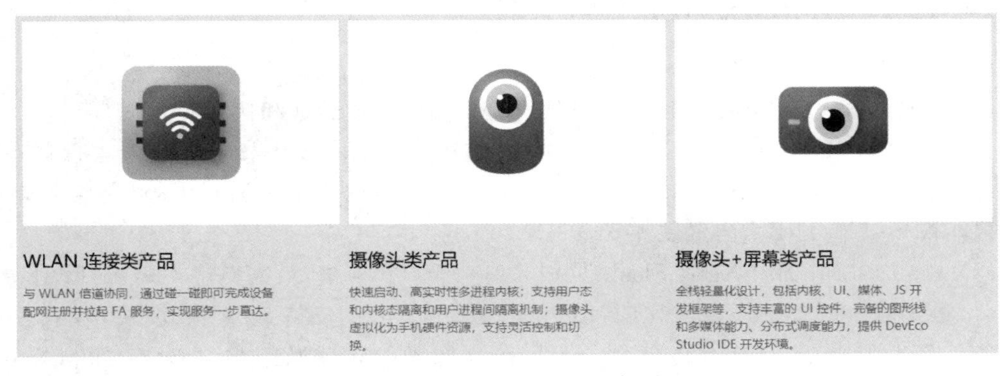

图 14-1 多种典型的开发场景

HiSpark 系列开发套件如图 14-2 所示。

14.1.1 HiSpark WiFi IoT 开发套件

HiSpark WiFi IoT 开发套件有以下特性。

(1) 支持鸿蒙 OS、LiteOS，方便进行物联网产品的原型验证和快速开发。

(2) 特性板搭载海思 Hi3861 芯片，最高运行频率为 160MHz，内置 352KB SRAM、288KB ROM，内置 2MB Flash，密码支持 IEEE 802.11 b/g/n，支持 STA 模式、AP 模式。

(3) 套件包含多个扩展板，包括 OLED 板、NFC 扩展板、环境监测板、红绿灯板、炫彩灯板、机器人板，并且集成了多种常见外设。

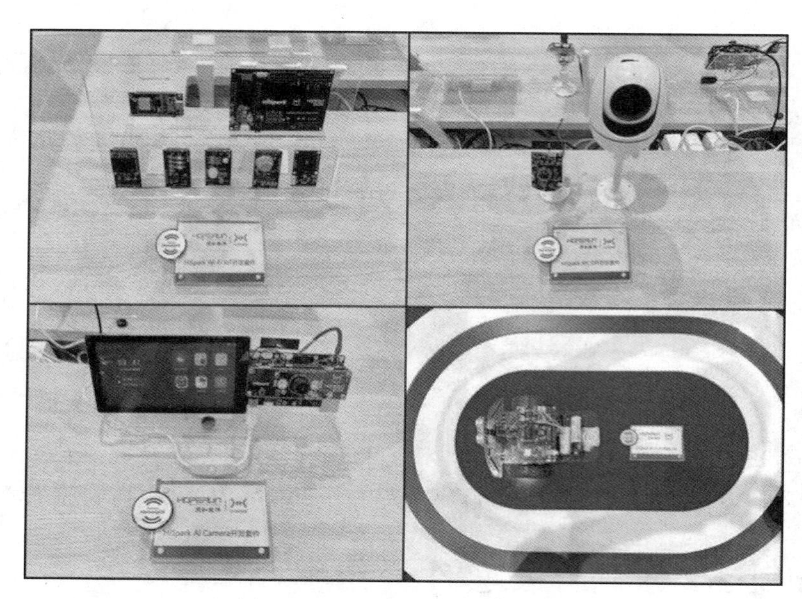

图 14-2 HiSpark 系列开发套件

14.1.2 HiSpark DIY IPC 套件

HiSpark DIY IPC 套件有以下特性。

(1) 支持鸿蒙 OS、LiteOS、Linux，方便进行产品的原型验证和快速开发。

(2) 板载海思 Hi3518EV300 芯片，内置 ARM Cortex A7 核心，最高运行频率为 900MHz，内置 64MB SDRAM，内置图像处理单元(ISP)，内置 H265/H264 硬件编解码器，内置智能视频引擎(IVE)，内置硬件安全引擎。

(3) 搭载 128Mb SPI NOR Flash，同时带有一个 TF 卡槽，最大支持 128GB TF 卡。

(4) 板载 Hi3881 WiFi 芯片，支持 IEEE 802.11 b/g/n，支持 STA 模式、AP 模式。

(5) 板载 JX-F23A 图像传感器，最大有效像素 2MB，视频最高支持 1920×1080 @25fps。

14.1.3 HiSpark AI Camera 套件

HiSpark AI Camera 套件有以下特性。

(1) 支持鸿蒙 OS、LiteOS、Linux，方便进行产品的原型验证和快速开发。

(2) 板载海思 Hi3516DV300 芯片，内置双核 Cortex-A7，最高运行频率为 900MHz，内置图像处理单元(ISP)，内置 H265/H264 硬件编解码器，内置智能视频引擎(IVE)，内置硬件安全引擎。

(3) 主控芯片内置神经网络推理引擎(NNIE)，8 位算力 1Tops，可进行端侧 AI 计算。

(4) 板载 1G DDR3 内存，最大传输速度为 1866Mb/s。

(5) 板载 8G eMMC,同时可外挂 2TB SDXC 卡。

(6) 板载 Hi3861 WiFi 模组,支持 IEEE 802.11 b/g/n,支持 STA 模式、AP 模式。

(7) 板载索尼 IMX 335 图像传感器,最大有效像素为 5.04M 像素,视屏最高支持 2592×1944@60fps。

14.2 HiSpark Hi3861 开发板

HiSpark WiFi IoT 智能家居套件基于华为海思 Hi3861 芯片,支持 LiteOS、鸿蒙 OS,实现 WiFi IoT 功能,广泛应用于常电智能家居(如白电、小家电、电工类)场景,如图 14-3 所示。

白电

小家电

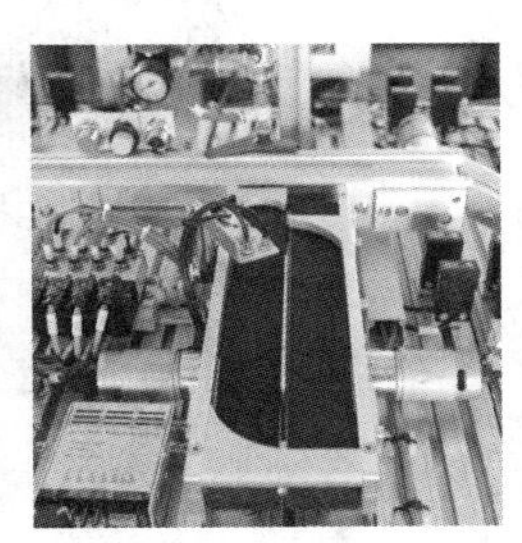

电工类

图 14-3 应用场景

14.2.1 开发板介绍

Hi3861 WLAN 模组是一片大约 2cm×5cm 大小的开发板,是一款高度集成的 2.4GHz WLAN SoC 芯片,集成 IEEE 802.11b/g/n 基带和 RF(Radio Frequency)电路。支持 HarmonyOS,并配套提供开放、易用的开发和调试运行环境,如图 14-4 所示。

另外,Hi3861 WLAN 模组还可以通过与 Hi3861 底板连接,扩充自身的外设能力,底板如图 14-5 所示,产品参数如图 14-6 所示。

14.2.2 烧录 HarmonyOS

烧录是指将编译后的程序文件下载到芯片的动作,为后续的程序调试提供基础。

Hi3861 系列开发板支持使用 Jlink 和 HiBurn 工具进行烧录。

(1) 使用 Jlink 工具烧录,开发者需自行下载并安装 Jlink 工具。

(2) 使用 HiBurn 工具烧录,DevEco Device Tool 已预置 HiBurn 工具。

使用 HiBurn 将.bin 文件烧录到 Hi3861 的步骤如下。

(1) Windows 10 系统执行前需要右击“属性”→解除锁定,否则系统默认会报安全警告,不允许执行。双击后,界面如图 14-7 所示。

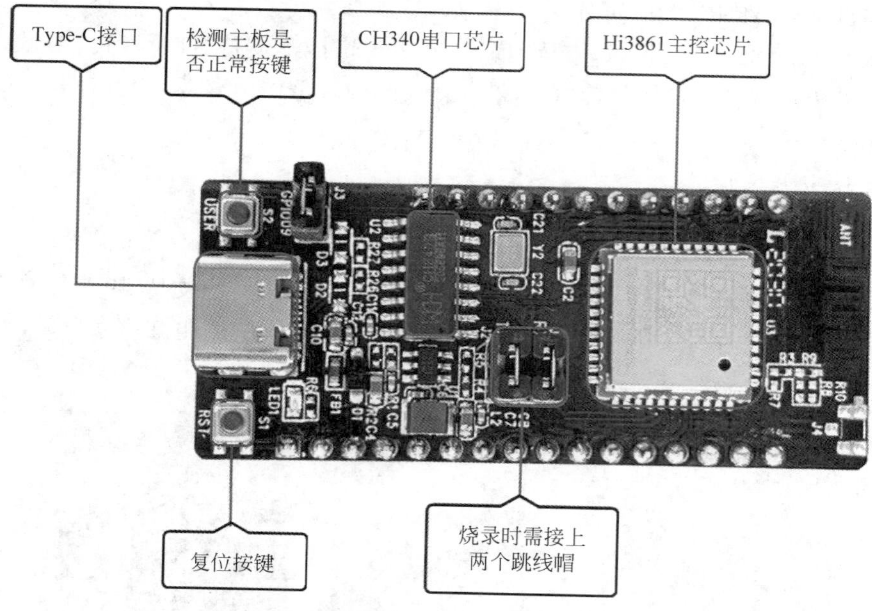

图 14-4　Hi3861 WLAN 模组外观图

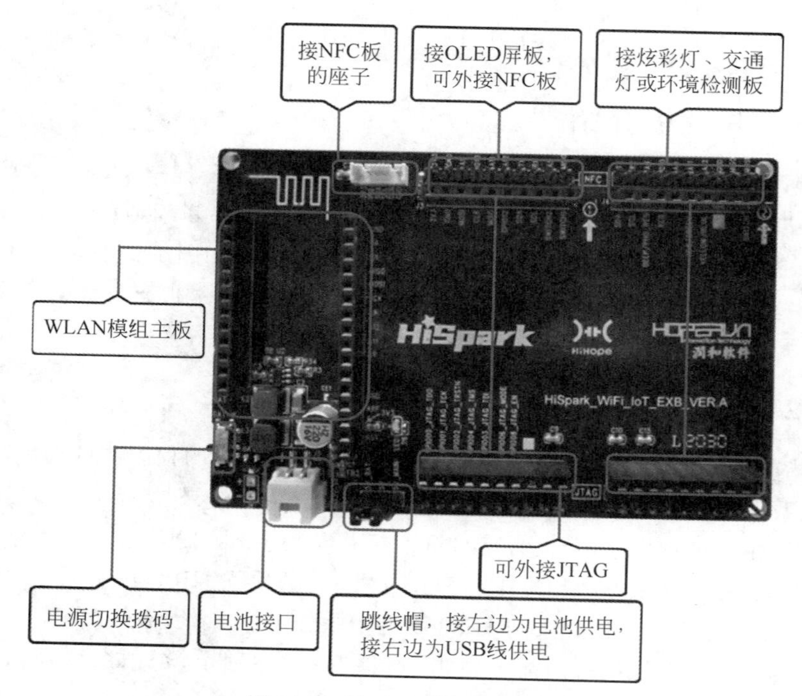

图 14-5　Hi3861 底板外观图

配置	参数
SoC	基于海思Hi3861/Hi3861L高度集成的2.4GHz Wi-Fi芯片 内部集成高性能32位微处理器、硬件安全引擎以及丰富的外设接口
操作系统	支持LiteOS、鸿蒙 OS和第三方组件，可与华为Hi-Link协同
通信能力	支持复杂环境下TPC、自动速率、弱干扰免疫等可靠性通信算法
组网能力	支持256节点Mesh组网 支持20MHz标准带宽和5MHz/10MHz窄带宽，提供最大72.2Mb/s物理层速率
网络能力	支持IPv4 /IPv6 /DHCPv4 /DHCPv6 Client /Server/DNS Client /mDNS /CoAP /MQTT /HTTP /JSON 集成IEEE 802.11b/g/n基带和RF电路 WiFi基带支持正交频分复用（OFDM）
安全能力	支持AES128/256加解密/HASH-SHA256/HMAC_SHA256 /RSA 支持ECC签名校验算法 真随机数生成，满足FIPS140-2随机测试标准 支持TLS/DTLS加速 内部集成EFUSE、安全存储、安全启动、硬件ID 集成MPU特性，支持内存隔离特性

图 14-6 Hi3861 产品参数

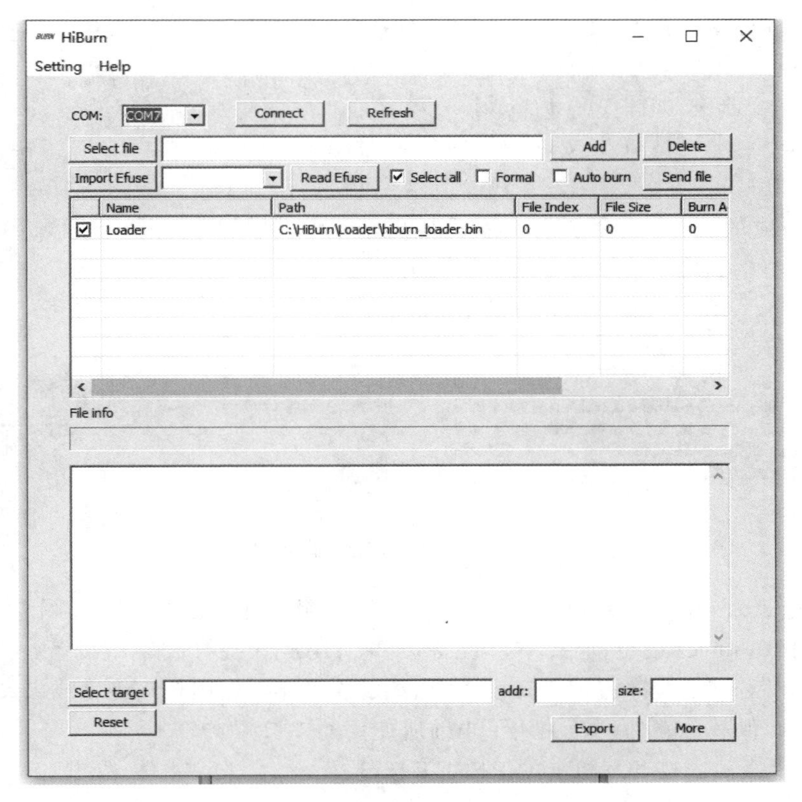

图 14-7 将.bin 文件烧录到 Hi3861

(2) 单击界面左上角的 Setting→Com settings 进入串口参数设置界面，如图 14-8 所示。

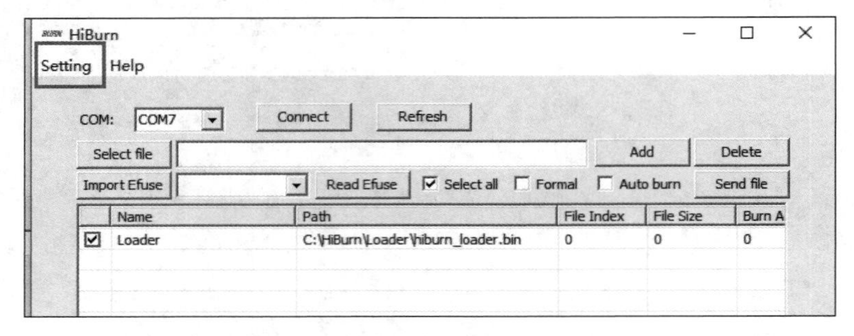

图 14-8　Hi3861 产品参数设置界面(1)

(3) 在串口参数设置界面，Baud 为波特率，默认为 115200，可以选择 921600、2000000，或者 3000000(实测最快支持的值)，其他参数保持默认，单击“确定”按钮进行保存，如图 14-9 所示。

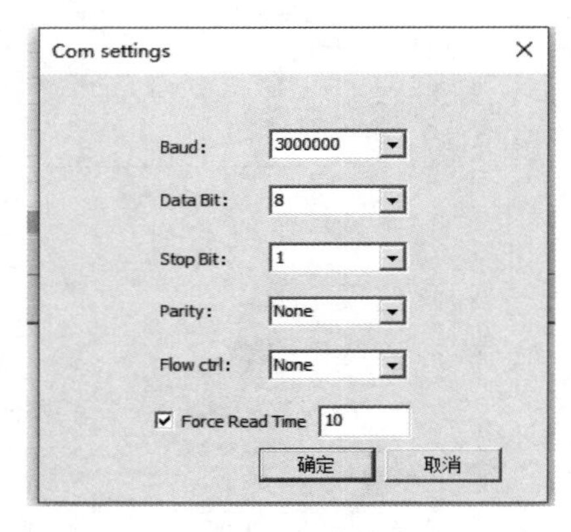

图 14-9　Hi3861 产品参数设置界面(2)

(4) 根据设备管理器，选择正确的 COM 口，例如笔者的开发板所选择的端口是 COM8，如果打开程序之后才插串口线，则可以单击 Refresh 按钮刷新串口下拉列表框的可选项，如图 14-10 所示。

(5) 单击 Select file 弹出文件选择对话框，选择编译生成的 allinone.bin 文件，这个 bin 其实是由多个 bin 合并而成的文件，从命名上也能看得出来，例如，笔者选择的是 z:\harmonyos\openharmony\out\wifiiot\Hi3861_wifiiot_app_allinone.bin。

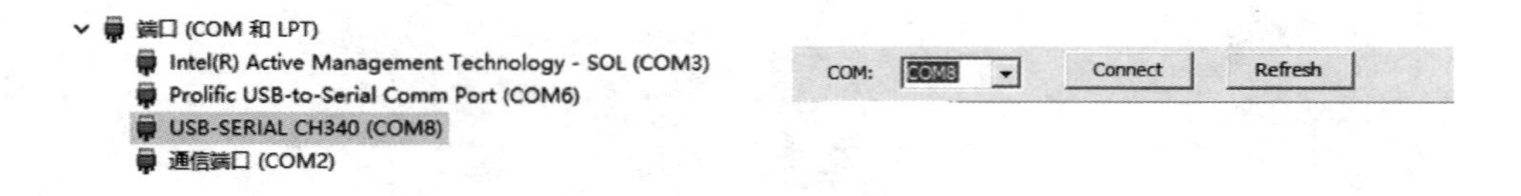

图 14-10　Hi3861 产品参数设置界面(3)

(6) 勾选 Auto burn，会自动下载多个 bin 文件，到这里，配置完毕，如图 14-11 所示。

(7) 单击 Connect 按钮，连接串口设备，这时 HiBurn 会打开串口设备，并尝试开始烧录，需要确保没有其他程序占用串口设备(烧录之前可能正在用超级终端或串口助手查看串口日志，需要确保其他软件已经关闭了当前所使用的串口)。

(8) 复位设备，需按开发板上的 RESET 按键。

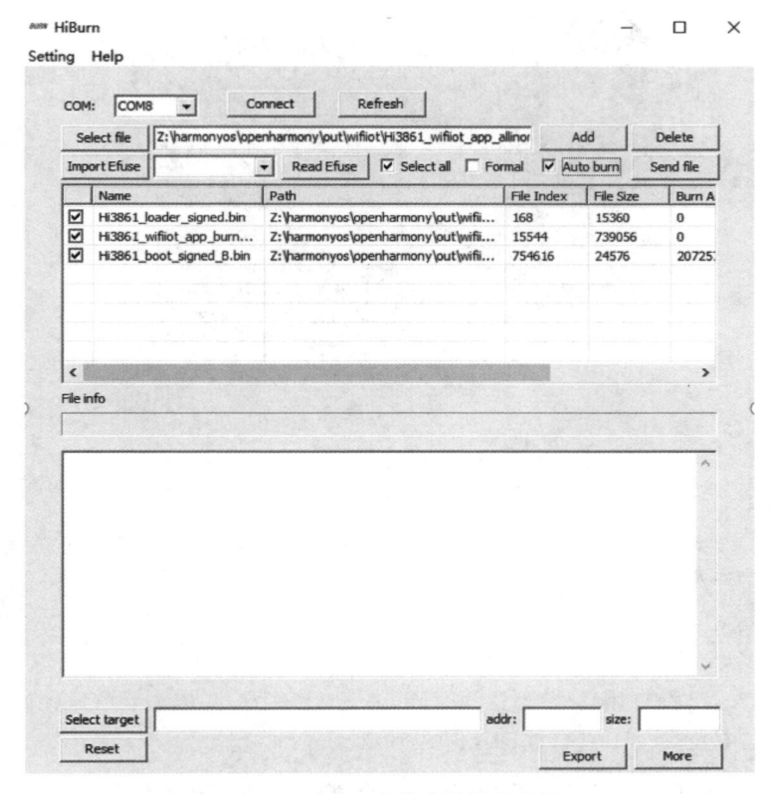

图 14-11 Hi3861 产品参数设置界面(4)

(9) 等待输出框出现 3 个“========”及上方均出现 successful,即说明烧录成功。

使用 HiBurn 和 DevEco Device Tool 方式进行烧录的优缺点如下。

使用 HiBurn 烧录相对于使用 DevEco Device Tool 烧录而言,主要有以下几个优点。

(1) 不依赖 VSCode,可以不用安装 VSCode、Node.js、JDK 及一些 npm 包。

(2) 下载速度更快,HiBurn.exe 最大波特率可以设置为 3000000,而 DevEco Device Tool 最大波特率只能设置为 921600,是它的 3 倍。

HiBurn 方式烧录有以下两个缺点。

(1) 需要手动单击 Disconnect 按钮主动断开连接,否则默认会重复下载。

(2) 烧录成功后,如果不断开串口,并且再次按了一下 RESET 按键,则会发现,它又烧录了一遍。

14.2.3 添加 Hi3861 显示屏驱动

HiSpark WiFi 开发套件提供了一个 OLED 屏幕,但是鸿蒙源代码中没有这个屏幕的驱动,需要自己进行移植,如图 14-12 所示。

图 14-12　鸿蒙源代码中没有这个屏幕的驱动

编写显示屏驱动代码步骤如下。

(1) 在 applications/sample/WiFi-iot/app 目录下增加应用目录 oled_demo，代码如下：

```
└── applications
    └── sample
        └── WiFi - iot
            └── app
```

(2) 设置 I2C 引脚复用：确定 I2C 引脚，查看原理图，可以看到 OLED 屏幕使用的是 I2C0，引脚是 GPIO13、GPIO14，如图 14-13 所示。

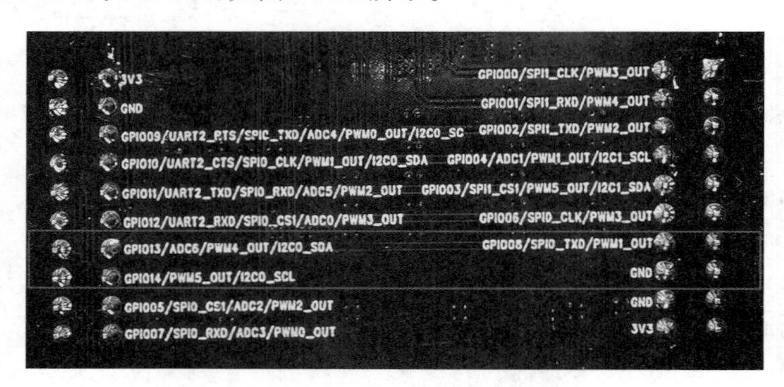

图 14-13　设置 I2C 引脚复用

所以需要修改 vendor\hisi\hi3861\hi3861\app\wifiiot_app\init\app_io_init.c 文件源代码，初始化 I2C 引脚的代码。修改后代码如下：

```
#ifdef CONFIG_I2C_SUPPORT
    /* I2C IO 复用也可以选择 3/4 或 9/10,根据产品设计选择 */
    hi_io_set_func(HI_IO_NAME_GPIO_13, HI_IO_FUNC_GPIO_13_I2C0_SDA);
    hi_io_set_func(HI_IO_NAME_GPIO_14, HI_IO_FUNC_GPIO_14_I2C0_SCL);
#endif
```

(3) 开启 I2C 功能。修改文件 vendor\hisi\hi3861\hi3861\build\config\usr_config.mk，增加 CONFIG_I2C_SUPPORT=y。修改完成后，重新编译即可成功驱动 OLED。

(4) OLED 屏幕驱动。入口函数如下：

```
void my_oled_demo(void)
{
    //初始化,此处使用的是 I2C0
    I2cInit(WIFI_IOT_I2C_IDX_0, 100000); /* baudrate: 100000 */
    led_init();
    OLED_ColorTurn(0);        //0 表示正常显示,1 表示反色显示
    OLED_displayTurn(0);      //0 表示正常显示,1 表示屏幕翻转显示
    OLED_ShowString(8,16,"hello world",16);
    OLED_Refresh();
}
```

使用 I2C 编写函数，代码如下：

```
u32 my_i2c_write(WifiIotI2cIdx id, u16 device_addr, u32 send_len)
{
    u32 status;
    WifiIotI2cData es8311_i2c_data = { 0 };

    es8311_i2c_data.sendBuf = g_send_data;
    es8311_i2c_data.sendLen = send_len;
    status = I2cWrite(id, device_addr, &es8311_i2c_data);
    if (status != 0) {
        printf("===== Error: I2C write status = 0x%x! =====\r\n", status);
        return status;
    }

    return 0;
}
```

14.3 HiSpark Hi3516 开发板

HiSpark AI Camera 基于华为海思 Hi3516DV300 芯片，支持 Linux、LiteOS、鸿蒙 OS，实现图像采集。识别、双屏显示、双向语音、红外夜视等功能广泛应用于智能摄像、安防监控、车载记录仪等，如图 14-14 所示。

14.3.1 开发板简介

Hi3516DV300 作为新一代行业专用 Smart HD IP 摄像机 SoC，集成了新一代 ISP、H.265 视频压缩编码器，同时集成了高性能 NNIE 引擎，使得 Hi3516DV300 在低码率、高

安防监控

工业监控

图 14-14　应用场景

画质、智能处理和分析、低功耗等方面领先行业水平，如图 14-15 所示。

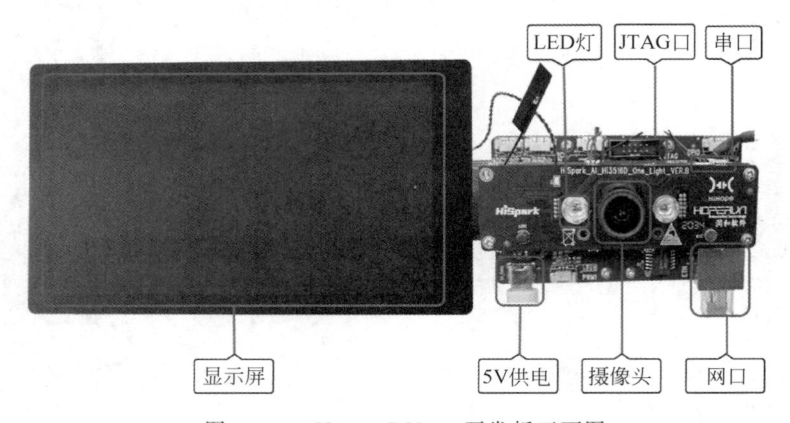

图 14-15　Hi3516DV300 开发板正面图

14.3.2　烧录 HarmonyOS

在 Hi3516DV300 开发版烧录 Open HarmonyOS，首先需要在 Linux 下编译出系统文件，再通过 Windows 平台进行烧录，如图 14-16 所示，编译命令如下：

```
python build.py ipcamera_hi3516dv300 -b debug
```

编译后的文件被存放在 out/ipcamera_hi3516dv300 目录下。烧录时需要用到 ipcamera_hi3516dv300 目录下的 OHOS_Image.bin、rootfs.img、userfs.img 3 个文件。

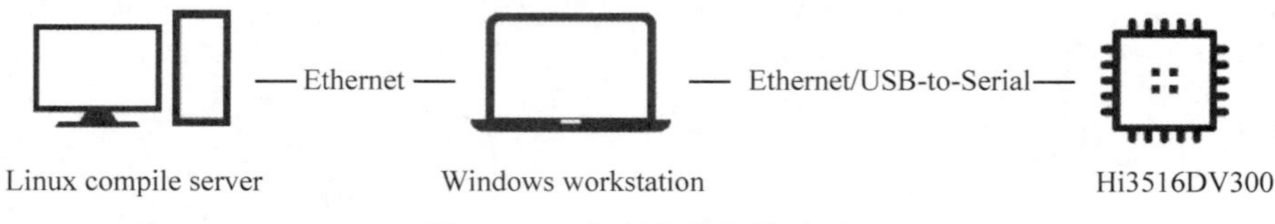

图 14-16　各硬件连接关系图

注意：这里需要注意，编译的时候需要加上-b debug 命令，否则无法通过串口进入鸿蒙 shell。

编译完成后，我们需要通过串口连接线把 Hi3516DV300 开发板连接到 PC 的 USB 接口上。这样 Windows 会自动生成一个串口号，我们打开计算机的设备管理器便可以查看了。

1. 串口烧录

Hi3516DV300 开发板烧录 Open HarmonyOS 最简单的方式是通过串口的方式进行烧录，但是通过串口烧录非常耗时，烧录通常需要 1.5h 左右的时间。下面具体看一看烧录的详细步骤。

步骤 1：查看串口号。

在 Windows 计算机上打开计算机的设备管理器，当串口线通过 USB 接口与计算机连接后，就可以查看并记录对应的串口号了。图 14-17 是计算机设备管理器中显示的 Hi3516DV300 板的串口号，如图 14-17 所示。

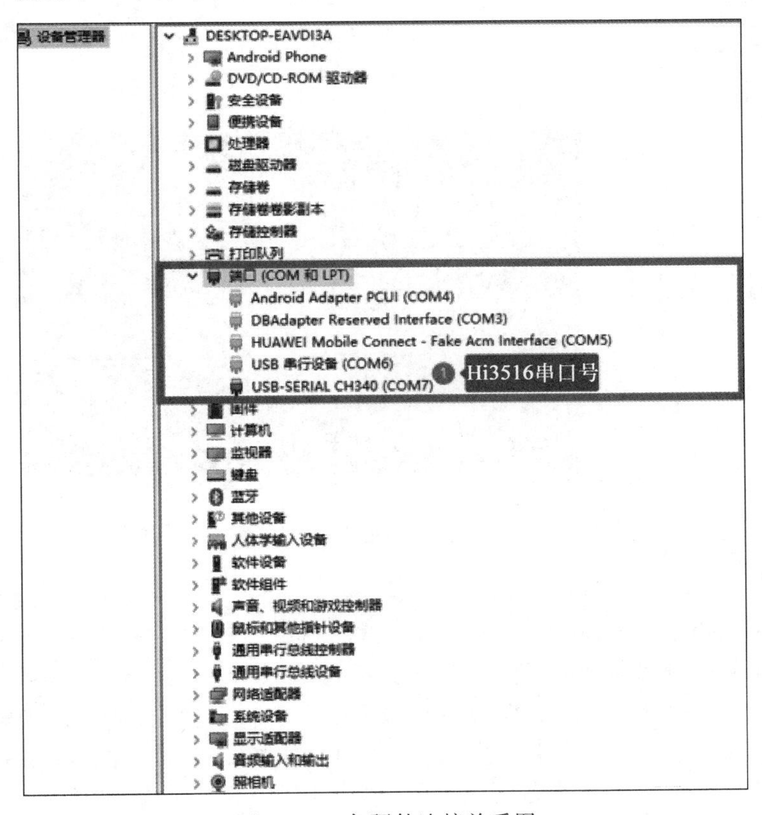

图 14-17 各硬件连接关系图

步骤 2：下载海思的烧录工具 HiTool。

HiTool 是海思半导体官方提供的一个海思芯片烧录工具。HiTool 工具可以到 HiHope 官网资源中心下载，如图 14-18 所示。下载网址为 http://www.hihope.org/download/AllDocuments。

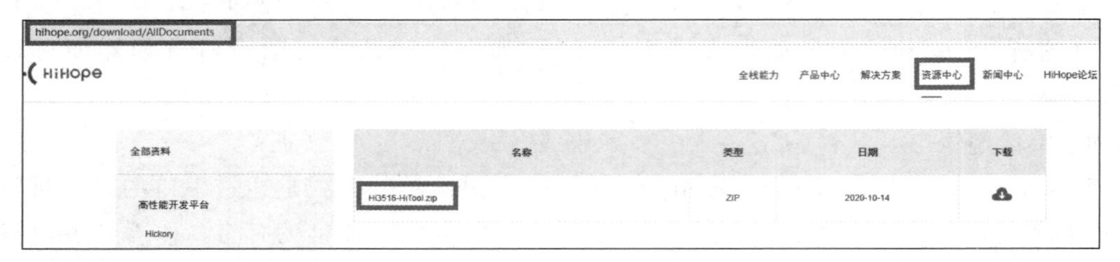

图 14-18 HiHope 官网资源中心下载

步骤 3：运行 HiTool 工具，选择芯片 Hi3516DV300。

将海思烧写工具全部解压出来，接着打开 HiTool.exe 文件，选择芯片 Hi3516DV300，如图 14-19 所示。

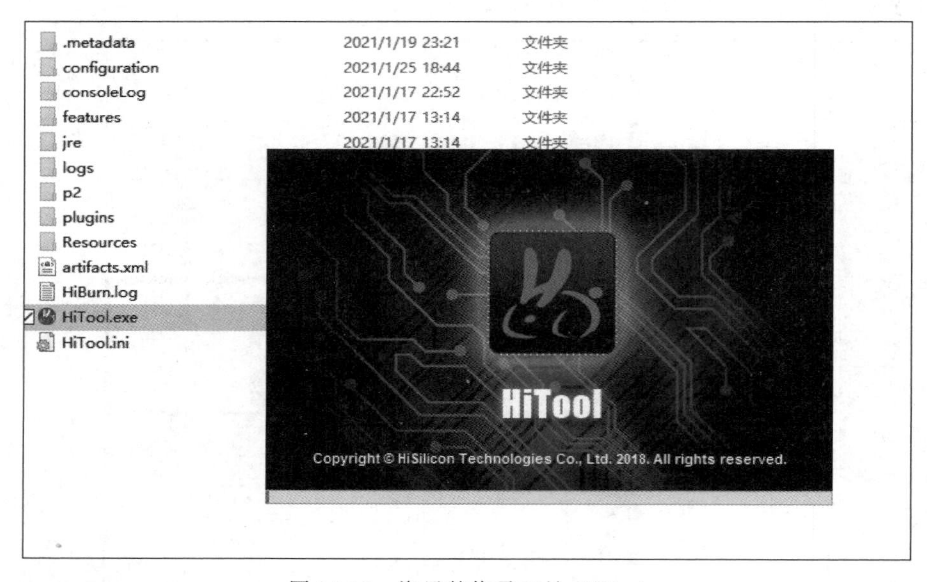

图 14-19 海思的烧录工具 HiTool

步骤 4：进入烧录界面，在右侧的“传输方式中”选择“串口”，如图 14-20 所示。

步骤 5：单击烧录界面中的“烧录 eMMC”按钮，如图 14-21 所示。

步骤 6：根据分区信息添加要烧录的文件，本次烧录只烧录内核 Kernel(OHOS_Image.bin)、根文件系统 rootfs(rootfs.img)、用户文件系统 userfs(userfs.img)，如图 14-22 所示。

步骤 7：单击烧录，如图 14-23 所示。

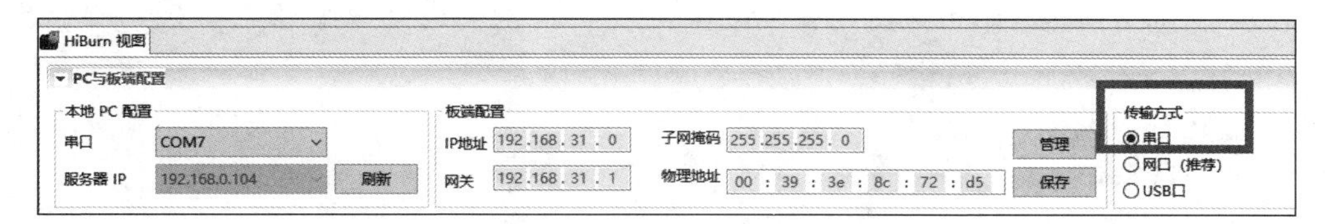

图 14-20 选择“串口”方式

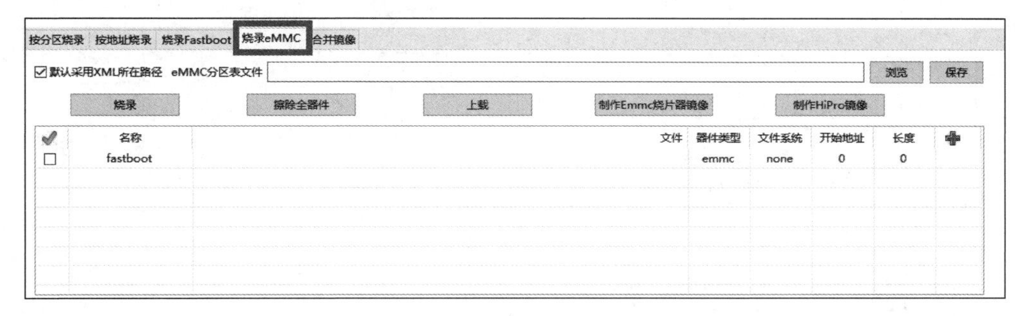

图 14-21 选择“烧录 eMMC”

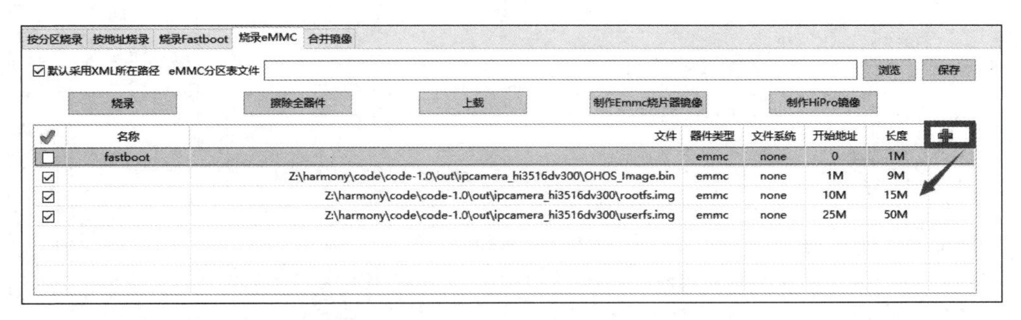

图 14-22 根据分区信息添加要烧录的文件

图 14-23 单击烧录

步骤 8：重启开发板，可以按动开关按钮，必须 15s 内完成该操作，如图 14-24 所示。

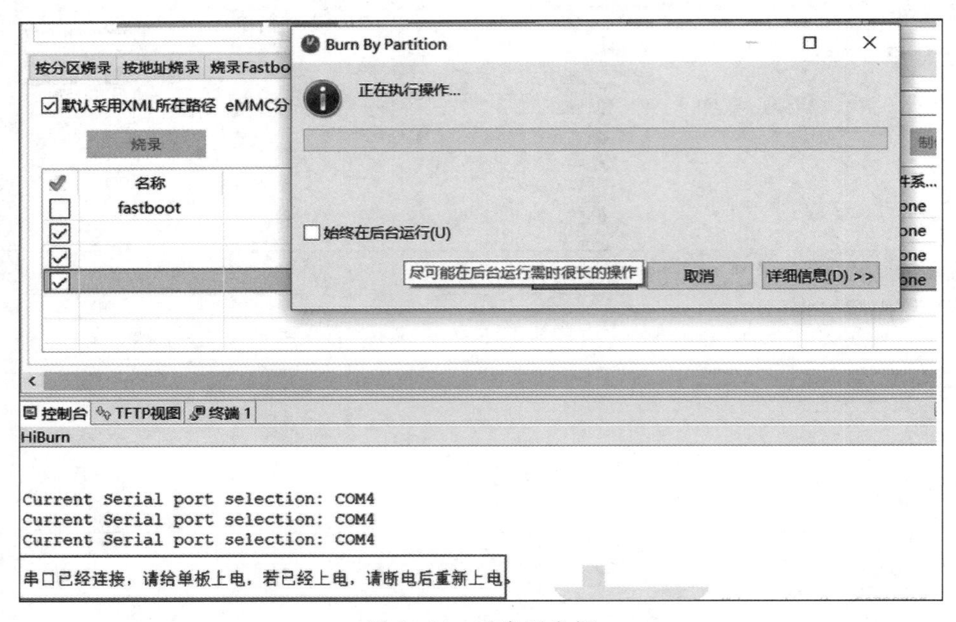

图 14-24　重启开发板

步骤 9：烧录成功，如图 14-25 所示。

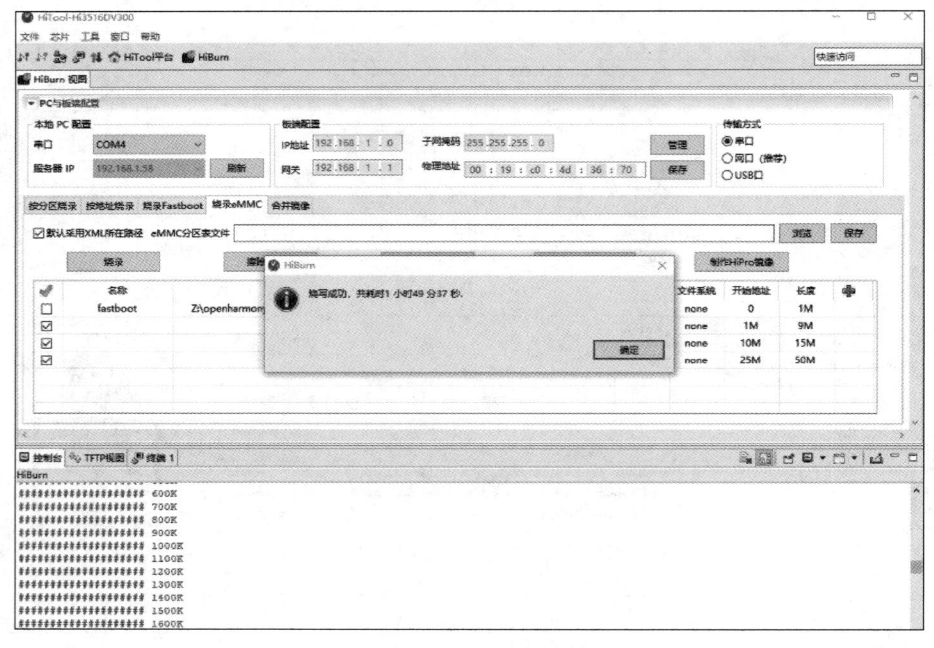

图 14-25　烧录完成

接下来，就需要等待 HiTool 工具烧录完成了，串口烧录时间比较长，一般需要 1.5h 左右。下面我们来看一看通过网口烧录的方式，这种方式要快很多，通常几分钟就可以完成。

2. 网口烧录

下面，通过网口的方式进行系统烧录。采用网口的方式首先需要配置好开发板与 PC 主机的网络连接，确保开发板与 PC 主板的网络是互通的状态，如图 14-26 所示。

图 14-26　连接好网线，串口

这里需注意，一定要同时插串口线和网线，因为即使采用网口烧录方式，网络配置命令和 fastboot 也是通过串口传输的。

步骤 1：配置开发板和 PC 主机网络。

检查防火墙是否关闭，防火墙可能阻止开发板访问计算机，如图 14-27 所示。

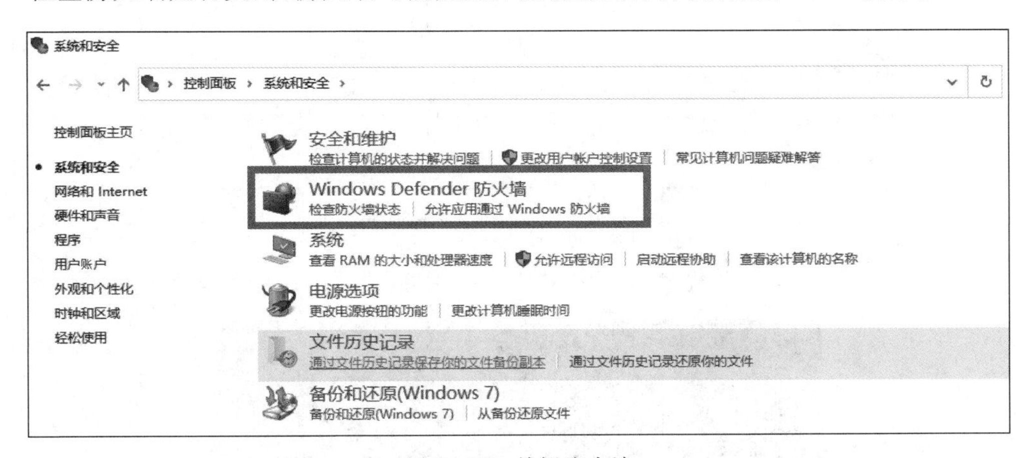

图 14-27　关闭防火墙

这里可以选择临时关闭防火墙，如图 14-28 所示。

在开发板的命令行中输入动态获取 IP 地址命令，如图 14-29 所示，代码如下：

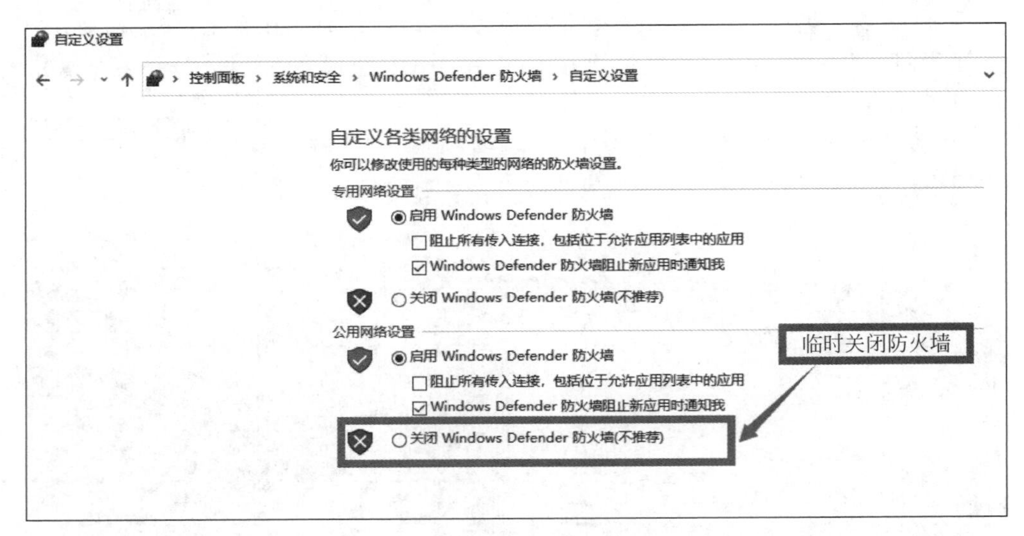

图 14-28　临时关闭 Windows Defender 防火墙

```
ifconfig eth0 up
dhclient eth0   //eth0 有线网卡
```

```
OHOS #
OHOS # ifconfig
eth0     ip:192.168.1.10 netmask:255.255.255.0 gateway:192.168.1.1
         HWaddr 86:dd:ef:2c:6c:34 MTU:1500 Running Default Link UP
lo       ip:127.0.0.1 netmask:255.0.0.0 gateway:127.0.0.1
         ip6: ::1/64
         HWaddr 00 MTU:0 Running Link UP
OHOS #
OHOS #
OHOS # dhclient eth0        动态IP地址
OHOS #
OHOS # ifconfig
eth0     ip:192.168.0.113 netmask:255.255.255.0 gateway:192.168.0.1
         HWaddr 86:dd:ef:2c:6c:34 MTU:1500 Running Default Link UP
lo       ip:127.0.0.1 netmask:255.0.0.0 gateway:127.0.0.1
         ip6: ::1/64
         HWaddr 00 MTU:0 Running Link UP
OHOS #
```

图 14-29　设置开发板 IP 地址

PC 测试是否可以连接上开发板，如图 14-30 所示。

```
C:\Users\      >ping 192.168.0.113

正在 Ping 192.168.0.113 具有 32 字节的数据:
来自 192.168.0.113 的回复: 字节=32 时间=4ms TTL=255
来自 192.168.0.113 的回复: 字节=32 时间=3ms TTL=255
来自 192.168.0.113 的回复: 字节=32 时间=2ms TTL=255
来自 192.168.0.113 的回复: 字节=32 时间=2ms TTL=255
```

图 14-30　PC 测试是否可以连接上开发板(一)

开发板测试是否可以连接 PC 端，首先确保防火墙处于临时关闭状态，如图 14-31 所示。

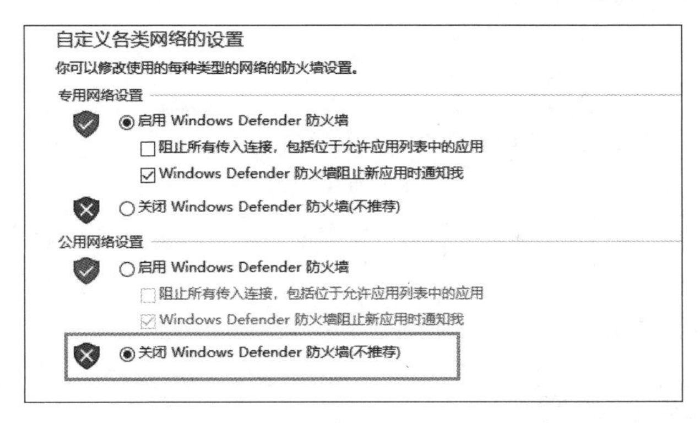

图 14-31　PC 测试是否可以连接上开发板(二)

测试开发板是否可以连接 PC,如图 14-32 所示。

```
OHOS # ping 192.168.0.104

[0]Reply from 192.168.0.104: time=10ms TTL=64
[1]Reply from 192.168.0.104: time=4ms TTL=64
[2]Reply from 192.168.0.104: time=2ms TTL=64
[3]Reply from 192.168.0.104: time=10ms TTL=64
--- 192.168.0.104 ping statistics ---
4 packets transmitted, 4 received, 0 loss
OHOS #
```

图 14-32　测试开发板是否可以连接 PC

步骤 2：配置海思 HiTool 烧录工具。

这里首先需要按以下 4 个步骤进行配置,如图 14-33 所示。

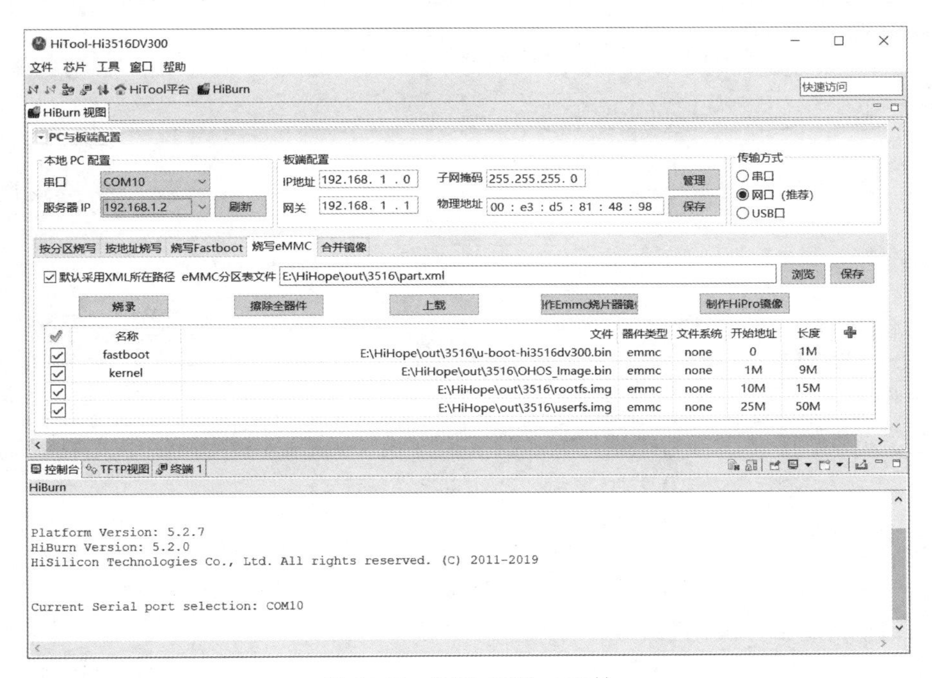

图 14-33　配置 HiTool 工具

(1) 选择烧录方式为网口烧录。

(2) 设置服务器 IP 地址,即 PC 主机的 IP 地址。

(3) 设置板端相关信息,即开发板的 IP 地址及物理地址等。

(4) 根据分区信息添加要烧录的文件,本次烧录只烧录内核 Kernel、根文件系统 rootfs、用户文件系统 userfs。

步骤 3:单击烧录,如图 14-34 所示。

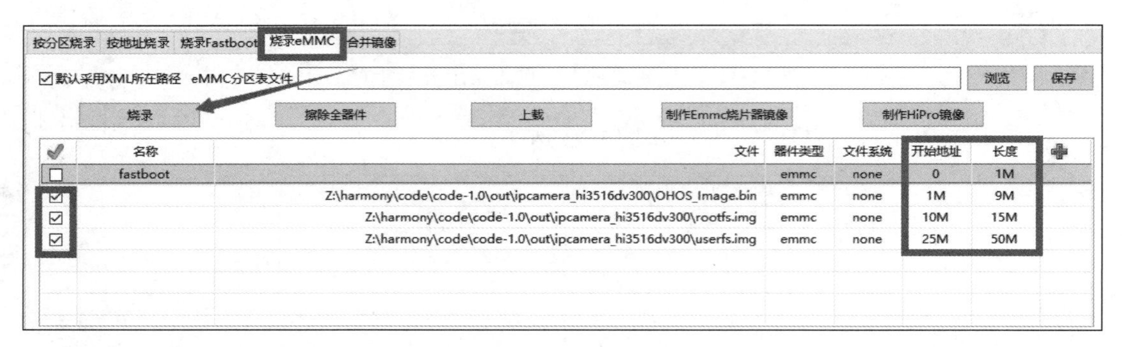

图 14-34 单击烧录

步骤 4:开始烧录系统,如图 14-35 所示。

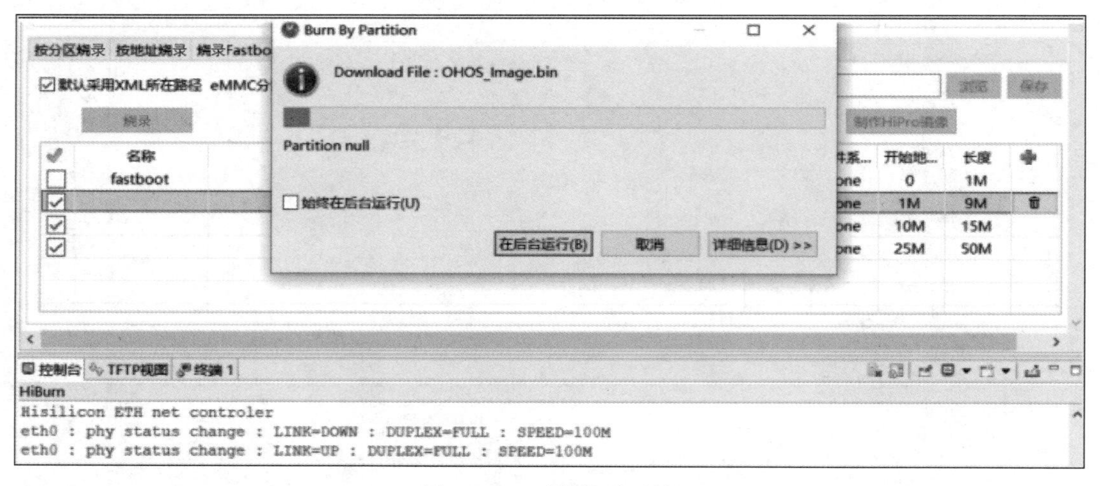

图 14-35 开始烧录系统

步骤 5:重启开发板,按动开关按钮,必须 15s 内完成该操作,如图 14-36 所示。

接下来会显示正在烧录,我们只需等待烧录完成的提示,如图 14-37 所示。

步骤 6:烧录成功,如图 14-38 所示。

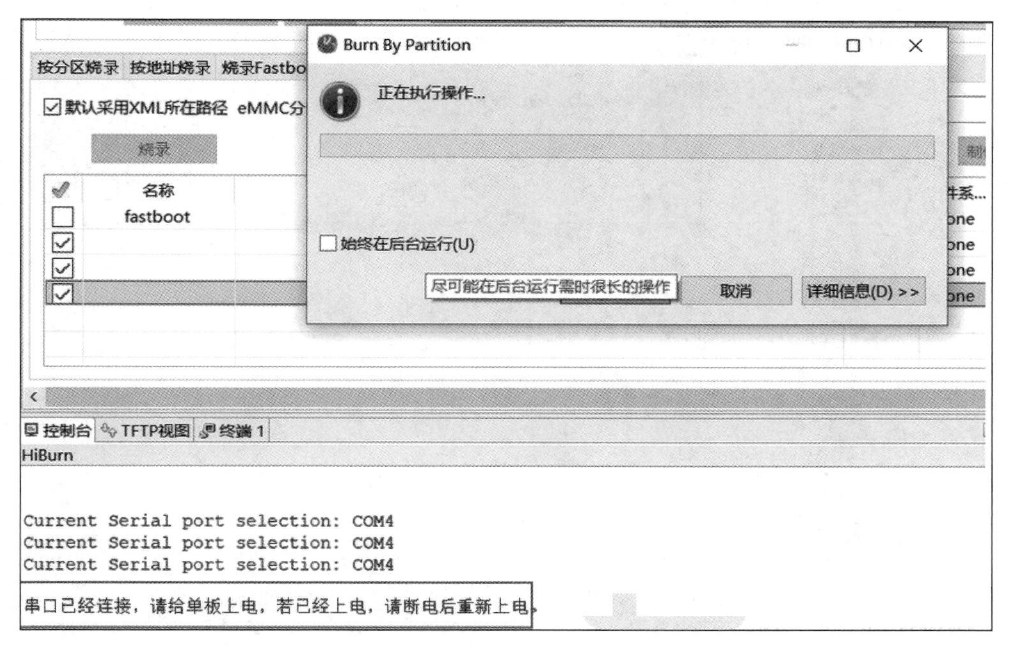

图 14-36　重启开发板

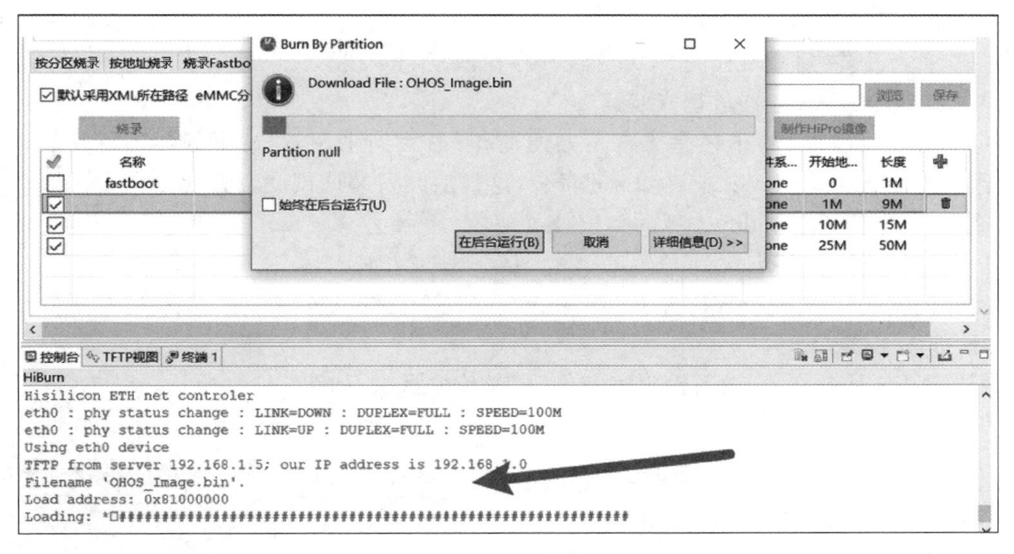

图 14-37　正在烧录

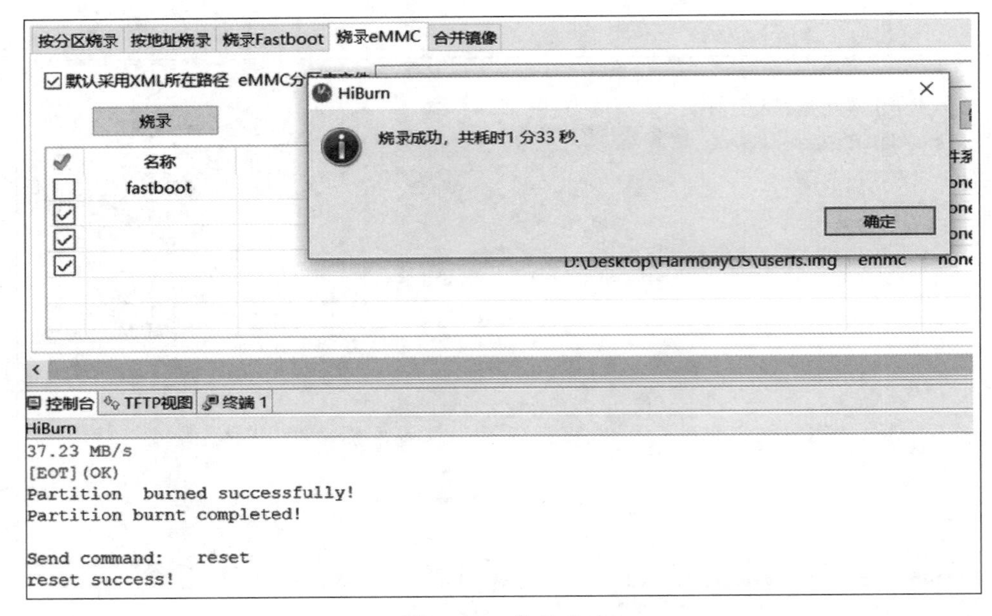

图 14-38 烧录成功

14.3.3 安装鸿蒙应用程序

目前把鸿蒙应用程序安装到 Hi3516DV3000 开发板上，有两种安装方法：

(1) 通过 SD 卡复制，在 PC 上把需要安装的 HAP 包及安装工具 bm、aa 复制到 SD 卡中，再把 SD 卡放置到开发板的 SD 卡插槽内，在鸿蒙 shell 目录下，执行安装。

(2) 通过 NFS 服务器进行安装。

每次安装时插拔 SD 卡还是非常不方便的，下面我们采用 NFS 服务器进行安装，让鸿蒙操作系统能直接访问 Windows 10 的目录，这样后续安装调试就会方便很多了。

这里，我们使用 Windows 版的 NFS Server。

1. 安装 NFS 服务器

我们先安装一个 haneWIN NFS 服务器，下载 nfs1169.exe 文件，一路单击"下一步"按钮即可，如图 14-39 所示。

配置 NFS 目录参数，编辑输出表文件，定义传输目录，如图 14-40 所示。

编辑内容如下：

```
#exports example
#c:\ftp -range 192.168.1.1 192.168.1.10
#c:\public -public -readonly
#c:\tools -readonly 192.168.1.4

d:\work2020\29-HarmonyOS\code\helloIot -public -name:nfs
```

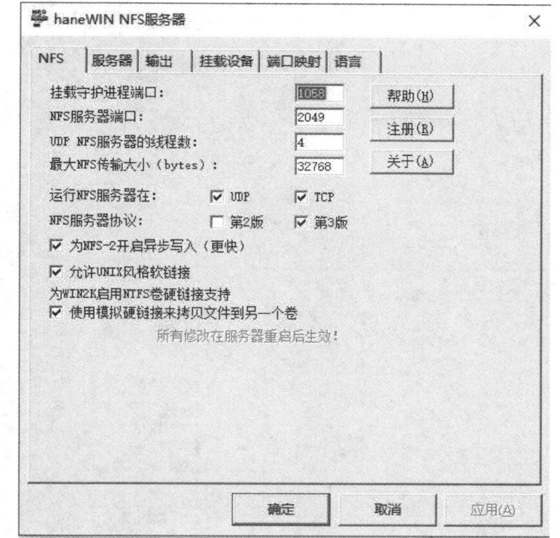

图 14-39　haneWIN NFS 服务器

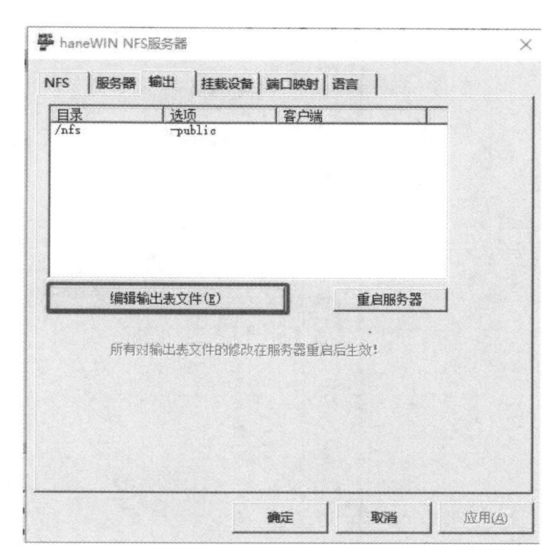

图 14-40　编辑输出表文件,定义传输目录

2. 重启 NFS 服务

右击管理员权限,重启所有服务,使配置生效,如图 14-41 所示。

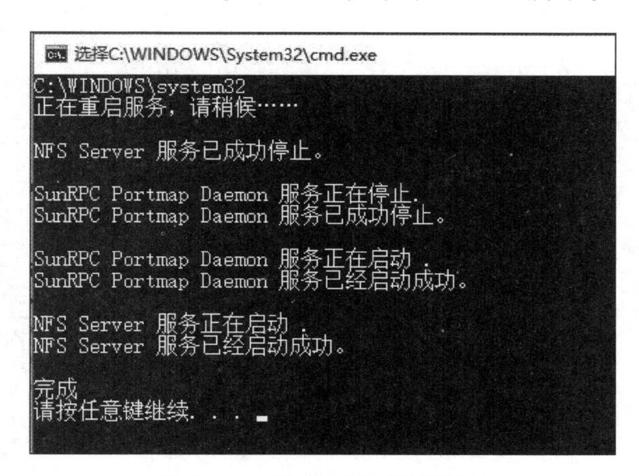

图 14-41　重启 NFS 所有服务

3. 设置防火墙

设置防火墙的 111、1058、2049 端口的 TCP 和 UDP,如图 14-42 所示。

4. 鸿蒙上挂载目录

主计算机的 IP 地址为 192.168.0.104,NFS 服务的别名为 nfs,对应的目录是 D:\work2020\29-HarmonyOS\code\helloIot。

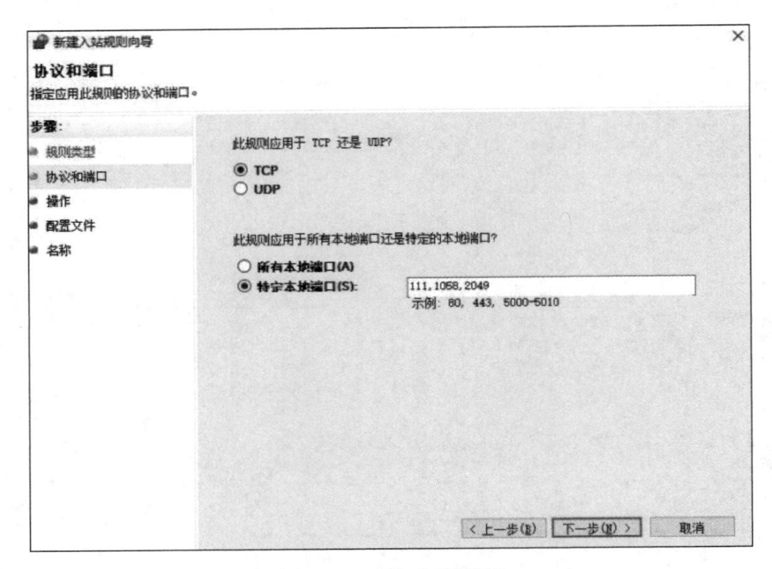

图 14-42　防火墙设置

```
mkdir nfs
mount 192.168.0.104:/nfs /nfs nfs
```

挂载到鸿蒙新建的/nfs目录下,我们可以复制安装包和安装工具,如图14-43所示。

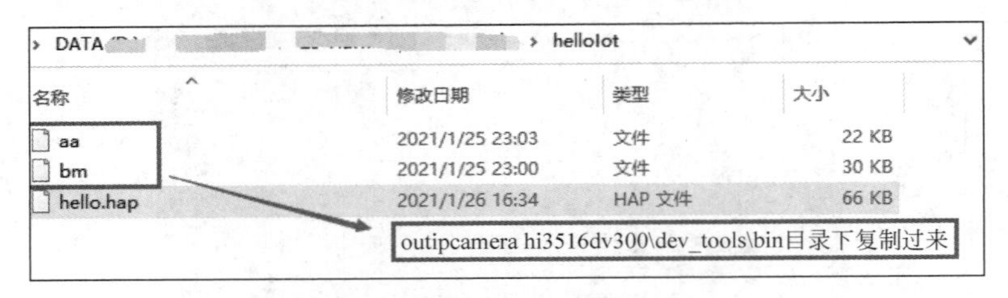

图 14-43　复制安装包和安装工具

在编译harmony sdk包的code out目录dev_tools下有两个安装工具文件aa和bm,将这两个文件和图14-43中的HAP包都放到SD卡中,如图14-44所示。

5. 安装应用

安装应用默认要校验签名,需要执行以下命令,关闭签名校验,命令如下:

```
cd nfs
./bm set -s disable
./bm install -p hello.hap
```

前面做了这么多铺垫工作,后续开发只要复制HAP安装包,直接执行一条命令便可安装,非常方便,如图14-45所示。

```
OHOS #
OHOS # ls
Directory /nfs:
-rw-r--r-- 22472     u:0      g:0      aa
-rw-r--r-- 29984     u:0      g:0      bm
-rw-r--r-- 66794     u:0      g:0      hello.hap
OHOS #
OHOS # ./bm install -p hello.hap
OHOS # 01-01 00:50:43.280 16 62 I 01800/Samgr: Initialize Registry!
01-01 00:50:43.280 5 32 D 01800/Samgr: Judge Auth<bundlems, BmsInnerFeature> ret:0
01-01 00:50:43.280 5 32 D 01800/Samgr: Find Feature<bundlems, BmsInnerFeature> id<38, 0> ret:0
01-01 00:50:43.281 16 62 I 01800/Samgr: Create remote sa proxy[0x249546f0]<bundlems, BmsInnerFeature> id<38,0>!
01-01 00:50:43.281 5 32 D 01800/Samgr: Judge Auth<bundlems, BmsInnerFeature> ret:0
01-01 00:50:43.281 5 32 D 01800/Samgr: Find Feature<bundlems, BmsInnerFeature> id<38, 0> ret:0
01-01 00:50:43.288 16 62 D 00000/(null): BMS client invoke install
01-01 00:50:43.288 16 62 I 00000/(null): BundleManager install or uninstall invoke return: 0
01-01 00:50:43.288 5 35 I 00000/(null): create ZipFile instance
01-01 00:50:43.288 5 35 I 00000/(null): open: /nfs/hello.hap
01-01 00:50:43.444 5 35 I 00000/(null): parse 16 central entries from /nfs/hello.hap
01-01 00:50:43.444 5 35 I 00000/(null): extract file config.json
01-01 00:50:43.444 5 35 I 00000/(null): get entry by name: config.json
01-01 00:50:43.444 5 35 D 00000/(null): get entry successed
01-01 00:50:43.452 5 35 I 00000/(null): unzip with store
01-01 00:50:43.452 5 35 I 00000/(null): seek to entry start 0x00000036
01-01 00:50:43.458 5 35 I 00000/(null): unzip with store success
01-01 00:50:43.458 5 35 I 00000/(null): close: /nfs/hello.hap
01-01 00:50:43.479 5 35 I 00000/(null): create ZipFile instance
01-01 00:50:43.479 5 35 I 00000/(null): open: /nfs/hello.hap
01-01 00:50:43.637 5 35 I 00000/(null): parse 16 central entries from /nfs/hello.hap
write file switch /storage/data/log/hilog2.txt
01-01 00:50:43.637 5 35 I 00000/(null): extract file config.json
01-01 00:50:43.637 5 35 I 00000/(null): get entry by name: config.json
01-01 00:50:43.637 5 35 D 00000/(null): get entry successed
01-01 00:50:43.644 5 35 I 00000/(null): unzip with store
01-01 00:50:43.644 5 35 I 00000/(null): seek to entry start 0x00000036
01-01 00:50:43.680 5 35 I 00000/(null): unzip with store success
01-01 00:50:43.680 5 35 I 00000/(null): close: /nfs/hello.hap
01-01 00:50:43.681 6 42 I 01100/bundle_deamon: BundleDaemon::Invoke(102): bundle_daemon invoke start 0
01-01 00:50:43.693 6 42 I 00000/(null): create ZipFile instance
01-01 00:50:43.693 6 42 I 00000/(null): open: /nfs/hello.hap
01-01 00:50:43.937 6 42 I 00000/(null): parse 16 central entries from /nfs/hello.hap
01-01 00:50:43.945 6 42 I 00000/(null): extract file config.json
01-01 00:50:43.945 6 42 I 00000/(null): get entry by name: config.json
01-01 00:50:43.945 6 42 D 00000/(null): get entry successed
01-01 00:50:43.962 6 42 I 00000/(null): unzip with store
01-01 00:50:43.963 6 42 I 00000/(null): seek to entry start 0x00000036
01-01 00:50:43.974 6 42 I 00000/(null): unzip with store success
01-01 00:50:43.985 6 42 I 00000/(null): extract file assets/entry/resources.index
01-01 00:50:43.985 6 42 I 00000/(null): get entry by name: assets/entry/resources.index
01-01 00:50:43.985 6 42 D 00000/(null): get entry successed
01-01 00:50:43.995 6 42 I 00000/(null): unzip with store
```

图 14-44　复制 HAP 安装包和安装工具(aa、bm)

图 14-45　App 安装效果

14.4　HiSpark Hi3518 开发板

HiSpark IPC DIY Camera 基于华为海思 Hi3518 芯片，支持 LiteOS、鸿蒙 OS，实现图像采集识别功能，广泛应用于智能摄像头、安防监控、车载记录仪等，如图 14-46 所示。

图 14-46 HiSpark IPC DIY Camera

应用场景如图 14-47 所示。

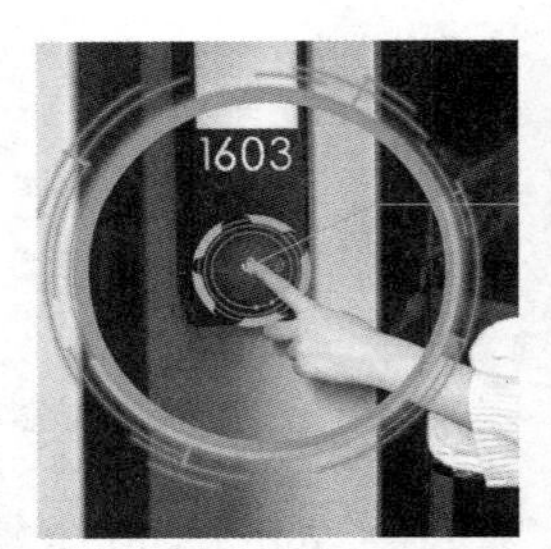

智能门铃

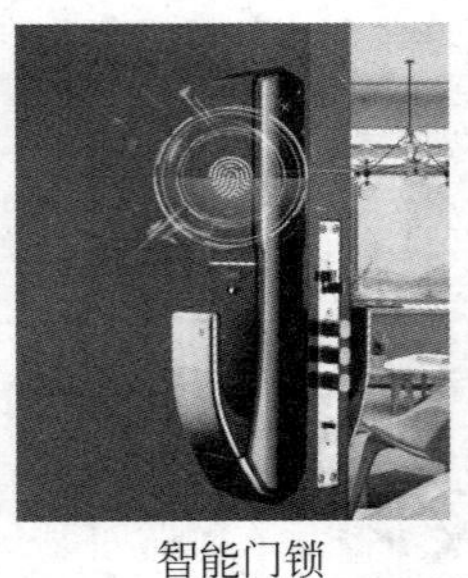

智能门锁

智能家居监控

图 14-47 应用场景

14.4.1 开发板简介

Hi3518EV300 作为新一代智慧视觉处理 SoC，集成了新一代 ISP（Image Signal Processor）及 H.265 视频压缩编码器，同时采用先进低功耗工艺和低功耗架构设计，使其在低码率、高画质、低功耗等方面引领行业水平，如图 14-48 所示。

产品参数如图 14-49 所示。

14.4.2 烧录 HarmonyOS

编译 Hi3518 目标平台的命令如下：

```
python build.py ipcamera_hi3516dv300 - b debug
```

编译生成的 Kernel、rootfs、userfs 映像文件会被保存到 out/ipcamera_hi3518ev300 目录下，u-boot 二进制会被保存在 vendor 目录（可以通过 find vendor -name u-boot *.bin 命

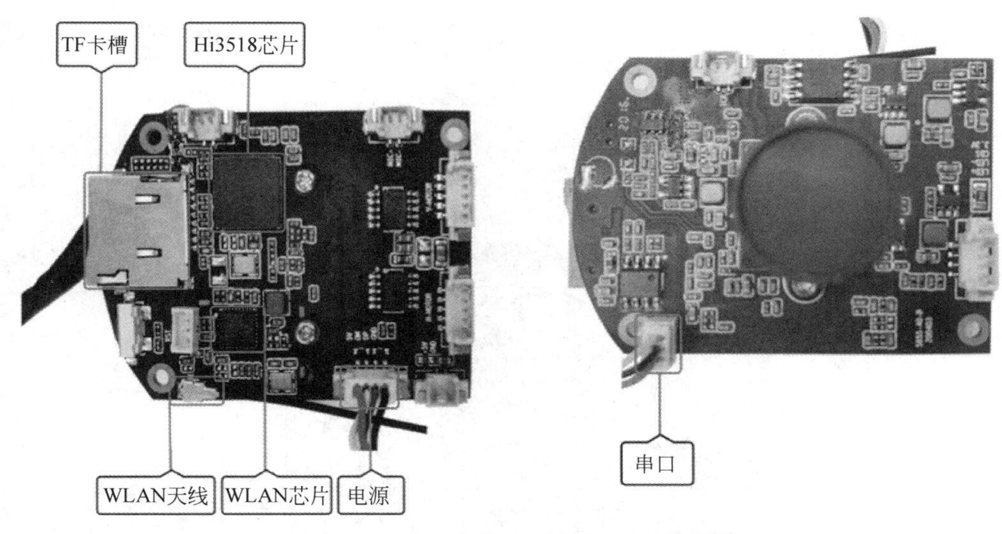

图 14-48 Hi3518EV300 单板正反面外观图

配置	参数
SoC	基于海思Hi3518EV300芯片 板载海思Hi3881 WiFi芯片
操作系统	LiteOS、鸿蒙 OS
网络自适应	设备能够根据当前网络状况，选择合适码流进行传输
视频能力	1920×1080/25fps高清实时监控，AI精准人形侦测与追踪
易用性	支持标准 UVC/UVA 协议做USB摄像头 使用支持跨网段自动出图功能（搭配 RS 系列 NVR） NVR 及 Easyvms 客户端集中管理
存储方式	支持多种：TF卡及手机本地存储/WiFi视频传输/云存储/AP热点模式直连
语言能力	设备支持7国语言语音播报（中文/英文/韩语/越南语/印尼语/葡萄牙语/西班牙语）；语音清晰，无回声噪声 App 支持16国语言（简体中文/繁体中文/英语/芬兰语/韩语/法语/葡萄牙语/泰语/西班牙语/波兰语/印度尼西亚语/德语/意大利语/俄语/日语/越南语/阿拉伯语); 支持手势放大，支持预置点设置，最大支持256个预置点
App 多功能	设备上线离线状态显示/视频码流帧率等设置/图像旋转(顶装模式)/定时抓图/报警设置（触发录像、触发 FTP、联动推送）/录像设置/OSD设置/查看WiFi信息等 支持监控时间戳显示/日夜模式设置/IR-CUT反向及切换时间设置

图 14-49 Hi3518EV300 产品参数

令进行查找)，各种硬件连接关系如图 14-50 所示。

1. DevEco Device Tool 烧录

Hi3518 开发板的代码烧录仅支持 USB 烧录方式。Hi3518 开发板的代码烧录，需要同时连接串口和 USB 口，确认好串口、USB 口的连接位置，如图 14-51 所示。

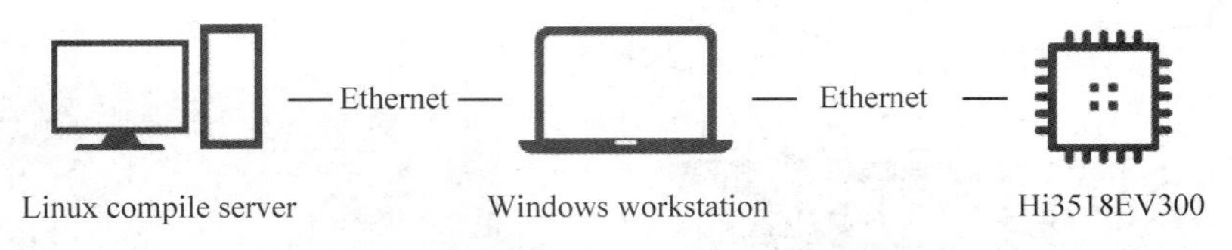

图 14-50　Hi3518EV300 各种硬件连接关系图

图 14-51　需要同时连接串口和 USB 口

打开计算机的设备管理器,查看并记录对应的串口号,如图 14-52 所示。

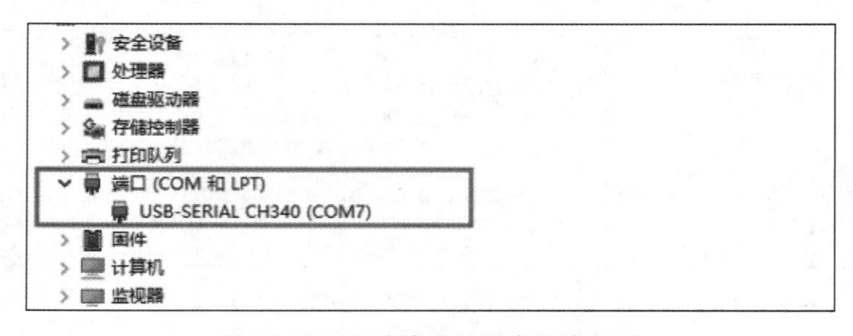

图 14-52　查看并记录对应的串口号

接下来,需要打开 Visual Studio,导入鸿蒙源代码项目,如图 14-53 所示。

选择 Device Tool→Configure→Burn 命令进入烧录配置界面,设置 Hi3518 系列开发板烧录信息。这里我们需要启动需要烧录的目标板,如图 14-54 所示。

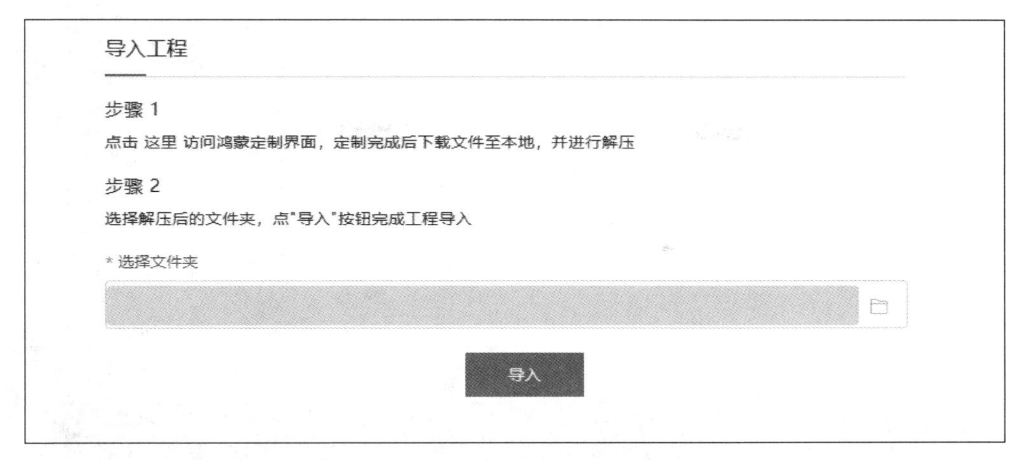

图 14-53　导入鸿蒙源代码项目

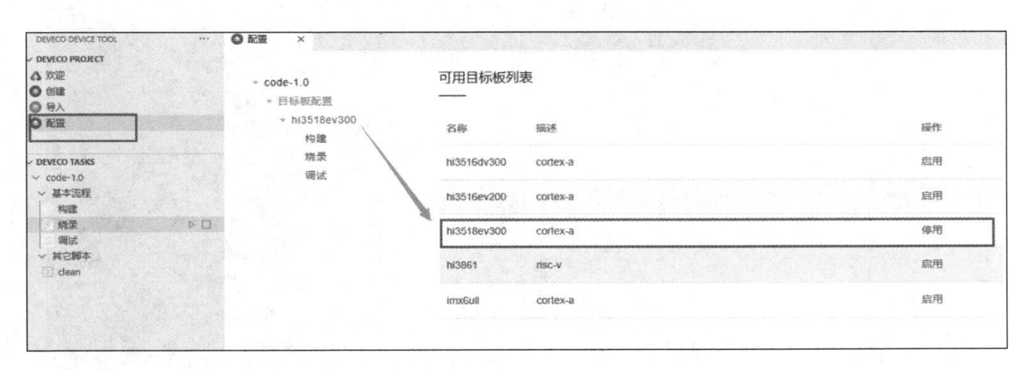

图 14-54　启用 Hi3518ev300

选择烧录方式(Burning Mode)，固定选择 usbport，如图 14-55 所示。

图 14-55　固定选择 usbport

设置 USB 烧录的烧录参数，如图 14-56 所示。设置 Port number，需选择已查询的串口号。设置 Baud Rate 和 Data Bits 参数，已根据开发板进行适配，保持默认值即可。设置器件类型(Memory Type)，Hi3518 系列开发板固定设置为 spi nor。需要根据如表 14-1 所示设置烧录文件。

表 14-1　设置烧录文件信息

待烧录文件	开始地址	文件大小	是 否 烧 录
u-boot-开发板名称. bin	0MB	1MB	是
OHOS_Image. bin	1MB	6MB	第一次使用开发板烧录时，必须同时烧录；在后续烧录过程中，如果修改了内核和驱动相关内容，才需要烧录
rootfs. img	7MB	8MB	第一次使用开发板烧录时，必须同时烧录；在后续烧录过程中，如果这两个文件未做修改，可以不烧录。建议每次烧录时，都烧录这两个文件
userfs. img	15MB	1MB	

Burn Files

Enable	File Name	Start Address	File Size
☑	ı-boot-hi3518ev300.bin	0	1MB
☑	ı300\OHOS_Image.bin	1MB	6MB
☑	ıi3518ev300\rootfs.img	7MB	8MB
☑	i3518ev300\userfs.img	15MB	1MB

New

图 14-56　设置烧录文件信息

修改了相关配置后，需单击最下方的 Save 按钮进行保存。在 DevEco Device Tool 中，单击▶按钮开始烧录，如图 14-57 所示。

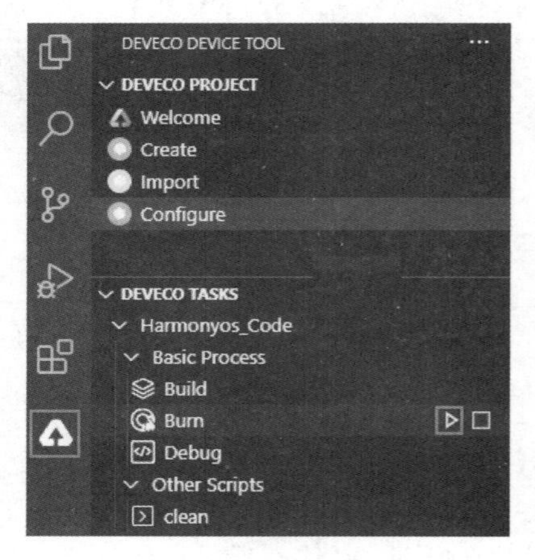

图 14-57　开始烧录

需在 15s 内手动重启开发板（下电再上电），如图 14-58 所示。

等待烧录完成，当控制台输出如下信息时，表示烧录成功，如图 14-59 所示。

注意，第一次烧录必选修改 U-boot 的 bootcmd 及 bootargs 内容，该步骤为固化操作，可保存执行结果，但如果 U-boot 重新烧入，则需要再次执行下述步骤，如表 14-2 所示。

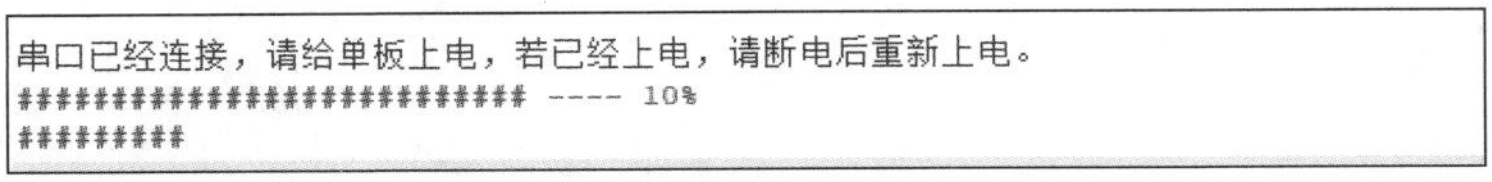

```
串口已经连接，请给单板上电，若已经上电，请断电后重新上电。
############################# ---- 10%
#########
```

图 14-58 重启开发板

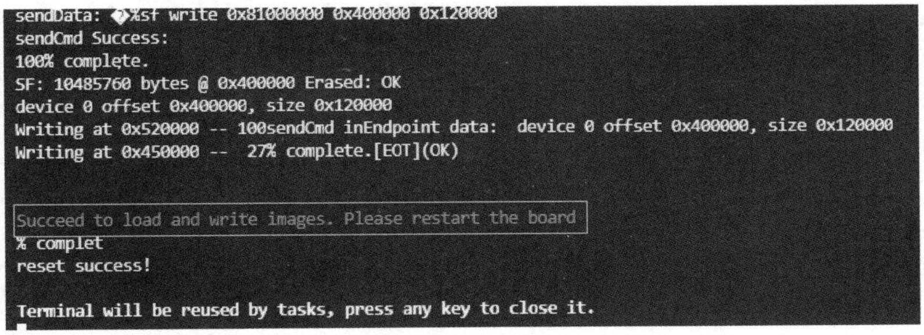

```
sendData: �%sf write 0x81000000 0x400000 0x120000
sendCmd Success:
100% complete.
SF: 10485760 bytes @ 0x400000 Erased: OK
device 0 offset 0x400000, size 0x120000
Writing at 0x520000 -- 100sendCmd inEndpoint data:  device 0 offset 0x400000, size 0x120000
Writing at 0x450000 --  27% complete.[EOT](OK)


Succeed to load and write images. Please restart the board
% complet
reset success!

Terminal will be reused by tasks, press any key to close it.
```

图 14-59 烧录成功提示

表 14-2 使用 HiTool 烧录

执行命令	命令解释
setenv bootcmd "sf probe 0;sf read 0x40000000 0x100000 0x600000;go 0x40000000";	设置 bootcmd 内容，选择 FLASH 器件 0，读取 FLASH 起始地址为 0x100000，大小为 0x600000 字节的内容到 0x40000000 的内存地址，此处 0x600000 为 6MB，与 IDE 中所填写的 OHOS_Image.bin 的文件大小必须相同
setenv bootargs "console=ttyAMA0,115200n8 root=flash fstype=jffs2 rw rootaddr=7M rootsize=8M";	表示将 bootargs 参数设置为串口输出，将波特率设置为 115200，将数据位设置为 8，rootfs 挂载于 FLASH 上，文件系统类型为 jffs2 rw，以支持可读写 JFFS2 文件系统。"rootaddr=7M rootsize=8M"处对应填入实际 rootfs.img 的烧录起始位置与长度，与 IDE 内所填大小必须相同
saveenv	表示保存当前配置
reset	表示复位单板
pri	表示查看显示参数

注意：go 0x40000000 为可选指令，默认配置已将该指令固化在启动参数中，单板复位后可自动启动。若想切换为手动启动，则可在 U-boot 启动倒数阶段按 Enter 键打断自动启动。

2. 使用 HiTool 烧录

也可以使用 HiTool 工具进行烧录，这里我们选择串口的方式进行烧录，如图 14-60 所示。

运行 HiTool 工具，选择芯片 Hi3518DV300。添加烧录文件，这里包括 u-boot、Kernel、rootfs、userfs 共 4 个文件，器件类型选择 spi nor 类型，开始位置和长度配置按图 14-60 所示进行配置。

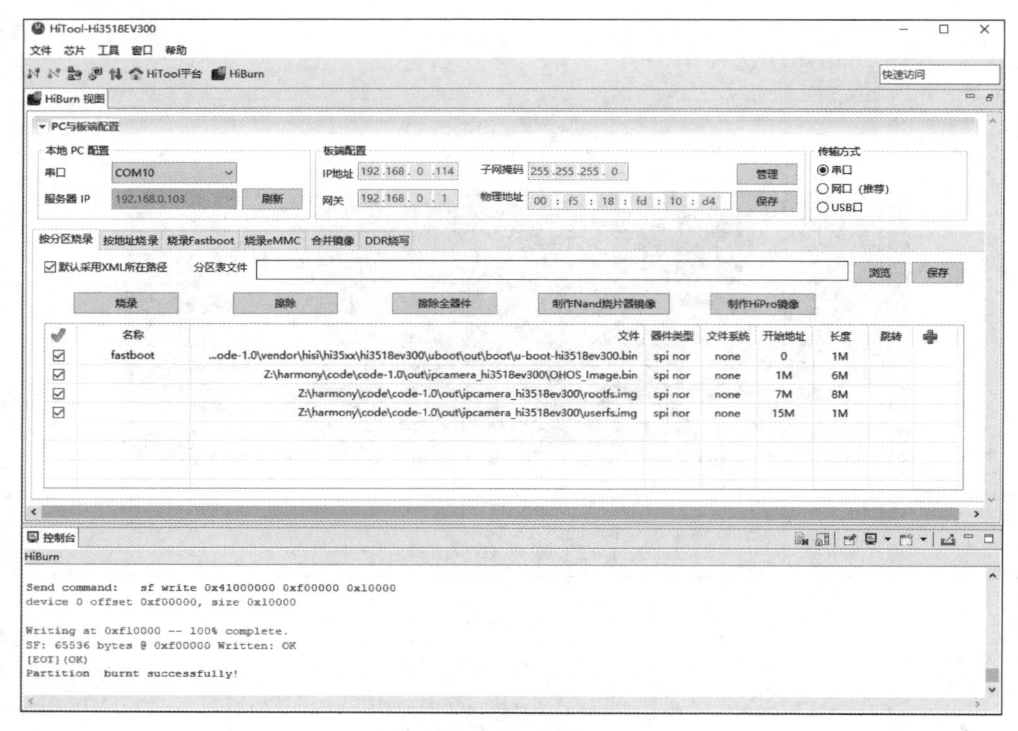

图 14-60 使用 HiTool 烧录

编译生成的 Kernel、rootfs、userfs 映像文件，在 out/ipcamera_hi3518ev300 目录下查找，u-boot 二进制文件会被保存在 vendor 目录（可以通过 find vendor -name u-boot *.bin 命令进行查找）。

单击烧录后，需要等待一段时间，等待固件下载完毕，如图 14-61 所示。

烧录成功后，重新复位设备。可能会出现未能启动鸿蒙操作系统的情况，此时只能启动 uboot，如图 14-62 所示。

如果提示 ##Error: "distro_bootcmd" not defined，则应该是 uboot 环境变量未设置好。输入命令 printenv 查看，如图 14-63 所示。

上面的错误问题，需要我们在 uboot 命令行下修改 U-boot 的 bootcmd 及 bootargs 内容，如图 14-64 所示，代码如下：

```
setenv bootcmd "sf probe 0;sf read 0x40000000 0x100000 0x600000;go 0x40000000";

setenv bootargs " console = ttyAMA0, 115200n8 root = flash fstype = jffs2 rw rootaddr = 7M
rootsize = 8M";

saveenv

printenv
```

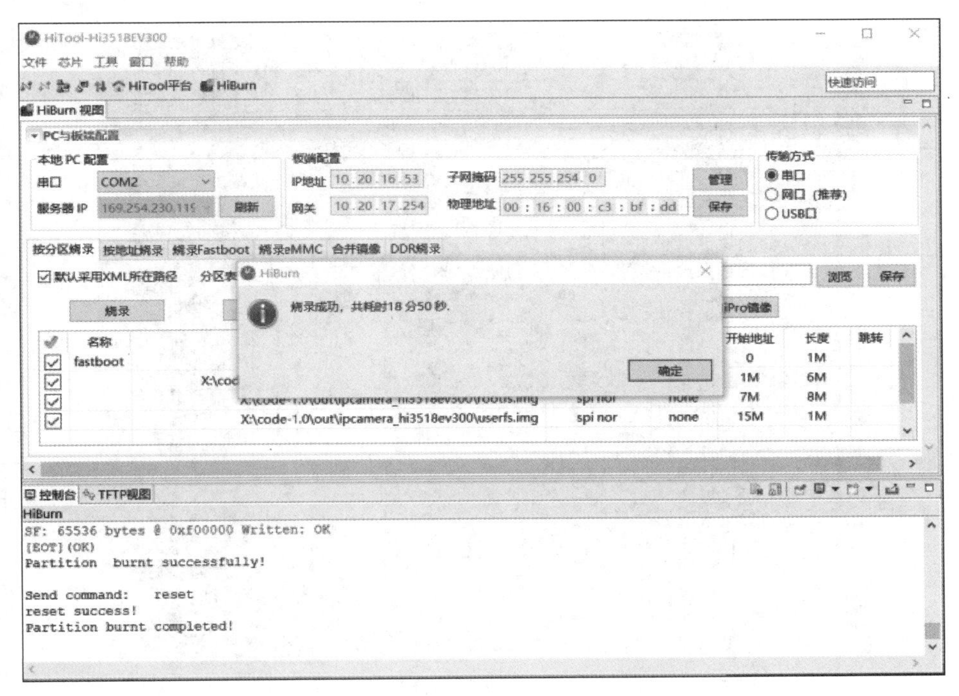

图 14-61　烧录成功

```
System startup

Uncompress Ok!

U-Boot 2016.11 (Jul 03 2020 - 19:39:05 +0800)hi3518ev300

Relocation Offset is: 03735000
Relocating to 43f35000, new gd at 43e94ef0, sp at 43e94ed0
SPI Nor:  hifmc_ip_ver_check(44): Check Flash Memory Controller v100 ...hifmc_ip
_ver_check(50):  Found
hifmc_spi_nor_probe(1664): SPI Nor ID Table Version 1.0
hifmc_spi_nor_probe(1689): SPI Nor(cs 0) ID: 0x20 0x70 0x18
hifmc_spi_nor_probe(1754): Block:64KB hifmc_spi_nor_probe(1755): Chip:16MB hifmc
_spi_nor_probe(1756): Name:"XM25QH128A"
hifmc100_spi_nor_probe(147): SPI Nor total size: 16MB
NAND:  0 MiB
MMC:
*** Warning - bad CRC, using default environment

In:    serial
Out:   serial
Err:   serial
Net:   No ethernet found.
Hit any key to stop autoboot:  0
## Error: "distro_bootcmd" not defined
hisilicon #
```

图 14-62　启动 uboot

```
protect - enable or disable FLASH write protection
pxe     - commands to get and boot from pxe files
reset   - Perform RESET of the CPU
run     - run commands in an environment variable
saveenv - save environment variables to persistent storage
setenv  - set environment variables
setexpr - set environment variable as the result of eval expression
sf      - SPI flash sub-system
showvar - print local hushshell variables
sleep   - delay execution for some time
source  - run script from memory
sysboot - command to get and boot from syslinux files
test    - minimal test like /bin/sh
tftpboot- boot image via network using TFTP protocol
true    - do nothing, successfully
ugzip   - Compress gzipfile with hardware IP
unzip   - unzip a memory region
usb     - USB sub-system
usbboot - boot from USB device
usbtftp - download or upload image using USB protocol
version - print monitor, compiler and linker version
hisilicon # printenv
arch=arm
baudrate=115200
board=hi3518ev300
board_name=hi3518ev300
bootargs=console=ttyAMA0,115200n8 root=/dev/mtdblock2 rw
bootcmd=run distro_bootcmd
bootdelay=2
cpu=armv7
soc=hi3518ev300
stderr=serial
stdin=serial
stdout=serial
vendor=hisilicon
verify=n

Environment size: 284/262140 bytes
hisilicon #
```

图 14-63 查看环境变量

```
hisilicon # setenv bootcmd "sf probe 0;sf read 0x40000000 0x100000 0x600000;go 0
hisilicon #
hisilicon #
hisilicon #
hisilicon # setenv bootargs "console=ttyAMA0,115200n8 root=flash fstype=jffs2 rw
hisilicon #  rootsize=8M";
hisilicon #
hisilicon #
hisilicon # setenv bootargs "console=ttyAMA0,115200n8 root=flash fstype=jffs2 rw
 rootaddr=7M rootsize=8M";
hisilicon # saveenv
Saving Environment to SPI Flash...
Erasing SPI flash...Writing to SPI flash...done
hisilicon # printenv
arch=arm
baudrate=115200
board=hi3518ev300
board_name=hi3518ev300
bootargs=console=ttyAMA0,115200n8 root=flash fstype=jffs2 rw rootaddr=7M rootsiz
e=8M
bootcmd=sf probe 0;sf read 0x40000000 0x100000 0x600000;go 0x40000000
bootdelay=2
cpu=armv7
soc=hi3518ev300
stderr=serial
stdin=serial
stdout=serial
vendor=hisilicon
verify=n

Environment size: 355/262140 bytes
hisilicon #
```

图 14-64 修改环境变量

执行修改配置后，再次复位系统。输入命令 reset。等待系统自启动进入系统，系统启动后，按 Enter 键，显示 OHOS 字样，如图 14-65 所示。

这个时候，我们就可以输入./bin/camera_app 后按 Enter 键，显示如图 14-66 所示。

```
[ERR]vfs_jffs_ioctl NOT SUPPORT
write file switch /storage/data/log/hilog1.txt
01-01 00:00:06.695 6 26 I 01800/Samgr: Initialized all core system services!
01-01 00:00:06.695 6 26 I 01800/Samgr: Goto next boot step failed! errno:-9
01-01 00:00:06.730 5 19 I 01800/Samgr: Create proxy[0x21996db0]<bundle_daemon, (
null), 27, 0>
01-01 00:00:06.730 6 26 I 01100/bundle_deamon: BundleDaemon::Invoke(102): bundle
_daemon invoke start 8
01-01 00:00:06.733 5 19 E 00000/(null): ScanSharedLibPath GetBundleInfos is erro
r
01-01 00:00:06.733 5 18 D 01300/abilityms: AbilityMgrHandler::OnServiceInited(12
2): start
01-01 00:00:06.734 5 18 E 01300/abilityms: bms query not found: query keep alive
 bundle info failure
[Init] start service media_server succeed, pid 7.
Media server initialize succeed.
[Init] start service appspawn succeed, pid 8.
[UnRegisteDeathCallback : 960]Wrong cbId:4294967295.
[UnRegisteDeathCallback : 960]Wrong cbId:4294967295.
01-01 00:00:08.712 8 32 D 01800/Samgr: RegisterRemoteFeatures<appspawn, (null)>
ret:0
01-01 00:00:08.712 8 32 I 01800/Samgr: Register endpoint<ipc client> and iunknow
n finished! remain<0> iunknown!
01-01 00:00:08.712 5 16 D 01800/Samgr: Register Feature<appspawn, (null)> pid<8>
, id<32, 0> ret:0
01-01 00:00:08.712 8 31 I 01800/Samgr: Register server sa<appspawn, (null)> id<4
294967295, 0> retry:1 ret:0!
01-01 00:00:08.712 8 31 I 00000/(null): [appspawn] get service name appspawn.
01-01 00:00:08.712 8 31 I 01800/Samgr: Init service appspawn <time: 5ms> success
!
01-01 00:00:08.712 8 31 I 01800/Samgr: Initialized all core system services!
01-01 00:00:08.712 8 31 I 00000/(null): [appspawn] get service name appspawn.
01-01 00:00:08.712 8 31 I 01800/Samgr: Goto next boot step failed! errno:-9
01-01 00:00:08.712 8 30 I 00000/(null): [appspawn] HOS_SystemInit is called!
01-01 00:00:08.712 8 30 I 00000/(null): [appspawn] main, entering wait.
[Init] DoChown, failed for 0 99 /dev/hdfwifi, err 2.
[Init] main, entering wait.

OHOS #
```

图 14-65 重启系统

```
01-01 00:00:08.712 8 31 I 01800/Samgr: Initialized all core system services!
01-01 00:00:08.712 8 31 I 00000/(null): [appspawn] get service name appspawn.
01-01 00:00:08.712 8 31 I 01800/Samgr: Goto next boot step failed! errno:-9
01-01 00:00:08.712 8 30 I 00000/(null): [appspawn] HOS_SystemInit is called!
01-01 00:00:08.712 8 30 I 00000/(null): [appspawn] main, entering wait.
[Init] DoChown, failed for 0 99 /dev/hdfwifi, err 2.
[Init] main, entering wait.

OHOS # ./bin/camera_app
OHOS #
**************************************************

               Hello OHOS!

**************************************************

This is a sample: Param = 81
```

图 14-66 重启应用

14.5 本章小结

本章介绍了 HiSpark 系列开发套件及鸿蒙操作系统的烧录流程，大家可以根据书上的提示进行固件烧录。开发板烧录完成后，就可以基于鸿蒙操作系统开发自己的产品了。